Truth Machine

Truth Machine

The Contentious History of
DNA Fingerprinting

MICHAEL LYNCH,

SIMON A. COLE,

RUTH MCNALLY, AND

KATHLEEN JORDAN

THE UNIVERSITY OF CHICAGO PRESS   CHICAGO AND LONDON

The University of Chicago Press, Chicago 60637
The University of Chicago Press, Ltd., London
© 2008 by The University of Chicago
All rights reserved. Published 2008.
Paperback edition 2011
Printed in the United States of America

20 19 18 17 16 15 14 13 12 11   2 3 4 5 6

ISBN-13: 978-0-226-49806-5 (cloth)
ISBN-10: 0-226-49806-9 (cloth)
ISBN-13: 978-0-226-49807-2 (paper)
ISBN-10: 0-226-49807-7 (paper)

Library of Congress Cataloging-in-Publication Data

Truth machine : the contentious history of DNA fingerprinting / Michael Lynch . . . [et al.].

    p. cm.

    Includes bibliographical references and index.
    ISBN-13: 978-0-226-49806-5 (cloth : alk. paper)
    ISBN-10: 0-226-49806-9 (cloth : alk. paper)

    I. DNA fingerprinting—History.   I. Michael Lynch, 1948–

RA1057.55T78 2008
614'.1—dc22

2008001902

TO THE MEMORY OF MELVIN POLLNER (1940–2007)

Contents

Preface

During the premier season of the immensely popular American television series *CSI: Crime Scene Investigation,* the program's hero Gil Grissom frequently made pithy remarks that conveyed a characteristic attitude toward forensic evidence:

> Grissom *(admonishing Warrick Brown, a coinvestigator)*: Forget about the assumptions. Forget about your promotion. These things will only confuse you. Concentrate on what cannot lie—the evidence. . . . There is no room for subjectivity in this department, Warrick. You know that. We handle each case objectively without presupposition regardless of race, color, creed, or bubble gum flavor. (Episode 1.01: "Pilot")

> Grissom: We're crime scene analysts. We're trained to ignore verbal accounts and rely instead on the evidence a scene sets before us. (Episode 1.02: "Cool Change")

> Grissom *(later in the same episode, to another CSI investigator)*: People leave us clues, Nick. They speak to us in thousands of different ways. It's our job to make sure that we've heard everything they've said. Anything less is reasonable doubt.

> Grissom *(in the same episode)*: I tend not to believe people. People lie. The evidence doesn't lie.

The ideal investigator personified by Grissom distrusts human testimony, and by heroically acting without prejudice, he allows physical evi-

dence to speak for itself.[1] People lie, but the evidence speaks the truth—
that is, as long as one listens to it "without presuppositions." In *CSI* and
its numerous spin-off shows, forensic analysts are depicted as highly
skilled scientists who distrust mere testimony and circumstantial evidence in favor of crucial bits of physical trace evidence. Whenever such
evidence is allowed to speak for itself, criminal identifications reach a
conclusive end. The heroes and heroines in *CSI* not only collect and analyze physical evidence, like police detectives they also risk their lives by
directly confronting criminals with the truth (Nolan, 2007; Mopas, 2007;
Valverde, 2006).

Experienced forensic scientists are quick to point out that such fic-
tional portrayals are notoriously inaccurate, and that the reality of forensic science is far less clear, certain, and glamorous than portrayed on television. For example, according to forensic scientist Thomas Mauriello,
"40 percent of the forensic science shown on *CSI* does not exist" (Klein,
2004). Another forensic scientist, Max Houck, adds that DNA evidence
is used in a minority of cases. He also points out that few forensic analysts have advanced degrees in a science; much forensic science involves
older forms of judgmental comparison; and even when "scientific" evidence is available and uncontested, ordinary forms of circumstantial evidence may be crucial for the judgment of guilt or innocence (Houck,
2006: 87).

The popular impact of these television shows is such that a term "the
CSI effect" has been coined to describe an alleged effect on jurors' expectations. Supposedly, jurors in criminal trials who had followed the
shows expect prosecutors to adduce clear physical evidence; otherwise
they will not convict. Although the media reports "the *CSI* effect" as
a full-fledged sociolegal problem, further examination yields no empirical evidence of any such effect (Cole and Dioso, 2005; Cole and Dioso-
Villa, 2007; Tyler, 2006; Podlas, 2006; Forthcoming). The evidence for
the alleged effect in most media accounts is actually evidence of the effect of prosecutor, not juror, behavior. "The *CSI* effect" appears to be
more of a media panic (a media panic about the pernicious effects of media) than anything else. At a deeper level, however, "the *CSI* effect" may
be viewed as indicative of awe and anxiety about the perceived power of

---

1. Grissom exemplifies a modern conception of scientific objectivity that ascribes inherent properties to an object that can be ascertained only by means of a disciplined struggle against the prejudices of sense and mind (Daston & Galison, 1992; Megill, 1994).

scientific evidence, particularly DNA evidence. Indeed, some scholars have suggested that the "effect" may not be, as "*CSI* effect" proponents claim, the effect of the distorted view of reality portrayed on television, but rather the effect of reality itself: the effect of the increasing use of science and technology in criminal justice for building legal cases. This broader "tech effect" (Shelton et al., 2006) may be viewed as a reflection of the assumption that DNA evidence has attained a level of "mathematical certainty" that transcends the merely subjective status of eyewitness testimony, confessions, and older forms of forensic evidence that purportedly generated "moral certainty."[2] Naïve jurors are by no means the only persons who hold such an assumption.

Although professional analysts remind us that the current state of the art in criminal investigation is far removed from the hi-tech science portrayed in *CSI,* they often make exception for DNA analysis. It is common to read that DNA fingerprinting is the "gold standard" in forensics; that it is a "truth machine"; that it represents "God's signature"; and that it produces "unassailable" or "infallible" evidence (McCartney, 2006: xii). These characterizations are not only made by enthusiasts for popular television shows in which crime scene investigators snare criminals with "science"; they are also made by science journalists, knowledgeable lawyers, and legal scholars. For example, in an article in the *New Yorker,* Jeffrey Toobin disabuses readers of the impression from *CSI* that trace evidence such as hairs or toenail clippings can be linked to individual suspects with absolute certainty. Toobin (2007: 33) mentions a forensic conference presentation in which Houck and FBI analyst Bruce Budowle reported results from a study that compared visual hair comparisons with the results from mitochondrial DNA (mtDNA) tests (Houck and Budowle, 2002). The mtDNA analysis indicated that in nine out of eighty cases in which forensic analysts using microscopes reported matching hairs, the samples could not have come from the same person. When Michael Risinger and Michael Saks (2003) first used that

---

2. Laudan (2003: 297) points out that the contrast between "mathematical" and "moral" certainty traces back to John Locke and John Wilkins (also see Shapiro, 1986). According to this distinction, mathematical certainty is subject to rigorous proof, beyond all doubt, whereas moral certainty *can* be doubted by a skeptic, but is nevertheless strongly supported by evidence and practically reliable. It is associated with an older, nonmathematical, conception of probability. Shapin (1994) elaborates how moral certainty was no less intrinsic to the role of witnessing in experimental natural philosophy than it was to the rise of standards of proof in law.

study to impugn the accuracy of microscopic hair comparison, Houck attempted to neutralize the criticism by refusing to accord epistemic priority to DNA: "Microscopical and mitochondrial DNA analyses of human hairs yield very different but complementary results, and one method should not be seen as 'screening for' or 'confirming' the other" (Houck, 2004). This response provoked some incredulity precisely because it has become virtually unthinkable to exhibit any confusion as to whether hair comparison is more trustworthy than DNA analysis.[3] For example, Toobin (2007: 34) quotes a comment about Houck and Budowle's study by law professor Margaret Berger suggesting that hair evidence should be excluded from criminal trials when there is "no mtDNA to back it up." Though used to counteract an alleged popular belief in the infallibility of forensic comparisons, such statements treat DNA analysis as itself infallible.

It is notable that such language is reminiscent of what commentators ascribed to the original form of fingerprinting nearly a century ago. At that time, the fingerprint was proclaimed to be "a God-given seal" (*Finger Print & Identification Magazine,* 1925: 10), "God's fingerprint language" (Brayley, 1909: 7), and "voiceless evidence" (International Association of Identification, 1920: 26). It also was declared that "until further research is made on the subject we shall assume that the Creator placed them there as a means of positive identification" (Murphy & Murphy, 1922: 9).[4] Such statements compound mathematical certainty with theological certitude. However, as we discuss below, the transcendent evidential quality once assigned to fingerprinting is now more often attributed to "DNA," as the older form of evidence has begun to lose its luster.

In this book, we question how DNA became a "truth machine." We chronicle a series of technological and legal changes starting in the mid-1980s, and focus on controversies in the courts and the science press. We also question the basis for the truth ascribed to the machine, and sug-

3. More recently, Houck and Budowle have invoked subtler arguments. They contend that the failure of hair comparison in these cases represents poor discrimination, not errors. This is perhaps an example of the "sociology of error" (Cole, 2005b): an account that presumes correct access to reality and reserves social and psychological explanations for instances in which that reality is obscured, misrecognized, or resisted by others (Bloor, 1976).

4. The religious terminology seems reminiscent of theological rather than mathematical certainty, but the two can be merged in a Platonic view of mathematics as a transcendent language of the gods.

gest that the probative value of DNA evidence rests on practices, circumstantial knowledge, and administrative assurances that are not essentially different from those that support other, less glamorous, forms of evidence. Finally, we examine the social and legal implications of the widespread belief that DNA evidence produces ground truth in the contentious world of criminal justice.

The Study

The research that went into this book took place over a fifteen-year period and involved a series of collaborations. The study began in the United States in the late 1980s and early 1990s, at a time when forensic DNA testing was being actively debated in the courts and in the pages of *Science* and *Nature*. It continued through the mid-1990s, at a time when the U.K. Home Office and Forensic Science Service were constructing a national DNA database (NDNAD). During that phase of the study, the televised O. J. Simpson trial in 1994–95 yielded a mother lode of videotaped testimony, transcripts, interviews, news coverage, and scholarly writing. Though the *Simpson* trial itself was less significant for our purposes than many of the other cases we covered, the mass of material greatly informed our understanding of the issues.[5] Our research continued through the late 1990s, when DNA testing was established as a new "gold standard" in forensic science. Finally, starting in 1999, we followed the story of how DNA testing provided leverage for challenging *other* forms of evidence, including fingerprinting.[6]

Because our research covered an extensive period while running concurrently with many of the events described, it combined historical and ethnographic methods of analysis. As historians of recent events, we collected and examined a substantial amount of documentary material, including published articles in forensic science and law journals and popular sources, unpublished records such as written forensic witness

5. For a series of studies on the O. J. Simpson trial and related issues, see the special issue of *Social Studies of Science* guest edited by Lynch & Jasanoff (1998).

6. Simon Cole, who had studied the history of fingerprinting, acted as an expert witness for the defense in several challenges to the admissibility of fingerprint evidence, and we draw upon his experience as a scholar and witness (see especially chapter 9). For a dialogue on the dilemmas raised by Cole's participation as an expert witness, see Lynch & Cole (2005).

statements and administrative reports, court transcripts, and written judicial summaries, and (in the *Simpson* case) videotapes of testimony.[7]

The ethnographic aspect of the study was "multi-sited" (Marcus &
Fischer, 1999) and "multi-timed," in the sense that it involved many sites
of investigation—principally in North America and Britain—over an
extended period of time. We relied upon interviews, observations, and
variable degrees of participation at laboratories and courtrooms, and
also performed a significant amount of documentary reconstruction.
Given limits of time, technical competence, and access to forensic laboratories, we relied upon interviews with laboratory staff at different
levels of administrative hierarchy, ranging from research administrators, midlevel officials, and staff scientists, to various grades of technician. We made repeat visits to selected facilities in the United States
and United Kingdom, where, in most cases, we were permitted to tape-
record and transcribe the interviews. For example, during several visits to the Forensic Science Service facility in Lambeth, South London,
we were permitted to take photographs (some of these are included in
chapter 7). We were given tours and demonstrations of equipment and
laboratory results, as well as reprints and references to technical publications. We also interviewed attorneys and expert witnesses involved
in the cases we studied, and we examined court transcripts, and (in the
Simpson case) video recordings of live television coverage. Some of our
analyses of trial transcripts and videotapes were informed by conversation analytic procedures, but we did not aim to make a technical contribution to that field.[8]

To a large extent, we tailored our methods to the specific themes presented in the chapters that follow. We borrowed liberally from our own

7. Some of these materials were collected with support from the National Science Foundation, for building an archive at Cornell University on the forensic issues surrounding the
O. J. Simpson case. The O. J. Simpson Murder Trial and DNA Typing Archive, 1988–1996
(collection number 53-12-3037), is housed at the Division of Rare and Manuscript Collec-
tions, Carl A. Kroch Library, Cornell University.

8. There is a large literature in conversation analysis. The two-volume set of transcribed
lectures by Harvey Sacks (1992), the founder of the field, remains the single best source of
insight into the workings of ordinary conversation. Still useful is the anthology by Atkinson and Heritage (1984). For conversation analytic studies of court discourse, see Atkinson & Drew (1979); Pomerantz (1987); and Drew (1992). For studies of credibility in tribu-
nals see Brannigan & Lynch (1987), and for studies on uses of textual materials to leverage
testimony, see Lynch & Bogen (1996), and Gonzalez (2006).

and others' earlier studies of laboratory discourse, the visualization and mathematization of scientific results, and the organization of testimony.[9] Although focused on specific legal decisions, legal briefs, trial transcripts, and selected videotapes of broadcasts of the O. J. Simpson trial, our study attempted to encompass a much broader scope than usually is the case for detailed studies of text and talk. The study focused on practices in different national contexts—principally the United States and United Kingdom, and to a lesser extent other European nations, the European Union, and Australia—over a twenty-year period.

We discuss cases from different national legal systems, and mention some differences between legal and criminal justice systems in connection with particular cases and technical developments. Our focus is mainly on the role of expert evidence in adversary legal systems, mainly in the United States and United Kingdom, but many of the developments we discuss are international in scope. For example, we noticed during the study that forensic scientists, criminal justice administrators, and lawyers were attuned to key cases and technical developments in other nations. International news coverage of notable cases and networks of forensic scientists transcend national borders. In addition, deliberate efforts were underway to "harmonize" DNA profiling systems in Europe and throughout the world. Consequently, while we have not made a systematic effort to compare national DNA profiling systems and practices, we believe that the developments described in our study apply broadly, if not universally.

The research required us to read technical publications, and to seek explanations from practitioners. One of us is trained in biology; two of us were given an elementary hands-on lesson in laboratory routines; one took part in a training session on crime scene investigations involving DNA evidence; and one of us was an expert witness in court cases involving fingerprint evidence.[10] Discussions among ourselves and with colleagues and informants helped upgrade our understanding of legal

9. See Lynch & Woolgar (1990) for a collection of studies on visualization in science, Goodwin (1994) on visual rhetoric used in court, and Jasanoff (1998) for an interpretation of courtroom materials informed by such studies.

10. Patrick Daly, an inspector with the British Transport Police, was a part-time PhD student at Brunel University when Lynch and McNally were there, and he was a good source of information and insight about police investigations and police views of DNA evidence.

and technical issues. The issues—particularly the probability issues—are widely acknowledged to be difficult, and indeed they were difficult for us as well as for judges and jurors. However, our aim in this book is not to give technical advice on legal, statistical, biological, or forensic matters, nor is it to present a comprehensive historical narrative of the controversy over forensic DNA profiling (for such a history, see Aronson, 2007); instead, our aim is to narrate a sociological history in which this controversy illuminates the fraught relationship between science and law. Each of the associated practices can be explored in much greater technical depth, but we shall not do so here.

Organization of the Book

The book is organized into ten chapters, which are interspersed with five "interludes" that present relatively compact accounts of some of the legal and technical issues featured at various points in the discussion. The overall structure is that of a narrative that runs through phases, starting in the mid-1980s when DNA "fingerprinting" was invented and continuing until the present time. At a more abstract level, the narrative is framed by phases of legal-scientific controversy about forensic DNA evidence. The controversy began in the late 1980s and ran for a few years before winding down in the mid-1990s and entering a "postclosure" phase in which "DNA" was invoked as a gold standard with which to challenge all other forms of criminal evidence, including fingerprinting.

Themes discussed in each chapter summarize issues that were brought into focus by particular cases and phases of controversy, as well as questions of general interest for science and technology studies (S&TS) and legal research:

- How do controversies open and close in a hybrid legal-scientific field?
- How do innovations disperse in time and place, and how is technical stability maintained?
- How does expert evidence relate to ordinary "commonsense" evidence?
- How are the credibility and scientific status of evidence presented and undermined in adversary discourse?
- How do quantitative measures influence the credibility of evidence?
- How is the probative value of evidence constructed and deconstructed?

These questions relate to topics that are often considered epistemological—objectivity, representation, quantification, and scientific truth—but
they are addressed in a case-specific way that draws our attention to historical contingencies and circumstantial judgments.[11]

The concluding chapter critically reflects on the story of how DNA
evidence was granted such extraordinary truth status. We recognize
the delicacy of such criticism: the currently popular view of "DNA" as
a "truth machine"—an unassailable basis for exposing the fallibility of
all other forms of criminal evidence—has been crucial for postconviction exonerations of death row inmates. While we applaud such exonerations, and recognize the immense effort that has gone into their achievement, we also believe that uncritical assumptions about the objectivity
and certainty of DNA evidence also are encouraged by those who would
lend scientific certainty and legal finality to death penalty convictions.
The chapters that follow underline a lesson that is, or should be, familiar to students of criminal justice if not to *CSI* enthusiasts: expert evidence does not *itself* determine guilt or innocence; instead, its probative
value depends upon circumstantial judgments that place the evidence
within a story of the case at hand. The exceptional credibility assigned
to DNA evidence does not exempt it from the judgments and contingencies that surround its use in criminal investigations. This is not to say
that DNA evidence is somehow less certain or more error-prone than
"real" science. Rather, the lesson from a large body of research on the
history and social organization of the natural sciences is that the credibility of research results in *every* empirical science rests upon judgments
and contingencies that can never be fully explicated. In addition to ex-
ploring how that lesson applies to a self-proclaimed science that is im-
planted within a legal domain, we believe that our study of protracted
disputes and deliberations about DNA evidence sheds light on how scientific credibility is achieved as a practical, rhetorical, and administrative matter.

11. Peter Dear (2001) uses the term "epistemography" to describe a detailed historio-
graphic or ethnographic orientation to the traditional topics of epistemology. This idea
has some kinship with a long-standing ethnomethodological program for "respecifying"
the recurrent themes in social theory and methodology (Garfinkel, 1991). Lynch speaks of
these themes as "epistopics" (recurrent topics of epistemological discussion and debate)
that tend to appear again and again in philosophy, history, and sociology of science (Lynch,
1993). Also see Garfinkel (2002: 181–82) for a discussion of "perspicuous settings."

Acknowledgments

This book would not have been possible without the help of many people. Sheila Jasanoff, Saul Halfon, and Arthur Daemmrich conducted interviews for a joint project with three of the coauthors of this book. In a related project, Jasanoff and Bruce Lewenstein helped compile textual and video resources for an NSF-supported project that produced an archive that is currently housed at Cornell University, Kroch Library. Rachel Dioso-Villa collaborated with Simon Cole on a study of the so-called *CSI* effect. Patrick Daly (who was both a graduate student in human sciences at Brunel University and an inspector with the Transport Police in London during a key phase of the research) provided valuable information and personal insight on police investigations, and Grant Donovan located a number of valuable media sources as an undergraduate assistant at Brunel University. We also are very grateful to Jay Aronson and Robin Williams for sharing draft papers and research materials with us from their own studies of DNA profiling. We are grateful to James Starrs for assistance in obtaining documents from the case *Thomas v. Trent.*

We also would like to express our gratitude to many of the individuals we interviewed—especially those with whom we had lengthy and repeated discussions. Our exchanges with them were more like discussions with colleagues than interviews with subjects, and we gained key insights as well as detailed information from them. We are especially grateful to William Thompson, who has written extensively about the technical and legal controversies in which he has been an active participant. Professor Thompson gave us many hours of illuminating conversation, as well as an extensive body of published and unpublished materials. Barry Scheck and Peter Neufeld took hours out of their very busy schedules for lively discussions with us about their prominent involvement in key cases and the Innocence Project. We also are grateful to Ian Evett, and Christophe Champod of the Forensic Science Service (FSS), for sharing information and insights with us, and to Chris Hadkiss of the FSS for giving us repeated tours of forensic laboratories in Lambeth, South London. Peter Donnelly provided us with extremely valuable verbal and written information about his involvement in cases discussed in chapters 5 and 6. Peter Martin, an independent consultant formerly with the FSS, met with us for hours to discuss European efforts to "harmonize" DNA profiling systems. Paul Gaughan, Paul Debenham, David Hartshorne, Nigel Hodge, Cathy Lee, and members of the Bedfordshire (U.K.) police provided us with valuable information about specific cases, technical developments, and trends in the United Kingdom. We also thank Peter D'Eustachio, Lawrence Kobilinski, and Victor Weedn for informing us about developments, problems, and cases in the United States. Members of the FBI labs graciously met with Kathleen Jordan for several hours on two occasions in the early 1990s, and gave us valuable information about early technical and administrative developments with DNA profiling. This list is incomplete, and we apologize to those we have not mentioned, or whom we have mentioned anonymously in the book.

We also would like to thank Catherine Rice, Christie Henry, and Pete Beatty of University of Chicago Press for guiding the manuscript through the review process, and the anonymous reviewers for their advice on how to improve the draft. Finally, we would like to thank Brian Lynch for invaluable help with preparing the graphics for the figures in chapter 4, and Andy Webb for figures A.1, A.4, and A.5.

The research that went into this book was partially supported by several grants: Michael Lynch and Kathleen Jordan's research received support from the National Science Foundation, Studies in Science, Technology & Society, "The polymerase chain reaction: The mainstreaming of a molecular biological tool" (Award no. 9122375, 1992–93); and a subcontract from the National Science Foundation, Studies in Science, Technology & Society, "DNA fingerprinting: Law and science in criminal proceedings" (Award no. 9312193, Sheila Jasanoff, PI, 1993–94). Michael Lynch and Simon Cole received support from a grant from the National Science Foundation, science and technology studies panel, "DNA profiling and fingerprinting: Relations between closure and controversy" (Award no. SES-0115305, 2001–04). Simon Cole also received support from National Science Foundation Grant no. IIS-0527729 and the National Institutes of Health Grant no. HG-03302. Michael Lynch and Ruth McNally received support from the Economic and Social Research Council (U.K.), "Science in a legal context: DNA Profiling, forensic practice, and the courts" (Award no. R000235853, 1995–98).

We gratefully acknowledge Orchid Cellmark, Ltd., U.K., for permission to reproduce figures A.2 and A.3, and LGC Ltd., U.K., for permission to reproduce figure A.6. We also acknowledge H. M. Collins and Sage Publications, Ltd., for permission to reproduce figure 2.1 from Collins (1999), "Tantalus and the aliens: Publications, audiences, and the search for gravitational waves," *Social Studies of Science* 29: 164; and *Nature* magazine for permission to reprint figure 2.2 from Eric Lander (1989), "DNA fingerprinting on trial," *Nature* 339 (14 June): 503. We also thank Wolters Kluwer for permission to reproduce figure 4.7 from Henry C. Lee et al. (2004), *American Journal of Forensic Medicine & Psychiatry* 15 (4): 273. We thank David Ashbaugh and CRC Press for permission to reprint figure E.2 from David R. Ashbaugh (1999), *Quantitative-Qualitative Friction Ridge Analysis: An Introduction to Basic and Advanced Ridgeology* (Boca Raton, FL); and David Stoney and the *Journal of Forensic Sciences* for permission to reprint figure E.3 from David A. Stoney & John I. Thornton (1986), "A critical analysis of quantitative fingerprint individuality models," *Journal of Forensic Sciences* 31 (4): 1187–1216. We also thank Chris Hadkiss of the FSS for permission to take and reproduce the photographs of the FSS Laboratory facilities in Lambeth in chapter 7; Mary Ann Kuharki, director of Prolife Across America for permission to publish the Prolife Minnesota Billboard in figure 9.1; and John Vanderkolk and the *Journal of Forensic Identification* for permission to reprint figure 9.2 from Vanderkolk (2001), "Levels of quality and quantity of detail," *Journal of Forensic Identification* 51 (5): 461–68.

Some of the interview excerpts, transcripts, and case materials discussed in this book are presented in previous publications by the authors. Materials and arguments in chapter 3 are presented in two papers: Kathleen Jordan & Michael Lynch (1998), "The dissemination, standardization, and routinization of a molecular biological technique," *Social Studies of Science,* 28 (5/6): 773–800; and Michael Lynch (2002), "Protocols, practices, and the reproduction of technique in molecular biology," *British Journal of Sociology,* 53 (2): 203–20. There is some overlap between chapter 4 and Michael Lynch & Ruth McNally (2005), "Chains of custody: Visualization, representation, and accountability in the processing of DNA evidence," *Communication & Cognition* 38 (3&4): 297–318. Some of the arguments and materials in chapter 6 appear in Michael Lynch & Ruth McNally (2003), "'Science,' 'common sense,' and DNA evidence: A legal controversy about the public understanding of science," *Public Understanding of Science,* 12 (1): 83–105. Arguments and some examples in chapters 8 and 9 appear in Michael Lynch (2004), "'Science above all else': The inversion of credibility between forensic DNA profiling and fingerprint evidence," 121–35 in Gary Edmond (ed.), *Expertise in Law and Regulation* (Aldershot, U.K.: Ashgate); and Simon Cole & Michael Lynch (2006), "The social and legal construction of suspects," in J. Hagan (ed.) *Annual Review of Law & Social Science* (Palo Alto, CA: Annual Reviews): 39–60.

CHAPTER ONE

A Revolution in Forensic Science?

This book is about a "scientific revolution" in forensic science. As with *the* scientific revolution of the seventeenth century, there may be some reason to doubt that it actually was a revolution or even that it ever occurred, but nevertheless it remains a compelling subject to write about (Shapin, 1996: 1). The revolution we chronicle has been widely heralded, and its banner is "DNA." In the context of forensic science, "DNA" is shorthand for a family of techniques and their analytical products. Common names for these techniques include DNA fingerprinting, DNA profiling, DNA typing, and DNA testing.[1] Technical names include multilocus probe (MLP), single-locus probe (SLP), and short tandem repeat (STR) systems. DNA fingerprinting, like its traditional namesake and other forensic identification techniques,[2] is used for comparing trace evidence found at crime scenes with a suspect's (or sometimes a victim's) evidence. But unlike comparisons of fingerprints, handwriting samples, threads, hairs, or bite marks, DNA profile comparisons use laboratory

---

1. The most widespread name, especially early in the history of the techniques, was "DNA fingerprinting." For reasons we shall explain, the analogy with fingerprinting later was downplayed, and other terms such as "DNA profiling," "DNA typing," and "DNA testing" were preferred. Sometimes, DNA profiling and DNA typing are distinguished, with the former referring to earlier techniques making use of restriction enzymes, Southern blotting, and autoradiography, and the latter referring to more recently developed techniques using the polymerase chain reaction as a key constituent. Several techniques have been used in the past twenty years, and there have been numerous variants on each technique. We briefly review some of the main techniques in interlude A.

2. Throughout this book, we use "forensic science" to describe practices performed by various public and private organizations, such as the Forensic Science Service in the United Kingdom, to assist police and, less often, defendants in criminal investigations. Forensic science is involved in other kinds of investigation, and more precise terms (such as "identification sciences") can be given to fingerprinting and DNA profiling.

methods and equipment that have widespread use in biological research and clinical testing. In public discourse about DNA evidence, laboratory science is associated with truth. In recent years "DNA" has become emblematic of a level of objectivity and certainty unmatched by any other mode of criminal evidence. In this book, we chronicle how forensic DNA testing attained such extraordinary status, and work out some of its implications for criminal justice. Of primary concern for us is how the attribution of scientific status to DNA evidence seems to have lifted it above and beyond the contestable status of other forms of expert and nonexpert evidence. By paying close attention to arguments by scientist-critics and lines of attack used by savvy defense lawyers during key controversies and legal disputes, we show that the credibility of DNA evidence largely rests on a fallible combination of technical, administrative, and legal practices. We argue that, in many respects, its credibility, as well as its occasional vulnerability to attack, arises from institutional practices that support or undermine the credibility of less glamorous forms of criminal evidence.

Although we focus on legal matters, our orientation largely draws from the field of science and technology studies. Sometimes called "science studies" or "science, technology and society," S&TS is a transdisciplinary field that combines the history, social study, and philosophy of science. In recent decades, research in the field has become notorious for its critical, often skeptical, approach to the "construction" of scientific facts, laws, and entities. Constructionist S&TS is frequently caricatured as a form of philosophical relativism that holds that everything under the sun (and, for that matter, the sun itself) is a figment of cultural imagination, not essentially different from fairies and goblins. Without going into the frequently rehashed arguments on the subject,[3] we can say, simply, that our orientation to the "deconstruction" of scientific evidence follows paths that have been blazed by scientists and lawyers who have had prominent roles in disputes about DNA evidence. In this respect, we pursue an ethnomethodological line of research (Garfinkel, 1967; 2002), meaning literally a line of research that investigates the varieties of "people's practices" (ethno-methods) that establish facts on the ground. Rather than seeking to establish that DNA evidence is, in the

3. Opinionated attacks on constructionism are found in Gross & Levitt (1994), Gross et al. (1996), and Sokal and Bricmont (1998). See Hacking (1999) for a broad and well-informed philosophical critique of constructionism, and Lynch (1993) for a more focused, critical examination of social constructionism in S&TS.

final analysis, a social construction, we examine the interactional and administrative practices through which lawyers, experts, and nonexperts build up and break down its credibility. In this introductory chapter we set up our chronicle of the recent history of forensic science by critically discussing a prominent conception of a "paradigm" shift in that field. This allows us to introduce a distinct view of the construction and deconstruction of expert knowledge in criminal justice systems.

A Paradigm Shift?

DNA fingerprinting has been said to mark a "paradigm shift" in forensic science. Though misleading in many respects, the analogy with a new scientific paradigm is interesting for those of us in the field of science and technology studies who grew up with Thomas Kuhn's (1970 [1962]) *Structure of Scientific Revolutions* as our canonical text. Kuhn's historicophilosophical account of scientific revolutions itself had historic, revolutionary significance for the S&TS field. Especially significant was his conception of paradigm shifts in the basic or "pure" natural sciences: grand conceptual sea changes in astronomy, physics, and chemistry. According to Kuhn, a new paradigm is no mere novelty: discoveries that cumulatively build upon prior discoveries are emblematic of "normal science," whereas a new paradigm requires a much more radical rupture with the past—a level of novelty that is at first resisted, because it defies conventional wisdom. A new paradigm overthrows the old: heliocentrism overthrows geocentrism; chemical theory overthrows phlogiston and caloric; relativity replaces classical mechanics. Importantly, and controversially, Kuhn (1970 [1962]) argued that competing paradigms were "incommensurable." Referring to the much-abused alternating figure of the duck-rabbit in gestalt psychology (also see Hanson [1961] for a similar argument using an antelope-bird alternating figure), Kuhn suggests that a scientist's alignment with one paradigm rather than another is not a rational choice; instead, it is more like the way a viewer sees the alternating figure *as* a duck or *as* a rabbit. But, unlike a viewer who can freely alternate between seeing the figure as a duck and then as a rabbit, scientists tend to have their feet firmly planted in a single paradigm. The nexus of theoretical commitments and the communal networks that make up and support the paradigm commit subscribers to a standpoint from which they simply cannot see the sense of the alternative universe.

The transition between paradigms is not a sudden "switch," but more of a gradual migration or generational change within a fractiously divided community of practitioners. Once established, a new paradigm generates its own "normal science" tradition, in which novelty is framed by no-longer-questioned presumptions that were controversial during the prior period of revolutionary change.

To speak of DNA profiling in forensic science as a new paradigm is hyperbolic. For all its advantages as a method of criminal identification, it does not represent a new paradigm in Kuhn's sense of a grand conceptual shift that overthrows existing theory and normal science. It does not surpass a prior normal science tradition with something completely different. Instead, as Saks and Koehler (2005: 893) would have it, the revolution is a matter of rising *to,* rather than rising *above,* normal scientific status. Speaking of pre-existing craft traditions of forensic science, they observe,

> In normal science, academically gifted students receive four or more years of doctoral training where much of the socialization into the culture of science takes place. This culture emphasizes methodological rigor, openness, and cautious interpretation of data. In forensic science, 96% of positions are held by persons with bachelor's degrees (or less), 3% master's degrees, and 1% Ph.D.s. . . . When individuals who are not steeped in the culture of science work in an adversarial, crime-fighting culture, there is a substantial risk that a different set of norms will prevail. (Saks & Koehler, 2005: 893)

By treating normal scientific status as a positive historical goal, Saks and Koehler imply that, until now, forensic science has been a *subnormal* science.[4] It has subnormal status, not so much in comparison with Kuhn's normal science, but with an idealized conception of "real" science according to which some forensic disciplines appear to be a loose array of police crafts graced with an aura of science.[5] According to a

4. Kuhnian normal science is an established, stable, and cumulative research tradition that develops in the aftermath of a scientific revolution. Saks and Koehler's normal science is a general status attained by a specialized practice. Their view is akin to general conceptions of science expressed by the U.S. Supreme Court in rulings about the admissibility of expert evidence. Normality in this sense is an aspect of public credibility (Shapin, 1995). It is also subject to legal stipulations and public administration.

5. Kuhn sometimes spoke of a preparadigm phase in the history of a science, and he noted that some disciplines (such as many in the social sciences) remain stuck in a pre-paradigmatic phase. These differ from the forensic sciences in that the latter are often

conventional (and rather whiggish) history,[6] for more than a century practitioners of various criminological crafts for identifying individuals (commonly called forensic sciences today) have called themselves "scientists," and novel techniques such as fingerprinting and tool mark and bite mark analysis were heralded as scientific breakthroughs. But, as Saks and Koehler argue, these traditional practices have not yet attained (and may never attain) the normal scientific standing enjoyed by DNA typing. Accordingly, DNA typing rose to the status of normal science because it applies knowledge "derived from core scientific disciplines," and offers "data-based, probabilistic assessments of the meaning of evidentiary 'matches'" (Saks & Koehler, 2005: 893).[7]

In support of Saks & Koehler's view, it can be said that, notwithstanding dramatic portrayals in currently popular television shows, forensic science has long held secondary, and even dubious, standing when compared with "high" science. More than forty years ago, Paul Kirk, a prominent forensic analyst, wrote (referring to "criminalistics," the specialized forensic practices used in criminal investigations), "It seems fair to state that criminalistics may now be considered a science in its own right, but that it lacks at this time the full development that will allow general recognition" (Kirk, 1963: 238). According to a much more recent account, this characterization still holds:

"mature," stable practices, and some of them (such as latent fingerprint examination) are characterized by a high degree of consensus (see Cole, 1998). From the point of view of university-based sciences, they are subnormal because of their lack of connection to scientific disciplines, and scientific culture more generally, and their strong association with police work and police organizations.

6. The "whig interpretation of history" (Butterfield, 1931) is a critical label affixed to historical accounts that treat the past as a condition for the emergence of present-day democratic institutions and religious tolerance. Kuhn and many others criticize histories of science that treat present-day textbook knowledge as a basis for understanding and evaluating earlier eras of science and natural philosophy. In this sense, Saks and Koehler present a whiggish (or, in Wilson & Ashplant's [1988a, b] terms, a "present-centered"]) interpretation of forensic science that forgets the extent to which claims to scientific truth and associations with established academic sciences have characterized many earlier eras in the history of forensic and identification science (Golan, 1999, 2004a).

7. This differs from saying that the older forensic sciences are preparadigmatic in Kuhn's sense of having yet to attain a consensual paradigm that unifies the discipline and overrides intractable conceptual difficulties and interminable disputes. Some forensic science professions (exemplified by latent fingerprint examiners [Cole, 1998]) have long enjoyed impressive internal consensus and credibility in the courts. Saks and Koehler use an idealized image of DNA profiling—its use of probability measures and its adoption of molecular biological techniques—as a basis for making invidious comparisons with older forensic sciences.

The situation, unfortunately, hasn't changed much since Kirk described it 40 years ago. Technical innovations, especially in instrumentation, have far out-distanced any attempts to establish a theoretical framework for criminalistics as an autonomous discipline. The dearth of, and recent alarming decrease in, academic programs in forensic science only serve to underscore the dangers of analysis without a framework for thoughtful interpretation. (Inman & Rudin, 2001: 64)

Unlike the discovering sciences (still epitomized by physics, though with molecular biology in the ascendancy), forensic science involves routine procedures, limited objectives, and little or no orientation to discovery. It is not rocket science. It is related to what Steven Shapin (2004) has called "sciences of the particular": practical research geared to tangible objectives set by other social institutions.[8] More conventionally understood, it is an "applied" science:

8. Routine clinical trials of pharmaceuticals can be viewed as a "science of the particular" that is subjected to regulatory "scientific" norms articulated, administered, and arguably enforced, by government and corporate organizations. Kuhn was sometimes criticized for portraying normal science as unexciting and routine work, but the routines of the clinical trial or forensic laboratory are far less creative than anything Kuhn deigned to include in his conception of normal science. Increasingly, clinical trials are performed by contract research organizations (Angell & Relman, 2002; Mirowski & Van Horn, 2005) and sponsored by pharmaceutical companies. They are regulated (perhaps not heavily enough) by government agencies, and their production is highly normalized. Even the scandalous practice of ghostwriting in pharmaceutical research—so vividly exposed by David Healy (2004)—points to a high degree of normalization through which the authorship function becomes reduced to an endorsement separate from the arts of writing (a perverse twist on the trends discussed by Biagioli & Galison, 2003). They are not internally governed normal sciences in Kuhn's sense, but are instead monitored and regulated by government agencies. Steven Epstein (1996) uses the term "impure science" to describe clinical trials, but he means more than that they represent a corruption of the ideals of "pure science." Epstein describes how AIDS activist groups effectively lobbied regulatory agencies and pharmaceutical firms to accelerate clinical trial protocols in the interest of getting experimental drugs to patients. The fact that standard methodological steps and controls were bypassed was not, for Epstein, indicative of shoddy science; rather, it pointed to how the methodology of clinical trials answered to multiple social interests and agendas. Epstein's conception of "impure science" is relevant to the way DNA evidence is handled, presented, and evaluated by a range of agents and agencies besides scientists. However, the "subnormal" status of (some) forensic science has to do with an invidious comparison with a "paradigm" that is supposedly rooted in accepted scientific concepts and procedures. Another partial cognate is Silvio Funtowicz and Jerome Ravetz's (1992) term "post-normal science" (see also Ravetz, 2006). This term describes sciences, such as those associated with assessing possible causes of global climate change, that address urgent problems under contentious and uncertain conditions. Although forensic sciences can involve uncer-

> Little of what goes on in forensic science resembles the classical description of how science develops theories, tests hypotheses and revises its ideas and understandings. This is partly because the scientific method is a description of pure, or basic, science (knowledge building), while forensic science is an *applied* science. (Thornton & Peterson, 2002: 13; emphasis in original)

However, even its status as an applied, as opposed to basic, science can be questioned. After all, which branches of basic research do the fingerprint examiner, tool mark examiner, and handwriting analyst apply? Although latent print examiners' visual comparisons of fingerprints may be informed by self-tutelage in embryology, it is difficult to see how "science" governs the comparison and evaluation process itself. At their best, these "sciences" are diligent practices of collecting, reading, and comparing trace evidence. In some disciplines forensic scientists have historically risen from the police ranks, and their higher education credentials are limited. Their skills are learned on the job, rather than acquired through advanced education.

Although its nominal status as science may give forensics some authority in courts of law, its association with the natural sciences also is a burden. More conspicuously than university research, or even some corporate research, forensic science lacks autonomy: it is answerable to initiatives and evaluations that are not controlled by the community of practitioners.[9] Administrative quality assurance/quality control (QA/QC) regimes are designed by associations of lawyers and scientists who advise government agencies on how to normalize a science that is not trusted to stand on its own feet. Layers of internal staff and various external agencies and institutions are empowered to review and evaluate the research and its products. Such QA/QC recommendations incorporate scientific ideals (double-blind testing, peer review, and the characteristic format of research reports) that are enacted and supervised

---

tainty in high-stakes situations of decision making, notwithstanding *CSI* drama they tend to be routinized and performed by anonymous agents.

9. Some forensic laboratory directors, for example, report to nonscientists in the law enforcement chain of command, a practice that can raise questions about how scientific autonomy is meant to function in such contexts. Forensic science is not heavily regulated when compared with industries such as clinical laboratory testing, or even pet food manufacture. With the increased salience of intellectual property concerns, human subjects committees, and protections against charges of intellectual misconduct, university scientists also are becoming subject to legal and regulatory norms, formal regulations, and reviews (Myers, 1995; Packer & Webster, 1996).

within bureaucratic administrations.[10] The regulations and standards often mechanically invoke popularized Popperian (pop-Popperian, we might say) notions of testing, testability, and falsifiability.[11] By implication, when forensic science adheres to such standards (peer review, reliability, testability, etc.) it becomes "normal," though as we shall see, the legitimating function of such standards in courts of law is independent of the extent to which they are actually followed, in practice.

Earlier "Paradigms"

The current tendency to treat DNA typing as an exceptional and strongly scientific form of evidence obscures the fact that to a large extent its institutionalization has followed a path set by earlier techniques which are now deemed to be subjective and prescientific. When Alphonse Bertillon's "signaletic" system was adopted by the Paris Prefecture of Police, it was used to profile criminals who were in police custody. A filing system was set up in which standardized information was presented on cards. It was possible to search the cards to find individuals with identical measurements. This property was held to be a crucial feature of campaigns to apprehend "recidivists"—repeat offenders who often literally and figuratively masked their identities. Bertillon argued that his measures were more trustworthy than photographs, and in France and many other nations the system he established lasted well into the twentieth century, decades after fingerprinting had become the preferred identifi-

10. Robert K. Merton's (1973) famous schema for the norms of science (universalism, disinterestedness, communal property rights, and organized skepticism) supported the argument that scientific objectivity can be optimized only under conditions of autonomy from political, religious, and economic interests and authorities. The schema is often criticized for offering rhetorical support for the view that science is cognitively exceptional, free from politics, and uncorrupted by commercial interests. What such criticisms tend to miss is that Merton's scheme borrows heavily from Weber's (1968: 956ff.) ideal type of bureaucracy, with its emphasis on conduct in accordance with rules (rationality), impersonal criteria of evaluation, specialization of function and circumscribed expertise. These ideal features support a conception of what might be called "administrative objectivity"— the claim that administrative decisions follow from legitimate rules rather than personal biases.

11. This is, of course, a reference to Sir Karl Popper's philosophy of science (see, for example, Popper [1963]), though as Susan Haack (2005) argues, the popularized versions of falsification that legal actors (including U.S. Supreme Court justices) invoke tend to be watered down and conflated with incompatible philosophical ideals.

cation method in Britain, India, and Argentina. Although Bertillonage is now archaic as a method of identification, it often is credited with initiating a scientific era of criminal forensics. Ginzburg (1989) links it to an emergent "evidential paradigm," which ranged across numerous fields of investigation and was characterized by an effort to reconstruct individual identities from mundane signs and traces, rather than merely inferring character from them. Nonetheless, efforts to discern character from somatic markers remained—and still remain today through popular and professional interest in a genetic basis for crime (Allen, 2004)—closely related to identification.

The crucial advantage of fingerprinting was not so much the individuality of fingerprints or the accuracy of the system, but rather its ability to lower and defer costs by enabling identification data to be recorded by relatively unskilled labor, while requiring skilled labor (the indexing of the prints) at a centralized "back end." Moreover, fingerprint examination not only could be used to counteract a recidivist's aliases and disguises, it also could be used to connect an individual to a crime scene. Newly professionalized "fingerprint experts" promoted the idea that matches between latent marks (recoverable traces left at a crime scene) and rolled prints (inked prints of each finger taken under controlled conditions) could be determined with absolute certainty.

Until very recently, dactyloscopy was not a digital technology, except in the punning sense of being about fingers.[12] However, the older meaning of "digit" also signaled an advantage, as many crimes involve acts of touching. Much in the way that the human face is a meaningful surface of identity in social interaction (as noted by sociologist Erving Goffman (1959), in his classic essay on "face work"), the ends of the fingers are an interface with the material surfaces with which a person interacts. As Ginzburg (1989) observes, fingerprints are usually inscribed inadvertently, as byproducts of action rather than expressions of intention, and this enhances their forensic credibility. In Goffman's (1959: 2ff.) terms, they are signs "given off" rather than "given," which are legible for an indefinite time after they have been inscribed. Prior to their use for purposes of criminal identification, fingerprints were sometimes used

12. Although fingerprints are now recorded and filed in digital form, latent fingerprint analysis continues to require a human examiner to assess possible matches turned up by a computerized database search.

as equivalent to signatures: direct bodily impressions onto a paper surface. What is so useful about fingerprints for purposes of criminal identification is that, for the most part, "finger writing" is inscribed invisibly and unintentionally, whenever the tips of the fingers contact a surface leaving oily traces of ridge and pore patterns. The technique of exposing and collecting the latent fingerprint sets up their codification and comparison. Footprints are an alternative form of bodily trace, and forensics makes ample use of them, but the practice of wearing shoes tends to blunt the discriminatory power of the footprint. In modern urban life a footprint is a trace of a removable manufactured object, which may show unique patterns of wear, but it is a different category of trace than a fingerprint or handprint.[13]

The technology of latent fingerprint analysis is aided immeasurably by features of the built environment in the modern world. Our fingers continually come into contact with flat surfaces composed of, or covered with, glossy paint, plastic, glass, polished metal, and other substrates that collect and maintain the integrity of the oily imprint. Moreover, our environments, and our ways of life within them, have standardized features that facilitate inferences about the places and circumstances of normal and suspicious entry and exit.

As Paul Rabinow (1996b) observes, Francis Galton, who coined the word and founded the program of "eugenics," initially hoped that fingerprint patterns would provide a meaningful index of hereditary lineages. Galton failed to find significant ethnic correlations for fingerprint patterns, although other researchers were able to establish such links. Despite his own disappointment, Galton, together with Edward R. Henry and Juan Vucetich, made a significant contribution to fingerprinting by developing a rudimentary classification system for fingerprint patterns, which proved useful not for tracing heredity but for individual identification. The widespread adoption of fingerprinting as a method of criminal investigation was based on the dogma that no two individuals have the same dermal ridge patterns, together with the assumption that professional examiners were capable of making error-free fingerprint identifications. Fingerprint evidence was contested in trials early in the twentieth century, and fingerprint experts sometimes had to conduct in-court

13. Shoe prints can be highly significant, as viewers of the O. J. Simpson trial may recall. The prosecution presented evidence of bloody footprints recovered from the crime scene that matched a rare kind of shoe. Simpson denied owning such shoes, but during the trial he was shown wearing them in a televised broadcast.

demonstrations to persuade juries of their ability to distinguish matching from nonmatching fingerprints (Cole, 1997; 1998), but by the 1930s
the practice had become nearly unassailable in the courtroom.

Fingerprint identification bureaus were developed during the 1890s
in India and Argentina, and in Britain in 1901. Scotland Yard made the
first latent identification in Europe in 1902 in the Denmark Hill burglary
case. It made its first fingerprint conviction in a murder case in 1905, fin-
gerprinted its first corpse in 1906, and established its first legal precedent
("stated case") in the *Castleton* case in 1909, in which the only evidence
offered by the prosecution was that fingerprints on a candle matched
those of the accused (Lambourne, 1984: 67–86). Fingerprint evidence
and the experts who vouched for such evidence were uncertain and contentious matters in these early trials (Joseph and Wilson, 1996), but by
the 1930s it was widely accepted that when a fingerprint expert declared
a match between two prints, such testimony provided unambiguous evidence of identity. Courts were willing to accept that no two sets of fingerprints are exactly alike, and they reasoned from this assumption that
latent print identification must be "reliable." Subsequently, juries were
willing to convict on the basis of fingerprint evidence alone. Fingerprinting became a routine part of police procedure, and administrative systems were devised for collecting, filing, and analyzing fingerprints. In
1948, the first mass fingerprint screen in Britain was undertaken, and
in the same year police powers to take fingerprints were extended under the Criminal Justice Act (section 40), which permitted the police to
take fingerprints under court order from a person charged before a magistrate with a criminal offence without his or her consent (Lambourne,
1984: 123–29). Police were granted the power take fingerprints from suspects without a court order under the Police and Criminal Evidence
Act 1984.

The Credibility of Latent Fingerprint Examination

Even if one assumes that no two fingerprints are alike, latent print identification is really more about likeness than about unlikeness. The question is not whether individuals have unique friction ridge skin on their
fingertips; the question is whether latent print examiners can accurately
determine whether two unlike prints (for example, a "latent print" or
"mark" found at the scene of a crime, and an inked print taken from a
suspect) derive from a common source finger. The two prints will not

be identical: the latent print is likely to be partial, less distinct, and distorted by variations in substrate and pressure. Similarly, even if one assumes that prints from different persons will never be alike, this does not guarantee that the differences will be correctly detected in every case. Consider, for example, handwriting: one may assume that no two people have exactly the same handwriting, but this does not necessarily lead us to trust the handwriting expert. It took a few decades for fingerprinting to be accepted in the justice systems in Europe, North America, South America, India, and elsewhere, but eventually the courts accepted the dogma that no two individuals (even identical twins) have the same fingerprint patterns, and that fingerprint analysis was a reliable expertise. This dogma alone was not sufficient for establishing the power of fingerprinting. At least as important was the development of a distinct forensic profession of latent fingerprint examiners whom the courts trusted to declare whether or not two prints matched.

In most countries, a "points system" was devised for declaring matches. Although quantitative, in the sense that a specific number of points (16, 12, 8, etc.) provided a threshold for declaring a match, the number was not probabilistic, since examiners claimed that it was *impossible,* and not highly *improbable,* that fingerprints from different persons would match on the requisite number of points. This threshold system would mark a significant difference between fingerprint analysis and forensic techniques such as blood group analysis and DNA profiling which employed probability estimates. Scotland Yard abandoned its points system in the year 2000, partly because of its lack of probabilistic meaning, and for some time before that points had not been consistently used in the United States. Fingerprint experts who appeared in court continued to declare whether or not two fingerprints matched, without giving probability calculations. This was an either-or judgment, excluding the middle. To avoid ambiguity, the fingerprint examiner profession publicized the idea that examiners would deem "inconclusive" any and all fingerprint evidence that lacked sufficient clarity to make an absolutely certain judgment. Whether or not they actually adhered to this practice, examiners had such credibility as expert witnesses that, in the words of Anne Rafferty QC, chairman of the Criminal Bar Association in the United Kingdom, "If fingerprint evidence emerges when you are defending a client, then you tend to put your head in your hands. There is not really a question mark over it" (quoted in Grey, 1997). The testimony of a fingerprint expert was rarely challenged.

An Inversion of Credibility

When DNA evidence first was used in criminal investigation in the late 1980s, analysts spoke enviously of the "absolute" identifications provided by fingerprints. DNA "fingerprinting" borrowed the name of the established technique, but in the courtroom DNA matches could not be declared with the same absolute certainty—it was necessary to give probability figures. However, in the past several years, fingerprinting has begun to undergo a crisis. Two trends seem to be responsible for this crisis: a kind of feedback effect from DNA profiling, and (in U.S. federal law) changes in recommended legal standards for admitting expert evidence (Mnookin, 2001).

The legal history of DNA profiling in England and the United States shows a pattern of rise, fall, and rise again: rapid acceptance in the late 1980s, followed by challenges in the courts and science press from 1989 through the mid-1990s, followed by renewed acceptance at a stronger level. The initial credibility of DNA profiling can be attributed to an uncritical acceptance by scientists and the courts, while the later acceptance is often justified by noting that DNA profiling had "passed the test" and become a reliable scientific technique.[14] "DNA" passed the test so well that it became a standard for assessing the probative value of all other forms of criminal evidence, including its former namesake, "fingerprinting." Borrowing terms for describing controversies in science and technology studies, we use the term "postclosure" to describe this late phase in the establishment of a once-controversial technical innovation. Not only was the controversy settled, it set up a sequel in which "DNA" was used as an unquestioned model for challenging the epistemic status of *other* previously settled forms of evidence (particularly, and perhaps most interestingly, fingerprinting).

In the mid-1990s, just as the controversy about DNA "fingerprinting" was winding down, fingerprinting began to seem less "scientific" or even "old-fashioned" (McCartney 2006: 185). The relative credibility of the two technologies seemed to be undergoing inversion. The unquestioned credibility once assigned to fingerprinting was now assigned to "DNA," and the credibility of fingerprinting suffered by comparison.[15]

14. For popular accounts that herald the acceptance of "DNA," see Levy (1997) and Dawkins (1998). The eventuality of acceptance was widely forecast, even before it occurred, and announced tendentiously, most famously by Lander & Budowle (1994).

15. *Credibility* is a key legal concept that has deep resonances with a sociological approach to truth and truthfulness (Shapin, 1995). Credibility is commonly distinguished

Law, Science, and Society

In recent years there has been a convergence between science and technology studies (S&TS) and law and society (L&S) studies. This convergence reflects the fact that scientific and technical experts are increasingly prominent in legal disputes about health, safety, and environmental regulation. New forms of biotechnology challenge existing legal distinctions in areas such as intellectual property law, raising questions about whether genetically engineered organisms, or genes themselves, can be patented. Scientific knowledge and novel technologies challenge legal definitions of life, death, and viability. Such innovations sometimes reinforce traditional distinctions and regulations, but they also can be used to challenge or circumvent them. Even the practice of science itself is deeply penetrated by legal norms, regulations, and sanctions (Myers, 1995; Packer & Webster, 1996; Kevles, 1998). At a more abstract level, law and science offer distinct, but historically intertwined, procedures and traditions for determining facts, testing truth claims, and resolving disagreements (Jasanoff, 1995; 2005). When law turns to science or science turns to law, we have the opportunity to examine how these two powerful systems work out their differences.

In addition to offering a fund of topics for investigating the entanglements of science, technology, and law in modern (or not-so-modern) societies, research on law in science and science in law offers distinct methodological advantages. Mariana Valverde (2003b) articulates one such advantage very well, when she observes that ethnographic "science studies" of day-to-day laboratory practices enable sociologists and cultural anthropologists to gain insight into the unsettled arena of "science in action" (Latour, 1987), and to develop a more dynamic, contingent picture of research practices than was ever conveyed in philosophical and methodological writings.[16] Valverde then observes that the formal, public record of legal disputes yields a fund of material that is comparable to what the ethnographer turns up through extensive on-site investigation:

---

from truth, but in actual situations of judgment there may be no practical means available to distinguish the "appearance" of truth and truthfulness from the actual truth of what is said or conveyed. Nevertheless, as Shapin observes, various "proxies" are used in actual situations as guides for assessing truth and truthfulness.

16. Ethnographic studies of laboratories include Latour & Woolgar (1986 [1979]); Knorr (1981); Lynch (1985a) and Traweek (1988).

First, questions of evidence and of authority are often explicitly contested, with the contestations often forming part of a court's public record and/or going on in the public setting of the courtroom. Thus, unlike science studies scholars, who must gain access to social interactions that are not mentioned in scientific papers and that do not take place in public view, legal studies scholars have vast amounts of material—affidavits, trial transcripts, etc.—that can readily be analyzed, and we have automatic access to at least some of the struggles about what counts as evidence and who counts as an authority waged in legal settings. While recognizing that interviews and ethnographic methods can offer very important insights, and acknowledging that "the public record" is the product of a whole set of prior practices which are either black boxed or simply invisible, nevertheless, it seems to me that sociologists and anthropologists of knowledge should not neglect to explore processes that are readily accessible either through court observation or through the written public record of legal proceedings. (Valverde, 2003b: 1)[17]

From the side of S&TS, legal arenas not only offer the opportunity to *apply* an existing fund of concepts and research strategies to a significant institutional site, they enable us to *explicate* central concepts (including "science" and "technology") that are distinctly problematic. The very establishment of a field of S&TS might suggest that its members have specialized expertise about science, technology, and expertise itself (Collins & Evans, 2002). No doubt, members of the field are likely to be widely read on those topics, and to have thought deeply about them. And, in some cases, they can produce unusual insights and engage in sophisticated arguments about what is or is not a science, how technology develops, and what counts as expertise. However, those of us who turn away from the dream of a (social) science of science have less interest in developing better glossaries of the key words in our discipline than of understanding why and how those words prove so troublesome, not only for us, but more importantly for the parties whose actions and interactions constitute the institutions we study. As we shall see throughout this book, "science" and "expert" are contested terms in the law courts. In this study we are not interested in settling what those words

17. This quotation was taken from a conference paper (Valverde, 2003b) that was published under the same title (Valverde, 2003c). The published version did not include the quoted passage. For related discussions, also see Valverde (2003a; 2005).

mean in a context-independent sense. Instead, we are interested in how participants in the adversary system of the Anglo-American courts construe science and expertise.[18] American and English courts have a long history of wrestling with how to handle expert evidence (Golan 2004a, b). Such pragmatic construals may or may not compare well with philosophical or sociological conceptions of science, but the important point is that they *constitute what counts* in the legal system as science and expertise. Such constitutive work and its pragmatic effects are phenomena for this study, as we chronicle how DNA profiling was established as a supralegal form of scientific evidence.

Deconstruction

In S&TS writings about legal disputes, the word "deconstruction" is sometimes used to convey a sense of what happens to technical claims in an adversary system. "Deconstruction" in this context has only a superficial resemblance to the literary-philosophical method originated by Jacques Derrida.[19] The S&TS version of deconstruction tends to be more mundane,[20] as it refers to argumentative efforts to undermine scientific evidence by questioning the competence of the relevant scientists and the adequacy of their methods. Such arguments can occur among co-

18. Arguments in favor of treating social science concepts and variables as, in the first (and sometimes last) instance, ordinary words used in the social situations studied have a long history that is associated with pragmatist and interactionist approaches. See Mills (1940), Winch (1958), and Sharrock & Watson (1984) on the concept of "motives" in social psychological explanations of human actions. Also see Garfinkel (1967; 1991), and Garfinkel & Sacks (1970), and Blumer (1969) for general outlines of the sociological implications of such a reorientation toward basic concepts.

19. Jasanoff (1992: 348) provides a succinct definition of the mundane version as "nothing more arcane than the pulling apart of socially constructed facts during a controversy." Latour & Woolgar (1986 [1979]: 179) first used the term "deconstruction" to describe the way disputing scientists tend to take apart rival factual claims and methodological procedures: "The 'reality out there' . . . melts back into a statement, the conditions of production of which are again made explicit." See Jasanoff (1995: 211–15) for an example of how the term tends to be used in S&TS studies of science in law; for a critical exchange about the relation (if any) between the more mundane sociological application and Derrida's original, see Fuchs & Ward (1994) and Agger (1994).

20. To say they are "mundane" is not to suggest that they are uninteresting. Melvin Pollner's (1974; 1975; 1987) conception of "mundane reason" in traffic courts, psychiatric clinics, and other organizational settings illuminates how ordinary discourse about what happened (or *can* have happened) in a mundane world provides an empirical counterpoint to erudite metaphysical debates about the "problem" of reality.

workers in a laboratory, between rival scientists at a conference, or through an exchange of letters in *Nature*. The arguments can become extremely elaborate when they delve into previously unmentioned details of laboratory technique and possible sources of noise or artifact. Harry Collins likens scientific controversies to "breaching experiments," such as those deployed by ethnomethodologist Harold Garfinkel (1963; 1967) to illuminate taken-for-granted features of ordinary social activities. Like a "breaching experiment" in which Garfinkel's (1967: 42ff.) students persistently question intimates about the meaning of their routine actions and expressions, a scientific controversy in which routine laboratory practices are called to account can momentarily "explode" the quiet agreement that operates in a settled field of action. For sociologists, a breaching experiment is a research tool and didactic device for exposing taken-for-granted norms, background knowledge, and tacit practices. Studying a scientific controversy does not require an intervention on the part of the sociologist to "explode" the quiet agreement characterizing a form of life; instead, in heated exchanges among participants minor flare-ups, and sometimes major explosions, enable us to gain a vivid appreciation of the extent to which contrasting assumptions, leaps of faith, previously undocumented skills, and competing interests are featured in scientific practice but in a way that tends to be hidden when members of a field present a unified front. Two, partly related, senses of the word "explode" are relevant in this context: one is a noisy spectacle that disrupts quiet routine and peaceful coexistence, and the other an "exploded view" of inner workings, akin to diagrammatic conventions for showing the parts of a machine or the anatomy of a body. Of the two, the second explosion, more analytical and less spectacular, is more salient for the present study.

The metaphor of "opening the black box" expresses one of the principal analytical aims in studying contentious science. The expression derives from cybernetics. In a technical variation on the gestalt theme of figure-ground, an engineer can draw a "black box" around certain components in a circuit diagram, in order to place the details of those components in the background. In the S&TS field, "black box" has become a common term of trade with diverse uses. Often it refers to taken-for-granted, presumed, or collectively "forgotten" aspects of the social history of a scientific or engineering innovation. A different sense of "black box" has to do with the transparency or opacity of a device for a specific community of users. The components under the hood of a contemporary automobile, for example, are far less transparent (both conceptually and mechanically) to an average owner, or even garage mechanic, than were, for example, the "guts" of a 1960s-vintage Volkswagen Beetle. Similarly, the electronic components of a contemporary radio not only are inaccessible to view, they are inaccessible to repair, unlike the tubes and wires of a vintage radio. Although black boxing holds many advantages for producers and users alike, it also can be problematic when things do not work as planned, or when facts and artifacts are incorporated into nonstandard settings.

Bruno Latour (1987) and his large following in S&TS have developed a theoretical agenda around the theme of opening up or deconstructing black boxes. This theoretical interest differs from the practical aim of reverse engineering in which an intact artifact is disassembled in order to reconstruct how it was put together. For Latour and others the scope of such reverse engineering is unusually broad and multifaceted. In addition to delving into how facts and artifacts are technically constructed, the S&TS deconstructionist explores broader horizons, interests, competencies, and networks. The didactic aim is to demonstrate how "heterogeneous engineering"[21] tightly weaves together social networks and material components, to the point of challenging the salience of the initial social-technical distinction. The lesson aims both to elucidate broader historical and social horizons of science and engineering, *and* to reveal how scientific and engineering innovations embody social actions and social forces in ways that escape the conceptual apparatus of standard social theories. In addition to being an analytical tool for S&TS, deconstruction also identifies what scientists and engineers *do* to their rivals' ontological claims during technical disputes, and thus it is a fine topic for social and historical studies of "technoscience."

In an esoteric field of science, the deconstruction of rival claims tends to occur in discourse among specialists. In an adversary system of interrogation, such as in the United States or United Kingdom, experts are explicitly (and often publicly) asked to explain, defend, and justify factual claims and technical choices.[22] It is a situation of "pure institutionalized mistrust" in Brian Wynne's (1989b: 33) memorable phrase. Social scientists rarely have the authority (or gall) to interrogate experts in

21. The theme of "heterogeneous engineering" was introduced by John Law (1986).

22. Just how such public inquiries are organized is itself an interesting issue. For an illuminating study of the play between public disclosure and closed inquiry in a science advisory panel, see Hilgartner (2000).

a sustained and adversary way, and so public interrogations of experts hold special interest for scholars who aim to delve into the "construction" of facts and artifacts.

It may be argued that "deconstruction" has become second nature to legal actors in adversarial systems of justice, and, indeed, the deconstruction of expert witnesses has a long and colorful history (Golan 2004a: 58). Unlike literary theorists and constructionist social scientists, courtroom adversaries rarely, if ever, call into question fundamental assumptions about truth, nature, and science. Courtroom deconstruction is practical and circumscribed. In jury trials in the Anglo-American tradition, the juror's "common sense"—nonspecialist understandings established as part of communal life—is assigned decisive importance. Contemporary courts resort to juries less and less frequently, but in trials and hearings in which no jury is present the presiding judge takes the role of the commonsense fact-finder. What is or is not a matter of "common sense" is itself subject to adversarial dispute and rhetorical play. "Common sense" is a wonderfully protean concept—sometimes upheld as a valuable public resource, and at other times condemned as a source of illusion and superstition—and in adversary courtrooms "science" and "common sense" are discursive categories that are actively worked into arguments and counterarguments.

Occasions in which expert evidence is deconstructed hold special interest for S&TS scholars,[23] because the adversary questioning skeptically explores the contingent production of the evidence. When successful, such questioning can explode just-so statements about objective facts to reveal hidden or actively suppressed sources of contingency and uncertainty. As Sheila Jasanoff (1995: 215) points out, these occasions can provide public tutorials (or "civic education") on the construction of expert evidence, thus encouraging reflection about expert authority and promoting a picture of science at odds with a realist or positivist version.[24] Jasanoff also notes, however, that courtroom "deconstruc-

---

23. What S&TS specialists *are*—(social) scientists, scholars, academics—is not easy to specify. The term "scholar" does not carry the weight (or baggage) of the term "scientist," and it can signal a lack of interest (or even disdain) for empirical detail. Our research is oriented to such detail, though it is not empiricist in the philosophical sense of claiming to derive propositions strictly from sensory observation (see Lynch [1993] for programmatic arguments).

24. A somewhat darker picture of the "civics lesson" promoted by the "applied deconstruction" of the State's evidence in a public tribunal is provided by Lynch & Bogen (1996: 14), in an analysis of testimony at the 1987 Iran-Contra Hearings.

tion" cannot be taken on its own terms as an instrument for exposing actual uncertainties and contingencies. Participants in a legal dispute rarely question the idealized conceptions of scientific method and fact to which they hold expert witnesses accountable,[25] and interrogators have been known to dramatize and exaggerate the consequences of minor departures from idealized conceptions of scientific knowledge and method (Oteri et al., 1982). Consequently, courtroom deconstruction cannot be taken as a reliable basis for probing and exposing the construction of expert evidence. Instead, consistent with the research policies in the field of ethnomethodology, it is necessary to view courtroom deconstruction as an interactional production, and not as a method for unmasking a reality that supposedly lies beneath the construction of expert evidence. In other words, courtroom "deconstruction" cannot be relied upon as a method for "laying bare" underlying contingencies and inherent uncertainties; instead, it is itself a contestable (and frequently contested) source of evidential claims and counterclaims.

The chapters that follow examine courtroom dialogues and judicial summaries of trials and admissibility hearings in which DNA profile evidence (and fingerprint evidence) was challenged. The challenges are of interest for the way the dialogues explicate unresolved tensions about science, expertise, and laboratory practice. The overall frame of the study is that of a history of a technical-legal controversy that began in the mid-1980s, was resolved a decade later, and was then followed by controversies about *other* forms of evidence. Analysis of selected excerpts from key cases will provide materials for specifying the phases and articulating the thematic elements of the controversy.

The Problem of Expertise

The problem of expertise is a thematic element that pervades the DNA fingerprinting controversy and also is a topic of lively debate in the S&TS field.[26] Briefly, it is a problem of how to reconcile the role of expert

25. For a critical account on the general theme of the idealization of science in law, see Caudill & LaRue (2006).

26. As Burney (2000: 8), has noted, "'Experts' are key figures in the history and historiography of the modern state." See also Golan (1999, 2004a). Turner (2001) provides a summary and critique of the "problem" as it is often posed in social and political theory. The problem of jury understanding of expert evidence (Edmond & Mercer, 1997) is a circumscribed variant of that problem. Empirical studies of "lay expertise" developed, for example, by patient and activist groups (Epstein, 1995; Rabeharisoa & Callon, 2002), or of local

knowledge (which is often assumed to be beyond the critical comprehension of the "lay" public) with the ideals of public participation in a liberal democracy. In a courtroom inquiry, the nonexperts have limited acquaintance with, and very little opportunity to learn about, the technical matters that expert witnesses are asked to present. Consequently, jurors (as well as many judges, prosecutors, and defense attorneys) can be at the mercy of experts who make uncontested assertions about what the evidence shows. And, when the experts disagree, the lay participants have no technical basis for deciding which claims to believe. Legal scholars and social scientists disagree about the extent to which jurors are capable of understanding expert evidence. Many argue that judges and jurors are ill equipped to distinguish genuine expertise from "junk science." However, others argue that the difficulties jurors experience with expert evidence can be relieved by more effective presentation of expert evidence in court. According to this view, juries are capable (or, at least, no less capable than judges are) of grasping what they need to know to decide a case.[27]

In the present study, we do not take up the empirical question of what jurors actually understand, or can be brought to understand. Instead, we address how actions and pronouncements by courtroom participants (principally, attorneys, judges, and expert witnesses) implicate what a silent jury can understand or appreciate. Although jury trials occur in a small, and decreasing, proportion of criminal and civil cases in the Anglo-American courts, the jury continues to stand proxy for the common citizen's place in the justice system.[28] The jury is also a "virtual" presence: even when they settle cases out of court, litigants frequently argue about how a jury would or could react to elements of the case at hand (Burns, 2001). Legal precedents and rules of evidence traditionally

---

knowledge grounded in practical experience (Wynne, 1989b), provide dramatic examples of the limits of expert knowledge and the leading role in knowledge production that is (or can be) assumed by groups with no claims to scientific credentials. In contrast, Collins and Evans (2002) credit the substantive reality of expertise, and try to distinguish among various degrees and kinds of expertise that have a role in social and political life.

27. For discussions of the general problems associated with the role of expert evidence in legal and legislative decisions, see Smith & Wynne (1989), Wynne (1989a), and Jasanoff (1995). Contrasting views of jury competence are expressed by Huber (1991), who emphasizes that jurors and other lay participants in trials cannot tell the difference between genuine and "junk" science, and Gigerenzer & Hoffrage (1995), who claim that jury understanding of probabilistic evidence depends upon how it is presented.

28. The *New York Times* reported that in the past twenty years, jury trials ending with verdicts in the United States had dropped from 2 to 1 percent of civil cases and from 10 to less than 5 percent of criminal trials (Glaberson, 2001: A-1, 17).

reserve the "ultimate issue"—the judgment about guilt, innocence, or
liability—to the trier of fact and restrict the role of experts to "informing"
the court.[29] Procedures of cross-examination and the institution of jury
deliberation build in social-interactional and documentary mechanisms
for "testing" evidentiary claims, exposing and dramatizing inconsistencies, and airing opposing views. In traditional jurisprudence, such procedures for openly presenting and testing evidence before a lay audience
are likened to powerful machineries for exposing "the truth."[30]

Court cases and public debates about DNA testing and other forensic techniques frequently raise questions about the scientific status of
expert evidence. By equating DNA evidence with science, proponents
draw upon a powerful source of authority for legitimating their stories
and arguments, while undermining the credibility of evidence presented
by their adversaries.[31] The possibility that such authority can be exaggerated or otherwise misused was not lost on the courts.

Conclusion

This chapter outlined how the history of technical/legal controversy
about DNA testing is an apt subject for studying how an important social institution relies upon, incorporates, and translates scientific evidence. The chapters and interludes that follow take up the story in its
various phases and use specific legal cases and technical disputes to il-

29. "Ultimate issue" restrictions have been greatly lessened in many jurisdictions today
(see, for example, U.S. Federal Rule of Evidence 704a). Cole (2001a: 209) discusses appeal
cases in the United States in the late 1920s and early '30s in which the issue of jury usurpation pertained to the testimony of fingerprint examiners. In some cases the appeal courts
held that fingerprint examiners encroached upon the jury's province when they described
matching latent and rolled prints as matters of "fact" rather than expert "opinion." See
Iowa v. Steffen (1930).

30. John Henry Wigmore (1940: 29) famously stated that the presentation and testing
of evidence in the trial court is "the greatest legal engine ever invented for the discovery
of truth."

31. The quotation marks we place around "science" are not meant to signal skepticism about science, as such, though we recognize that scare quotes are frequently used
in the science studies field to "neutralize success words" (Stove, 1984: 23) such as "discovery," "fact," and "truth." We simply mean to denote that the word "science" (along with its
grammatical variants and synonyms) was deployed and appealed to in texts and speeches
by judges, lawyers, legal scholars, advisory panels, and the science press. The quotation
marks are intended to signal that the quoted item is a significant term with recurrent use in
the institutions we are studying.

luminate the issues and problems at stake. Our chronicle begins in the mid-1980s, when DNA fingerprinting was first developed for forensic investigations. We then move to the late 1980s and early '90s, when forensic DNA testing was challenged in the courts, the international forensic community, and the science press. The hyperbolic label "DNA wars" headlined a particularly intense flare-up that attracted widespread media attention.[32] Opposing views were aired in popular articles, technical publications, and legal testimony. Public displays of hostility among stereotypically sedate and studious scientists and the "culture clash" between scientists and lawyers provided newsworthy topics for the science press.[33] Aside from the entertainment value of the spectacle, the controversy highlighted key issues for scholarship on the role of science in law. The public discord helped to undermine the quiet agreement among legal and scientific experts about the effectiveness of a new criminal justice tool, and the trials, admissibility hearings, and official evaluations of DNA testing provided insight into a contentious relationship between scientific and forensic practices. More generally, the trials and investigations raised questions about the relationship between legal and scientific conceptions of evidence and methodological adequacy.

The controversy began to wind down in the mid-1990s, but then a very interesting phase began in which an "inversion of credibility" occurred. DNA evidence became an unquestioned ground for challenging all other forms of criminal evidence. We focus particularly on fingerprinting, because as noted earlier it had long held a status as a nearly unassailable form of criminal evidence, but starting in 1999, that status was challenged in a series of admissibility hearings. This postclosure phase of controversies about forensic DNA testing takes us to the present time, when "DNA" is equated with truth, and DNA databases grow ever larger. Recalling a theme that runs throughout this book, the concluding chapter emphasizes that this seemingly unassailable, transcendent form of criminal evidence remains bound up with stories that are infused with contingent judgments about the mundane meaning and significance of evidence.

32. For use of the expression "DNA war," see Thompson (1993). War analogies were also used by science journalist Leslie Roberts (1991) in her coverage of the *New York v. Castro.*

33. The theme of "culture clash" was headlined in an article by Roberts (1992).

DNA Profiling Techniques

According to Evelyn Fox Keller (2000), the twentieth century was the century of the gene, a period neatly flanked by the "rediscovery" of Mendel's principles of heredity at the beginning and the completion of the first draft of the human genome at the end. During the second half of this epoch it became widely accepted that DNA was the molecular basis of the gene.

DNA is famously described as a "double helix"—a molecule composed of two twisting strands. As a carrier of genetic information, its key feature is the ordering of four chemical units called "bases" known as adenine (A), thymine (T), cytosine (C) and guanine (G). The two strands of DNA's double helix are held together through the complementary pairing of base A on one strand with T on the other, and C on one strand with G on the other. This "complementarity" between pairs of bases explains the faithful replication of the DNA molecule (and genetic information) from one generation to the next. According to the principles of molecular genetics, the sequential ordering of these four bases encodes genetic information, and differences in their arrangement are the basis of genetic differences between species and between individuals of the same species. Each human cell contains about two meters of DNA located in a compartment called the nucleus. Here it is tightly packaged into twenty-three pairs of chromosomes, each of which contains a single DNA molecule of, on average, roughly 150 million base pairs. The totality of nuclear DNA in a cell—which in most people is virtually identical in almost every cell in the body—is popularly known as a "genome."[1]

---

1. The genome of a person's reproductive cells (eggs or sperm) is different from that of (somatic) cells in other body parts. "Chimeric" individuals and individuals with "mo-

In 1992, Walter Gilbert, one of the early proponents of the Human Genome Project (HGP), proclaimed that one day it would be possible to hold up a CD of one's own genomic DNA sequence and say, "Here is a human being; it's me!" (Gilbert, 1992: 84–85). The mapping of the concept of the gene onto the biomolecular properties of DNA underpins one of the most persuasive arguments for undertaking the HGP, namely, that the DNA genome is the "Book of Life" or, to modernize the metaphor, "Life's CD." According to this metaphor, biology, destiny, and identity are encoded within the molecular structure of the DNA genome.

DNA analysis is widely regarded as the "ultimate method of biological individualization" (Krawczak and Schmidtke, 1994: 1), hence its place in forensic science. However, there are two key differences between DNA analysis for forensics and DNA analysis for the HGP. One is that forensic analysis does not examine the entire genome of individuals, but a much smaller number of highly variable regions, comprising approximately one millionth of a genome. Secondly, forensic science is not primarily interested in those parts of genomic DNA that comprise genes.[2] On the contrary, forensic analysis has favored DNA sequences that are *external* to the genes, and are not believed to encode biological traits.[3]

Approximately 99.5 percent of the human genome has been found to be the same for everyone. Forensic techniques exploit those places in the genome, known as "polymorphic loci," which exhibit detectable variations. These detectable variations (called "allelomorphs" or "alleles") can be of two types: variation in the *sequence* of DNA bases; and *length* variation arising from differences in the number of DNA bases between two defined end points.

RFLP Systems

The first forensic DNA typing systems were based on a kind of length variation called "restriction fragment length polymorphism" (RFLP).

---

saicism" have more than one genome. For a fascinating discussion of these conditions see Martin (2007).

2. Conventionally understood, genes are those parts of the genome that are expressed as biological traits.

3. These are called "extragenic regions." The DQ-alpha system described below is a historical exception. There are technical and ethical reasons behind the choice of noncoding regions for forensic profiling systems. Technically the extreme variability at these

"Restriction fragments" are produced by treating a DNA molecule with a "restriction enzyme," which acts like a pair of "molecular scissors," making a precise cut every time it finds a specific short sequence of bases which are distributed throughout the DNA.

"VNTR profiling" uses restriction enzymes to identify highly polymorphic loci in variable number tandem repeat (VNTR) regions. A typical VNTR locus is composed of a short sequence of DNA bases (the "repeat unit") which is repeated a variable number of times.[4] Alleles at a VNTR locus differ from one another in length depending on how many repeat units they contain. A restriction enzyme that cuts DNA on either side of a VNTR will generate restriction fragments of lengths that are dependent on which VNTR alleles it contains. To detect these allelic differences, the fragmented DNA is loaded onto an indentation at one end of a flat plate made of agarose gel, and an electric current is applied across the gel. Under the influence of the electric field, the DNA fragments are pulled out of the well and through the jelly-like substance in a straight line (called a "lane"). The smaller the fragments, the faster they travel (see fig. A.1c). At the end of the "electrophoresis" process the fragments are separated from each other and arranged in length order along the lane.

Fragments containing loci of interest are picked out from all the other DNA fragments using "probes" labeled with a radioactive tag (fig. A.1d). Each probe is a short sequence of DNA that is "complementary" to the sequence in a VNTR unit. The probe seeks out and "hybridizes" with (i.e., binds to) all the VNTR fragments on the membrane that contain the complementary sequence.[5] The gel is then pressed against a nylon membrane to transfer the ordered DNA fragments onto the membrane for analysis. When the image ("autoradiograph") of the membrane is developed, each bound probe appears as a band whose position reflects the size of the allele on the fragment (see figs. A.1e, A.2, A.3).

---

noncoding regions makes them good choices for discriminating between individuals. Ethically, their selection is a privacy safeguard against the derivation of information about personal health or physical appearance from a forensic DNA profile.

4. Repeat units are typically from fifteen to thirty-five base pairs long.

5. This is an idealized description. As we shall see in chapters 5 and 6, forensic analysts sometimes account for apparent mismatches between DNA profiles by citing such artifacts as the "partial digestion" of DNA by restriction enzymes or the incomplete hybridization of probes with some, but not all, of the complementary sequences in a given DNA sample.

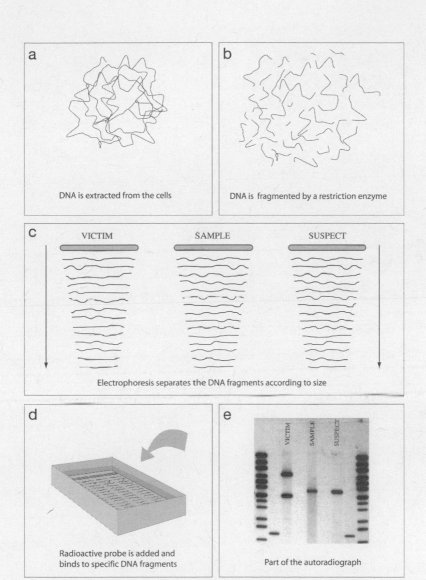

a DNA is extracted from the cells

b DNA is fragmented by a restriction enzyme

c VICTIM SAMPLE SUSPECT

Electrophoresis separates the DNA fragments according to size

d Radioactive probe is added and binds to specific DNA fragments

e VICTIM SAMPLE SUSPECT

Part of the autoradiograph

FIGURE A.I. Schematic representation of restriction fragment length polymorphism analysis using a single locus probe. Prepared by Andy Webb, Maple Print, England.

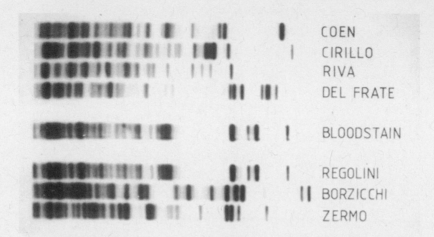

FIGURE A.2. Autoradiograph of "DNA fingerprints." Comparison between a crime blood-stain and seven suspect profiles. Each multilocus probe (MLP) detects multiple polymorphic positions in the genome. The result is that each profile shows an indefinite number of bands, some of which overlap in size. Source: Orchid Cellmark Ltd., Abingdon, U.K. Reproduced with permission.

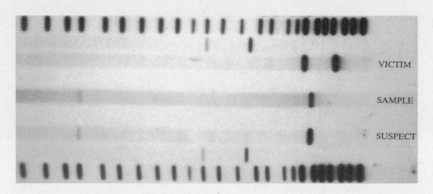

FIGURE A.3. Autoradiograph of SLP profiles. Results of analysing samples from a rape victim, the crime scene, and the suspect using a single-locus probe (SLP), which generates a maximum of two bands from each sample. Visual comparison of the profiles suggests a match between the crime scene sample and the suspect. The many bands in the outer lanes are the size standard used to calibrate the sample profiles. The lanes just inside the size standard lanes are where the control sample has been run. Source: Orchid Cellmark Ltd., Abingdon, U.K. Reproduced with permission.

Multilocus Probes

The type of probe Alec Jeffreys used when he and his associates invented "DNA fingerprinting" in 1984 is called a "multilocus probe" (MLP) (Jeffreys et al., 1985a). Each MLP is complementary to a sequence of bases common to repeat units in a "family" of VNTR loci located throughout the genome. Typically, one MLP would bind to approximately thirty VNTR fragments, resulting in the appearance of about thirty bands on an autoradiograph. Using several different MLPs to probe the same membrane produces a profile of multiple bands with the appearance of a barcode (see fig. A.2). In principle, MLPs can be used to produce profiles that are unique to an individual, apart from identical twins, hence the moniker "DNA fingerprint."

Single-Locus Probes

From 1986 to 1989, MLP analysis of VNTRs was the main DNA analysis system used by the U.K. Home Office Forensic Science Laboratories and the London Metropolitan Police Forensic Science Laboratory, and was the method used in the first mass screen that eventually identified Colin Pitchfork as the "Black Pad" murderer. The MLP system, however, was not adopted in the United States. The standard system adopted there when DNA testing was first introduced in 1986 was based on the use of "single-locus probes" (SLPs). From 1989 until the mid-1990s, the United Kingdom also used an SLP system for forensic analysis, based on probes developed by Alec Jeffreys.

Each SLP is complementary to only one VNTR locus, and so it generates a profile of just one or two bands per person (see victim, sample, and suspect lanes in figure A.3). If a person has inherited the same VNTR allele from both parents, then his or her SLP profile will contain just one band; if he or she has inherited two different alleles, then the profile will contain two bands. If five different SLPs are used, then a profile with up to ten bands can be generated, and the probability of two unrelated people having the same profile diminishes.

The VNTR loci used in forensic systems are highly variable, with each having hundreds of different restriction fragment length variants (alleles). Electrophoresis is, however, unable to discriminate between all of the different fragment lengths, and so for forensic purposes each locus is artificially divided into a smaller number (twenty to eighty) of discrete alleles through a process called "binning," whereby all fragments falling within certain lengths are assigned to the same bin and labeled as the same allele.

Forensic analysis typically involves the comparison of an SLP profile of a reference sample (a sample taken from a known individual) with a profile of an evidence sample from an unknown individual (often a "crime stain" recovered through investigation of a crime scene). If visual comparison of the position of the bands suggests that the two profiles match each other, then the lengths (in base pairs) of the DNA fragments in each profile are estimated by using the position of the bands on the gel in relation to a size standard. The "match criterion" for declaring a match between two bands from different samples does not require that they measure precisely the same lengths. Instead, two bands are said to match if their "uncertainty windows" overlap, that is, if their measurements are within a certain percentage of each other (see Rudin and Inman, 2002: 141; NRC, 1996: chap. 5).

The evidential weight assigned to a match between a reference sample and an evidence sample depends on how common the profile is in the population. For example, in *R. v. Adams,* which we discuss in chapter 6, a nine-band profile visualized using five SLPs was estimated to occur in one out of 200 million unrelated people in a relevant population. Estimates such as these are not based on empirical knowledge of the profiles of 200 million people. Instead, the frequency estimate is calculated from a very much smaller reference population believed to contain every *allele* in the typing system used, rather than every *combination* of every allele in the system. DNA from the reference population is typed, and the frequency of each allele is counted. The frequency of a given *profile* in the population is then estimated from the frequency of each of the contributing *alleles* in the reference population. This calculation is justified only if the population conforms to two population genetics principles called the "Hardy-Weinberg equilibrium" (HWE) and "linkage equilibrium" (LE). For a population approaching HWE, the frequency of the different pair-wise combinations of alleles ("genotypes") at a given locus can be calculated from the frequency of all the alleles at that locus. For a population at LE, the frequency of a profile can be calculated by multiplying together the frequency of all of the genotypes at each of the loci that make up the profile. Both HWE and LE are only approached in large populations, where migration is negligible and where mating is random. In practice, most human populations do experience migra-

tion and contain subpopulations of individuals who preferentially couple
with each other rather than with individuals from the wider population,
resulting in profile frequencies that depart from those predicted using
HWE and LE. When this is suspected, the amount of deviation from
HWE and LE is calculated and accommodated using statistical correction factors, a practice which initially attracted some controversy (see
Rudin and Inman, 2002: 142–47).

Like other bodily materials found at a crime scene, DNA molecules
are subject to degradation over time, breaking down into ever smaller
fragments due to exposure to heat, humidity, light, and various chemi-
cals. When such decayed samples are analyzed with forensic typing systems, the longer alleles are likely to be missing, and the resulting profiles
with missing bands are called "partial" profiles. The need to obtain "full"
profiles from very small samples and the desire for high-throughput,
automated profiling were served by the shift from RFLP-based DNA
profiling systems to systems based upon the polymerase chain reaction (PCR).

PCR Systems

PCR was invented in 1986 by Kary Mullis at Cetus Corporation in
California (Mullis et al., 1986).[6] Sometimes referred to as "molecular
Xeroxing," the PCR technique uses an enzyme called a "polymerase"
to replicate (or "amplify") a DNA sequence 100–2,000 base pairs long.
PCR produces millions of copies of the initial DNA sequence through a
chain reaction in which the products of one round of replication become
templates for the next (see fig. A.4). In each round, the precise DNA sequence replicated is defined by two synthetic "primers" (special probes)
that bind to either end, one to initiate the replication reaction and the
other to terminate it. A PCR system can generate profiles from samples
as small as 0.3–0.5 nanograms of DNA, about a hundred times smaller
than required for RFLP, and corresponding to the amount of DNA present in a few hundred sperm cells or a blood spot the size of a large pin-
head (Rudin and Inman, 2002: 17).

6. See Rabinow (1996a) for an anthropological account of the invention of PCR, and
Mullis (1998) for an autobiographical story by the self-styled maverick credited with the
invention.

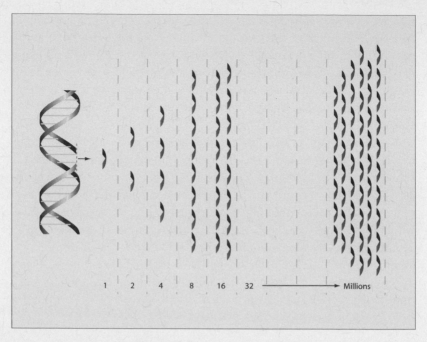

FIGURE A.4. Schematic representation of the PCR "chain reaction." Prepared by Andy Webb, Maple Print, England.

HLA DQA1 and Polymarker Typing Strips

The first commercial PCR kit used in forensic analysis was Cetus's "HLA DQ-alpha." This kit, and the subsequent "HLA DQA1" kit by Roche Molecular Systems, amplifies and types four alleles at the HLA DQA1 locus.[7] Rapid to use and requiring very little sample, the DQA1 kit analyses the "PCR product" (the DNA amplified by the kit) through a "reverse dot blot" process on nylon "typing strips" (see fig. A.5). Each typing strip contains an ordered row of probes.[8] Wherever an amplified DNA fragment binds to a probe, a blue dot appears on the typing strip. Comparison of the dot patterns from different samples indicates whether they have the same HLA DQA1 genotype and could, poten-

7. HLA DQA1 is a locus in the human leukocyte antigen (HLA) system, which plays a role in immune responses.

8. Nine probes for DQ-alpha; eleven for DQA1.

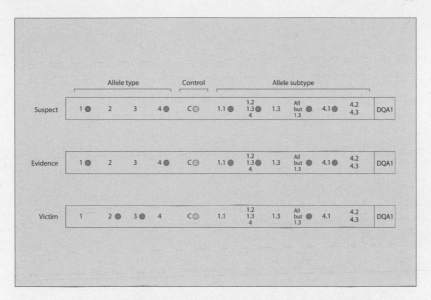

FIGURE A.5. Diagram of DQA1 reverse dot blot typing strips. Eleven probes are fixed to a single strip and DNA amplified by the kit is hybridized to them. The four probes on the left-hand side designate which of the four alleles are present; the control (c) dot should always appear, regardless of DQA1 type; and the probes to the right of the control designate the subtype of alleles 1 and 4. In the example above, the suspect and evidence samples are both type 1.1, 4.1, and the victim is type 2, 3. Prepared by Andy Webb, Maple Print, England.

tially, have come from the same source. Subsequently, the discriminating power of the DQA1 system was improved in the Perkin Elmer "Amplitype PM+DQA1" kit, which combined analysis of the DQA1 locus with the PCR amplification and analysis of five more loci, giving an overall discriminating power of approximately 1 in 2,000 (Rudin and Inman, 2002: 43–49). Referred to as the "polymarker" system, this was superseded by the STR system described below, although cases in which it was used in the past continue to work their way through the courts.

AFLPs

"Amplified fragment length polymorphisms" (AFLPs) are DNA fragments of differing lengths that are generated by amplifying a polymorphic DNA region with PCR. For a brief period a few forensic DNA laboratories used AFLPs of a VNTR locus called D1S80. In this system,

PCR was used to generate millions of copies of D1S80 alleles in the sample. These were then sorted by size using electrophoresis and visualized as a profile of one or two bands representing each person's genotype at this locus. [9]

STRs

The system used for most new suspect and case samples nowadays is PCR amplification of "short tandem repeat" (STR) loci. Although based on the same principles as RFLP analysis of VNTRs, the shift to PCR-based analysis of STRs enabled the development of high-throughput, computerized typing of degraded DNA samples with very high discriminating power.

STR loci comprise repeated short units of between two and five base pairs. Each allele is assigned a numerical "name" (such as "8" or "10") based on the number of repeat units it contains. Shorter than VNTR loci, STR loci can be used to profile older, partially degraded DNA samples. With only five to twenty alleles per locus, the discriminating power of each individual STR locus is limited. However, this deficit is compensated for by "multiplexing," which is the simultaneous PCR amplification of more than one locus by adding multiple sets of PCR primers to the sample. The polymarker system mentioned above was an early example of multiplexing. Examples of STR multiplex kits are Applied Biosystems' "Profiler Plus" and "COfiler," and Promega's "Power-Plex 16 BIO," which type the thirteen loci of the standard for the Combined DNA Index System (CODIS) convicted offender database in the United States, and Applied Biosystems' "Second Generation Multiplex Plus" ("SGM Plus"), which is the United Kingdom National DNA Database standard. Introduced in 1999, SGM Plus multiplexes ten STR loci plus another locus, called "amelogenin" (which is used to assign sex). [10] The probability of a match between full SGM Plus profiles from two un-

9. The Sutton case described in chapter 8 used the D1S80 and the polymarker systems. The case involved a mixed sample, and according to William Thompson in an interview (October 2005), the probability figures given by the prosecution failed to take proper account of the extent to which the mixture reduced the probative value of the evidence.

10. Although not an STR locus, this locus exhibits a stable sex-linked fragment length polymorphism. The amelogenin locus has just two alleles, a longer one on the Y chromosome and a shorter one on the X. This means that a DNA profile with just one band at this locus indicates that the sample comes from a person with two X chromosomes and is al-

related individuals is conservatively estimated to be of the order of 1 in a billion (10^9), although matches involving partial profiles, or profiles of close relatives have less probative value.

Initially, STR PCR product was loaded onto a gel, separated according to size by electrophoresis, stained, and analyzed. The gel was then dried and stored for future reference. Since the late 1990s, STR multiplexing kits have combined the processes of amplification and labeling through the use of fluorescent primers. In the latest platforms, the PCR product is separated in the liquid phase by capillary electrophoresis rather than on solid gel, allowing many more samples to be separated and analyzed in a single run.

The fluorescently labeled PCR fragments are detected in real time by laser as they travel past a fixed point under electrophoresis. Computer software (e.g., Applied Biosystems' "GeneScan" and "GenoTyper") measures the size of fragments by comparison with labeled fragments in an internal size standard in the same lane as the samples and then nominates their allelic identities by comparison with the "allelic ladder" which is included in every run. Software represents the profile on a computer screen in various ways, including a graph called an "electropherogram" (see fig. A.6).

Y-Chromosome STRs

The "sex" chromosomes X and Y contain STR loci that can be PCR-amplified. Because women are assumed not to have a Y chromosome, Y-STR profiling is used to distinguish between male and female "donors" of mixed crime stains, for example in sexual assault cases.

Mitochondrial DNA Analysis and Other Approaches

Another source of DNA that is PCR-amplified is from "mitochondria," which are small "organelles" inside cells. Although each mitochondrial genome is very small (16.6 kilobases), there are several thousand of them per cell. Its small size, relative abundance, and high mutation rate make the mitochondrial genome invaluable for forensic analysis, especially of

most certainly female, whereas a DNA profile with two bands indicates a person with one X and one Y chromosome who is almost certainly male.

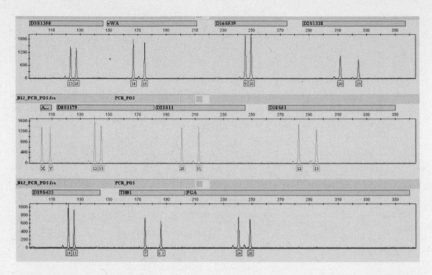

FIGURE A.6. Electropherogram of an SGM Plus STR profile. An electropherogram represents the profile as a histogram in which the positions of the peaks correspond to the sizes of the PCR-amplified DNA fragments, with the smallest on the left-hand side. The strength of the signal—represented by the height and area of the peaks—is a measure of the quantity of DNA present in the sample, a feature that can be informative when trying to distinguish individual profiles within a mixed sample. This electropherogram is of a full SGM Plus profile from one sample. There are two different alleles at each of the loci, including the amelogenin (A . . .) locus indicating that the sample is from a male. The names of the loci appear in the boxes above the peaks, and allele assignments are given underneath. Provided the order of the loci is clearly defined, this profile can be written for database storage as the following simple string text: "15, 16, 14, 16, 9, 10, 20, 23, X, Y, 12, 13, 28, 31, 12, 15, 14, 15, 7, 9.3, 24, 26." Source: LGC Ltd., U.K. Reproduced with permission.

samples so old and degraded that nuclear DNA typing is impossible. For example, it is possible to recover typeable mitochondrial DNA from dead cells in hair shafts, bones, and teeth (Rudin and Inman, 2002: 79–83).

Other approaches to the analysis of small and degraded DNA samples include increasing the number of PCR cycles, as in the "Low Copy Number" system developed by the Forensic Science Service (FSS), and the "miniSTR-multiplex" which amplifies very small lengths of DNA. However, as the amount of suspect DNA in the sample analyzed decreases, the risk of amplifying and profiling *contaminating* DNA increases (see Raymond et al., 2004). The amplification of contaminating DNA could mistakenly eliminate a suspect by giving a nonmatch that was a false negative result. Alternatively, a profile from a contaminant that matched a reference profile on a database could suggest a spurious link between a known individual and the crime.

Future Trends

Historically, the DNA variations used in forensic analysis have been deliberately selected from "extragenic" regions, that is, regions of the genome that are not expressed as detectable, physical traits.[11] This was in keeping with the mission of forensic DNA databases as evidence-building tools rather than investigatory ones. However, this policy is in tension with the forensically tantalizing prospect of reverse engineering a "DNA photofit" of the appearance of the perpetrator of a crime from analysis of DNA left at the crime scene (Jeffreys, 1993a). The temptation to exploit the potential to predict appearance from DNA patterns has proved too hard to resist, although the achievements to date have been more mundane and involve the nomination of target suspect groups for specific crimes based on physical traits or social relations that are assumed to be statistically linked to particular DNA profiles. Specifically, FSS offers a DNA analysis service that predicts whether or not the donor of a crime scene sample has red hair, and both DNAPrint Genomics (a company that performs genetic analysis) and FSS make ethnic inferences about a crime scene DNA sample donor's geographic or ethnic ancestry (Lowe et al., 2001; Williams et al., 2004; Nuffield Council on Bioethics, 2006; Nelson, forthcoming), in what has been characterized as a high-tech variant of racial profiling (Duster, 2004).

With the commercialization of the provision of forensic science services in the United Kingdom, the government-owned company FSS Ltd. and private companies, such as LGC Forensics and Orchid Cellmark Inc., are in competition with each other for DNA profiling contracts. In this competitive marketplace, commercial advantage is sought by developing DNA profiling systems that offer police forces improved crime detection rates, for example, through enhanced sensitivity. While diversification of suppliers enables defence lawyers to purchase reanalysis of DNA evidence in a different laboratory from the one used by the

11. Although it has been found that one of the loci used in SGM Plus is linked to type 1 diabetes (Williams et al., 2004).

prosecution, protection of commercial advantage through confidentiality could lead to insufficient technical disclosure between competing laboratories for the independent replication of profiles made using proprietary platforms, thereby effectively black boxing parts of forensic evidence from skeptical scrutiny. How the balance between competition and cooperation in the provision of forensic DNA services to the criminal justice system is to be managed remains to be seen. The commercial environment for the provision of forensic services, including DNA services, in the United Kingdom is clear from FSS Ltd.'s mission, which is "to retain and reinforce our leading position as the principal provider of forensic science to the U.K. criminal justice system and use this platform to become the leading provider world-wide, thereby enhancing long-term shareholder value."[12] The absence of any reference to *justice* in this mission statement is a worrying indication of the difference in priorities between forensic science as a fully commercialized industrial sector, and forensic science as a public service.

12. http://www.forensic.gov.uk/forensic_t/inside/about/docs/18715_Report_&_Accounts .pdf.

CHAPTER TWO

A Techno-Legal Controversy

Technical controversy is an established topic for research in the history and sociology of science. In addition to exhibiting divergent theoretical assumptions, rival experimental designs, and contrary evidential interpretations, controversies exhibit collective procedures for reaching "closure" and restoring a sense of continuity and consensus. The expectation that scientific communities should be able to resolve their disputes through rational procedures and crucial tests, rather than through brute force or majority rule, dates back to the early-modern origins of experimental science (Shapin & Schaffer, 1985). Although key scientific terms—evidence, tests, matters of fact—parallel, and even derive from, legal uses of those terms (Shapiro, 1983), science is now widely held to be a source of more certain, and less arbitrary, procedures for making factual judgments and ending disputes. The role of experts, and especially "scientific" experts, has become increasingly prominent in legal, governmental, and administrative circles (Smith & Wynne, 1989; Jasanoff, 1990; 1992; 1995). At the same time, the role of the jury has diminished in the United Kingdom, the United States, and other nations that grant "ordinary" citizens and their "common sense" a primary role in the fact-finding tribunal.[1] As we describe in chapter 8, with the ascendancy of expert evidence (and particularly DNA evidence), other, nonexpert forms of criminal evidence such as confessions and eyewitness testimony have undergone critical scrutiny. Nevertheless, in spite of these trends, judges often express skepticism about expert evidence and they do not readily

---

1. A *New York Times* article on trends in the United States (Glaberson, 2001) reported a steep decline in the proportion of cases settled by jury trials in both civil and criminal courts.

yield legal authority to experts. As we shall see when reviewing decisions about DNA profile evidence, the courts continue to place limits
on expert evidence so that expert witnesses do not usurp the traditional
province of judges and juries. The extent to which such traditional limits
should be maintained in the face of the extraordinary power ascribed to
DNA evidence is itself a major topic of controversy.

This chapter reviews a controversy about forensic DNA testing that
ran for several years in the late 1980s and early 1990s. As we shall elaborate, this controversy had a number of distinctive features. First, while
the *scientific* status of DNA evidence was a prominent issue, the controversy was brought to a head in the legal domain, and especially in contested admissibility hearings in U.S. state and federal courts. Disputes
between scientists took place in courtrooms, as well as in scientific journals and advisory panels, and the existence of the controversy—including
its parsing into phases (when it opened, and when it was more or less
closed) and specifications of what it was about—were framed by legal
procedures and criteria. Second, the most active phases of dispute included nonscientists (particularly judges and lawyers), as well as scientists and technical specialists from different fields. The identity of some
participants as scientists or nonscientists involved a degree of ambiguity and contestation. And, third, the controversy was not only about the
"scientific" status of a particular innovation (its acceptance as reliable
and as a source of valid results within bona fide scientific fields), but also
about whether that innovation retained its scientific status when it was
transferred to the domain of criminal investigation.

Controversy Studies

Studies of controversies in the sociology and history of science challenge
the idea that scientific disputes are resolved strictly through experimental testing and rational consideration of evidence. Without denying the
central place of empirical experiments and observations, controversy
studies emphasize that scientific disputes are more like political and legal disputes than is often assumed.

Controversy studies were inspired by Thomas Kuhn's influential
Structure of Scientific Revolutions (1970 [1962]). Kuhn's title and central thesis drew strong parallels between political and scientific revolutions. Particularly significant for sociological purposes was Kuhn's argu-

ment that revolutionary transformations of naturalistic understandings (for examples, the Copernican revolution in the sixteenth century, the chemical revolution at the end of the eighteenth century, and the rise of relativity in early twentieth-century physics) involved the overthrow of an established paradigm by an incommensurable matrix of theory and practice. Kuhn emphasized that incommensurable paradigms involved asymmetrical commitments: to work within one paradigm invariably skewed one's vision of the competing paradigm or paradigms. Although arguments during revolutionary transitions focus on theory and experimental evidence, historical scientists embedded in controversy had no recourse to historical hindsight or transcendental rationality when interpreting evidence and reaching consensus. Although Kuhn (1991) later put distance between himself and his sociological and political interpreters,[2] his writings were used to support arguments to the effect that a scientist's "choice" between competing paradigms (and, by extension, "choice" among competing theories in controversies of lesser scope) was no choice at all, because historical scientists did not simply compare the competing paradigms from a neutral standpoint. Instead, any choice already was embedded within a paradigm: a nexus of existential commitments, including naturalistic assumptions, a training regime, a network of colleagues and patrons, and a way of working with research materials.

Among the best-known controversy studies are H. M. Collins's analysis of disputes over gravity wave experiments and Andrew Pickering's history of the ascendancy of what he calls the "quark-gauge world view" in particle physics.[3] Collins's studies were particularly influential for

2. Kuhn was sparing in his comments on "Kuhnians" in social studies of science, and for the most part addressed criticisms arising from more traditional quarters in philosophy of science. In a conference presentation on "The road since 'Structure,'" Kuhn (1991) makes a brief and disdainful remark about sociologists of science associated with the Edinburgh School "Strong Programme" in the sociology of scientific knowledge (e.g., Bloor, 1976; Barnes, 1977). This remark is not included in the published essay by the same title in Kuhn (2000).

3. Collins has investigated two different phases of controversy, separated by twenty-five years (see Collins, 1975; 1985, chap. 4, for discussion of the early research, and Collins, 2004, for the later phase). The first phase, in the late 1960s and early '70s, focused on Joseph Weber's claims to have detected gravity waves with a relatively simple apparatus. Weber's claims were widely dismissed by other physicists, though Collins argues that there was never a crucial test or a definitive disproof. The second phase involves a far more credible (and expensive) effort to detect gravity waves. The main publication of Pickering's study of controversies in particle physics is his *Constructing Quarks* (1984).

S&TS research. His studies were based on interviews with active participants in the controversy and on-site visits to relevant experimental facilities. Collins identified a "core set" of scientists in the gravity wave field who performed the experiments that prosecuted the controversies and pursued their resolution. A key medium of interaction among members of the core set was furnished by published and unpublished reports of experiments, though members also engaged in more personal forms of conduct through which they established trust, communicated tacit knowledge, and built coalitions.

Collins noted how experimental designs and results varied remarkably, and yet predictably, among core-set rivals. He argued that, instead of being sources of independent evidence for resolving controversies, experiments tended to be *extensions* of the partisan arguments that sustained controversy. He observed that "closure"—the practical ending of controversy in a scientific community—was not based on any single crucial experiment. The eventual winners wrote retrospective histories of the dispute, while die-hard losers, some of whom left the field or became marginal figures, often remained dissatisfied with the alleged disproof of their claims. In other words, evaluations of experimental evidence were bound together with assignments of credibility and relevance to particular experiments and experimentalists. Over the lifetime of the controversy, the scientific evidence and the boundaries of the relevant scientific community were intertwined and coproduced.

For Collins and others who study scientific controversies (often associated with "constructionism" in the social sciences [Hacking, 1999]), closure does not result *entirely* from the accumulation of evidence, control of sources of error, or rational agreement about previously disputed facts. Although participants, and the sociologists who study them, cannot ignore possible sources of evidence and error, it sometimes happens that closure is announced, and even acted upon, despite the persistence of disagreement. Indeed, forecasting closure and even announcing it as a fait accompli is a common rhetorical tactic used by participants in ongoing controversies. A small, or in some cases a substantial, minority faction may continue to cite evidence supporting its views, while complaining of "political" machinations employed by the triumphant faction (Gilbert & Mulkay, 1984). A sociological analysis of such disputes is complicated by the fact that judgments about relevance and credibility—which disciplines are relevant, and who counts as a credible spokesperson for the consensus—can themselves be subjects of controversy. The cartographical metaphor of "boundary work" (Gieryn, 1983; 1995; 1999) and an analogy with political gerrymandering (Woolgar & Pawluch, 1985) are widely used in case studies of controversy and closure. In part, these sociological concepts point to the way controversies about facts are bound up in disputes about who is competent to "represent" the facts.[4] As consensus develops over the course of a controversy, communal alignments form and boundaries are drawn between who is deemed competent and credible and who is dismissed incompetent or incredible.[5]

Hybrid Controversy

In his study of the gravity waves controversy, Collins (1999: 164) employs a diagram of the core set and various peripheral groups who consume the written products of the core set's experiments (fig. 2.1). The diagram resembles a target, with the most heavily involved "core" investigators at the center, and less directly involved scientists further out from the center. Further out are policymakers, journalists, and other interested onlookers and commentators, and way out on the edge are members of "the public."[6]

    The DNA fingerprint controversy took place on rougher terrain, and

4. Both the political sense of "representation" (speaking or acting on behalf of a community or group) and the referential sense of "representation" (a sign or picture standing for an object or idea) come into play here (Latour, 2004b; Pitkin, 1967: 221–22).

5. Collins distinguishes between a more coherent "core group" that emerges from the core set during the active period of controversy. Some participants in the controversy become increasingly marginalized as consensus develops among members of the core group—a group largely defined by such consensus—and thus become members of the "set" but not the "group," and those at the very edge of the circle of participants (including some who may once have been central) may be denounced as cranks, purveyors of fraud, and pseudoscientists.

6. Superficially, the target diagram resembles a schema developed by phenomenologist Alfred Schutz (1964a) to describe the structure of social relationships centered around an individual, with close intimates occupying the center, and expanding outward to less and less intimate associates and consociates, ending finally with complete strangers at the outer edge. The major difference is that Schutz's scheme is a map of personal relations from a first-person standpoint, while Collins's scheme is meant to describe a social network characterized by greater expertise at the center, and decreasing expertise as one moves outward. Borrowing from Schutz, we can imagine that a peripheral player in Collins's scheme can occupy the center of a network in which that player's scheme of relevance takes precedence over the technical skills of the core set that Collins identifies and privileges. And, if we endow this new scheme of relevance with institutional authority (rather than simple subject-relevance), we can get some idea of what is at stake in judicial administration of expert evidence.

Target Diagram of Consumers of Scientific Papers

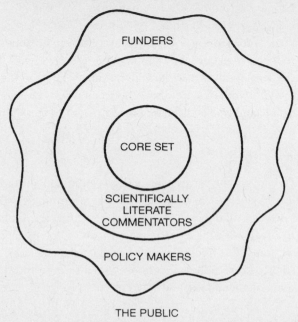

FIGURE 2.1. Core set diagram (target diagram of consumers of scientific papers). Source: H. M. Collins (1999: 164). Reprinted with permission from Sage Publications Ltd.

with less settled divisions among the participants. It was not a pure case of a scientific controversy, analogous to the gravity waves case. During the gravity waves controversy a relatively small number of experimental physicists in the core set battled over discrepant findings and interpretations. Although the core set was not a hermetically sealed group, a restricted group of laboratories had a central role in generating and resolving the controversy. Some other scientists and nonscientists may have become aware of the controversy, and a few may have had a peripheral role in it, but most had little interest or direct engagement in the esoteric dispute.[7]

The DNA fingerprint controversy also involved disputes among experts, but it was not confined to a clearly identified network of natural

7. In later work, Collins and Robert Evans (2002) developed a conception of expertise—and of the study of expertise—that extends the notion of the core set to apply to broader public controversies. For a critical exchange on Collins and Evans's theory, see Jasanoff (2003), Wynne (2003), Rip (2003), and Collins & Evans (2003).

scientists. The technical content of the dispute extended well beyond any single discipline, as key participants included scientists and mathematicians from several different specialties, including molecular biology, forensic science, population genetics, and statistics. Active participants also included lawyers, judges, police employees, legal scholars, government officials, and science writers. The core set thus was joined by a "law set" (Edmond, 2001) that participated in key legal decisions, as well as by an "administrative set" of review panels and advisory groups and a "literary set" of legal scholars, science journalists, and other scribes and chroniclers (including the authors of this book). This ecology of overlapping and sometimes contending sets was complicated by the fact that members of some sets (and to an extent all sets) faced the task of specifying who had a relevant and legitimate role in the dispute. Law courts, journalists, and scientific review panels faced the task of deciding which fields were relevant and which members counted as bona fide spokespersons for those fields. In other words, they mediated the dispute, translated its terms, and adjudicated its boundaries. Which specialties occupied the "core" of the dispute was not given from the outset, nor was it always clear which scientists and administrators counted as spokespersons for the relevant fields.[8]

The debates among the members of these sets covered technical questions about molecular biology, population genetics, and statistical procedure, but they also covered practical, legal, and ethical questions about

8. Because of its public visibility, the DNA fingerprint controversy was more like the "cold fusion" affair, which drew massive international publicity for several months after Stanley Pons and Martin Fleischmann's announcement of the discovery (see Gieryn, 1992; Lewenstein, 1995; Collins & Pinch, 1998). One of the prominent questions during the cold fusion affair concerned which subfields of physics and chemistry were relevant for testing the contested claims. Nuclear physicists claimed the high ground in the hierarchy of science, and the press tended to accept their judgments. The fact that Pons and Fleischmann were chemists tended to count against their credibility when prominent physicists dismissed their discovery. Which scientists and fields of science counted as part of the core set was contingent upon public acceptance of the credibility of particular fields, laboratories, and spokespersons. It was not the case that a small set of experts who possessed the technical competence to examine and criticize Pons and Fleischmann's claims settled the dispute. Partly because of the low cost of the experimental apparatus, a large and confusing array of replications was attempted. The press, Internet newsgroups, and other media sources were used to monitor and disseminate the results, and the public mediation of the experiments and results fed into the performance of the experiments and the accountability of the results. See the introduction to the Cold Fusion Archive by Bruce Lewenstein, available at www.wpi.edu/Academics/Depts/Chemistry/Courses/CH215X/coldfusion.html.

the way police and forensic organizations handle criminal evidence and implement laboratory protocols. At the heart of the dispute were legal procedures for deciding the admissibility of evidence. Closure was as much a legal and administrative matter as it was a technical or scientific issue. Technical "fixes" and the presentation of expert evidence were important for bringing about the eventual acceptance of DNA testing in criminal justice, but equally important were efforts to devise administrative standards for assuring the courts that DNA evidence was correctly handled and analyzed. Some of the most interesting and dramatic confrontations occurred when expert witnesses were cross-examined by attorneys in front of judges, juries and (during the O. J. Simpson trial) massive media audiences. Law, science, and the public understanding of science were deeply intertwined in these confrontations.

The Legal Framing of Scientific Controversy

The key players in the DNA fingerprinting controversy included specialized "expert lawyers" and scientists who actively participated in legal challenges and public policy debates. Legal standards for deciding the admissibility of evidence framed the controversy, and some of the landmark events and turning points in the history of DNA testing (and, later, fingerprinting) took place in the courtroom. The controversy was an episode in legal history as much as it was a chapter in the history of science. The courts, scientific review panels, and the media investigated and publicly discussed the controversy. Many of the themes that Collins and other social historians have used for analyzing controversies were themselves used by participants whose actions produced, or sought to end or foreclose, controversy. Commentators spoke openly about the controversy: they quoted contradictory claims about experimental results and the competence of the experimenters; and they made forecasts, and disputed others' forecasts, about the eventual resolution of controversy. These contentious commentaries were not side issues in the "real" controversy, because they were important for framing, conducting, and settling the dispute. The commentaries, syntheses, and speculations about closure were crucial for establishing public (and government) interest and disinterest in the continuing saga.

The technico-legal dispute about DNA fingerprinting addressed fundamental social and political questions about the role of technical expertise in a democratic system.[9] Participants raised questions about the meaning and authority of numbers, and about the participation of citizens and the role of "common sense" in legal decision-making. The courts were asked to resolve, in a case-by-case way, the problematic relationship between idealized norms of science (abstract principles, rules, and protocols) and day-to-day practices in the laboratory and at the crime scene. In effect, the courts provided a forum for publicly discussing topics of central importance for science and technology studies (Lynch, 1998). These debates had little input from S&TS research, and they were restricted in scope, as the courts sought to devise practical and administrative solutions rather than to develop novel ideas and sophisticated arguments. The common themes, if not common ground, between particular court cases and research in the history, philosophy, and social studies of science present us with a fascinating, if difficult and challenging, phenomenon. To put a name on it, we may call it "legal metascience": legally embedded conceptions of the general nature of science, and of the scientific status of particular practices and forms of evidence.

A study of legal metascience cuts both ways: it can deploy insight from philosophy, history, and social studies of science to analyze and criticize legal discourse about science (Jasanoff, 1991; 1992), but it also can lend critical insight into some of the conceptual problems that come into play in S&TS debates about science and its place in modern societies (Lynch, 1998; Lynch & Cole, 2005). The very themes of controversy, consensus, and closure, which historians and sociologists address in studies of technical controversy, are explicit topics of deliberation and debate in courtroom hearings about the admissibility of DNA profile evidence. The U.S. courts have codified specific standards for assessing novel forms of expert evidence. As we elaborate in interlude B, the "general acceptance" standard, which was articulated in a federal court decision *Frye v. U.S.* (1923), presents a judge with the social-historical task of deciding whether or not a novel form of (alleged) scientific evidence is generally accepted in the relevant fields of science. In its landmark decision *Daubert v. Merrell Dow Pharmaceuticals, Inc.* (1993), the U.S. Supreme Court advised federal courts to use a broader set of standards when assessing the reliability of expert scientific evidence, but many state courts

9. See Galbraith (1967) for the role of "technostructure" in industry, and Winner (1977) for a critical review of theories of technocratic politics.

continue to use the *Frye* standard. Criminal court systems in other nations place less stress on admissibility hearings, but debates about the
"scientific" status of forensic evidence also arise in trials and appeals, as
well as in science advisory panel inquiries. Our study of these debates allows us to treat controversy, consensus, and closure not only as substantive phases in the social history of innovation, but also as themes that are
used by key players who produce and assess the current state of controversy and closure.

Admissibility Hearings and the Unraveling of DNA Evidence

Two landmark U.S. Supreme Court rulings in the 1990s, *Daubert v. Merrell Dow Pharmaceuticals, Inc.* (1993), and *Kumho Tire Co. v. Carmichael* (1999), articulated standards for the admissibility of evidence in
U.S. federal courts. The Supreme Court's conception of science directly
influenced U.S. federal courts and many state courts during the DNA
fingerprinting controversy, and it had less direct influence on courts in
states (and even other countries) that do not formally adhere to U.S. federal guidelines for admissibility.[10] However, many of the U.S. cases discussed in this book occurred before the *Daubert* ruling, or occurred in
state courts that continued to used variants of the *Frye* "general acceptance" standard after 1993 to assess the admissibility of expert evidence.
Related questions about the "expert" and "scientific" status of evidence
are also aired in trial and appeal courts, not only in the United States
but also in other countries such as the United Kingdom. In the remainder of this chapter we focus on the emergence of controversy in connection with the most notable admissibility hearing in the history of DNA
profiling: the 1989 case *New York v. Castro*. First, however, we provide a
sketch of some earlier cases.[11]

DNA fingerprinting was used in criminal investigations soon after the
published announcement of its invention by Alec Jeffreys (who was later

10. For example, even though New York continues to adhere to the *Frye* standard, the
judge in *New York v. Hyatt* (a 2001 case involving fingerprint evidence discussed in Lynch
& Cole [2005]) explicitly mentions the *Daubert* factors even while acknowledging that they
do not formally apply. International influence is more difficult to trace, but our discussions
with attorneys and forensic scientists in the United Kingdom indicated that they closely
followed notable U.S. cases.

11. For a much more detailed historical account of *Castro* and other early challenges to
the admissibility of DNA evidence, see Aronson (2007).

knighted for the achievement) and associates at the University of Leicestershire, England (Jeffreys et al., 1985b). In 1986 and 1987, Jeffreys
assisted a police investigation of the notorious "Black Pad murders"—
two rape-murder cases in nearby Leicestershire villages.[12] The first of
the murders took place in 1983 along a footpath in a location known as
"Black Pad" in the village of Narborough, and the second occurred in
1986 in the village of Enderby. Both involved the rape and strangulation
of teenaged girls by an unknown assailant. The investigation was chronicled by crime writer Joseph Wambaugh in his book *The Blooding* (1989),
the title of which refers to the mass screening of "voluntary" blood samples taken by the police from thousands of men in the local area. A mentally imbalanced kitchen porter was initially suspected of the murders,
and even gave an ambiguous confession, but he was excluded when his
DNA evidence did not match the crime samples. A key part of the story
was that Colin Pitchfork, who eventually was convicted of the crime, initially evaded detection by submitting a blood sample given to him by a
friend.

In 1987, DNA evidence was used in the Florida murder trial of Tom-
mie Lee Andrews (see *Andrews v. State,* 1988), and many other cases
soon followed. The defense in *Andrews* and some other early trials such
as *New Jersey v. Williams* (1991) challenged the admissibility of DNA
evidence, but all of these early challenges were unsuccessful.[13] In some

12. Also see Office of Technology Assessment (1990: 8) for a brief summary.

13. Richard Charles Williams was indicted in 1983 and charged along with codefen-
dant Thomas Manning with the 1981 murder of a New Jersey state policeman, Philip Lamonaco. According to the prosecution's brief, the defendants were both sitting in a blue
Chevy Nova which had been stopped on Interstate Route 80 by Trooper Lamonaco. The
prosecution alleged that Williams shot Lamonaco, but that the trooper was able to dis-
charge his own weapon after being fatally wounded. The Nova was found abandoned a few
hours later, and blood was recovered from the passenger's seat, headrest, and door panel.
Ballistics evidence identified a gun that was recovered as the murder weapon, and other
evidence indicated that it had been purchased by Williams on the same day as the murder. His fingerprints were found on items left in the car which also were purchased that
day. Williams and Manning remained fugitives before being arrested in 1984 and 1985, re-
spectively, and they were jointly tried in 1986–87. Tests for blood type and enzyme markers presented at the trial indicated that the blood in the Nova could have come from either
defendant, but not from the victim. The trial resulted in a hung jury, and Williams was then
tried separately from Manning in 1991. Prior to Williams's retrial, the prosecution commissioned a new set of tests on the blood samples. One test employed a newly developed
method using the PCR DQ-alpha system, while others used older methods of blood analysis. The RFLP method was not used, because the blood samples taken from the Nova in
1981 were judged to be of insufficient quality.

of the early trials, the prosecution mobilized an impressive roster of experts, ranging from prominent molecular biologists, who endorsed the scientific status of the techniques, to forensic case specialists, who performed and supervised laboratory work for Cellmark, Lifecodes, the FBI laboratories, and other forensic organizations.[14] The defense in many of the early trials and admissibility hearings did not call any experts. According to Neufeld & Colman (1990: 25), in some instances, the presiding judge refused to authorize funds to retain expert witnesses for the (court appointed) defense: "A critical factor in the defense's successful challenge was the participation of several leading scientific experts—most of whom agreed to testify without a fee."[15]

Neufeld and Colman also mention that even when defense counsel were able to use expert witnesses, they found it difficult to find experts who would agree to appear: "The defense counsel in one case explained that he had asked dozens of molecular biologists to testify but all had refused. Interviews with some of the scientists revealed that most of them, being familiar with scientific research involving DNA typing, assumed the forensic application of the technique would be equally reliable" (Neufeld & Colman, 1990: 24). The assumption of reliability, and even infallibility, was encouraged by early statements by Jeffreys stressing that the technique (referring to the multilocus probe technique— see interlude A) produced near certain results: "The pattern is so varied (hypervariable) that any particular combination of the segments is as

14. When DNA evidence was first introduced as evidence in criminal trials, it was not yet established which field, or fields, would furnish expert witnesses. In early cases, prosecutors called a combination of expert witnesses, ranging from spokespersons for forensic labs to "independent" experts from academic fields of molecular biology and population genetics. Aronson (2007: 42–54) provides a detailed account of the first admissibility hearing in which the defense called expert witnesses to challenge the prosecution: *People of New York v. Wesley* (1988). In that case, the presiding judge accepted the prosecution's experts from the various fields, but placed restrictions on the defense's experts (and specifically, on the testimony of molecular biologist Neville Colman) on the grounds that their expertise was not directly related to forensic DNA testing. Despite such restrictions, the case set a precedent for treating expertise in molecular biology and population genetics as *relevant* to the evaluation of DNA evidence.

15. Both prosecution and defense can be deterred by the expense of DNA testing, but with the involvement of the FBI and the National Institute of Standards and Technology (NIST), and the allocation of large amounts of government funding to developing and testing DNA typing, forensics expertise is routinely available to prosecutors, but available only to defendants with sufficient funds and well-informed and energetic attorneys.

unique as a fingerprint"[16]; "Suppose we could test a million people every second. How long would it take to find one exactly the same? The answer is, the universe itself would die before we found one the same. It is simply an incomprehensible number" (Jeffreys, quoted in Grove, 1989). A Home Office spokesman stated, "The procedure is very complicated but it provides the scientists with a DNA fingerprint which has been shown to be specific to a particular individual."[17] Media statements through the 1980s also used the analogy with fingerprints to stress the (near) certainty of individual identification: "the perfect fingerprint: unfakeable, unique, and running in families"[18]; "This is the most important innovation in the fight against crime since the discovery of fingerprints."[19]

The first report of the National Research Council noted, in retrospect, that "in the publications in 1985 by Jeffreys and colleagues, the term 'DNA fingerprint' carried the connotation of absolute identification. The mass-media coverage that accompanied the publications fixed in the general public's minds the idea that DNA typing could be used for absolute identification. Thus, the traditional forensic paradigm of genetic testing as a tool for exclusion was in a linguistic stroke changed to a paradigm of identification" (NRC, 1992: 27). Another retrospective account also noted the early emphasis on error-free procedure and virtually certain frequency estimates:

In the popular mind the test became confused with the mapping of human genes, whereas in fact it probes only a handful of points on the chromosomes and ones which have no known function in determining physical makeup. The often faint, fuzzy, and distorted bands produced on autoradiographs were likened to the precise and unambiguous patterns of supermarket bar-codes. The chances of an innocent, coincidental match were touted at figures as low as 738,000,000,000,000,000 to one. The process was stated to be incapable of yielding a false match. (McLeod, 1991: 583)

16. Alec Jeffreys, quoted in the *Times,* London, 12 March 1985. In *New York v. Wesley* (1988), a similar statement was made by Michael Baird, a forensic scientist working for Lifecodes (the major competitor to Cellmark—the company with which Jeffreys was associated). Baird testified in direct examination "that the DNA patterns or prints that we get are very individualistic in that the pattern of DNA that you can obtain by using a series of DNA probes is as unique as a fingerprint" (quoted in Aronson, 2007: 46).

17. *Times,* London, 27 November 1985.

18. *Economist,* "Cherchez la gene," January 1986: 68–69.

19. M. McCarthy, quoted in the *Times,* 13 November 1987.

In the first few years after the introduction of DNA typing, defense lawyers were ill equipped to cross-examine the prosecution's experts on the
subject, and to challenge the extraordinary frequency estimates they
gave.[20] In *Andrews v. State* (1988) the defense challenged the admissibility of the DNA evidence, but all of the expert witnesses who testified
during the pretrial admissibility hearing were called by the prosecution.
The defense attorney cross-examined the expert witnesses, but the questions often seemed ill informed. For example, in the cross-examination
of Lifecodes witness Michael Baird, the defense attorney asked open-
ended questions that relied upon Baird to specify possible problems, and
did not press him with specific questions about documented sources of
error or doubtful population genetic and statistical assumptions. Not
surprisingly, Baird did not come to the aid of his interrogator.

> Q. Are there things other than the PH and the conductivity by which the re-
> agent can cause the test to go afoul?
> A. Those are the basic points that need to be in place in order for the test to
> work.
> Q. Would any type of foreign substance contamination within the substance
> make a difference?
> A. Not in our experience.[21]

The cross-examiner *sought,* rather than *used,* knowledge about how the
procedures in question worked, and the questions sometimes pursued
contingencies (such as variations in voltage applied to gels) that seemed
not to trouble the witness. The questions sometimes furnished the witness
with the opportunity to expose the questioner's ignorance, such as when
Baird corrected the defense attorney for assuming that the restriction
fragments used in Lifecodes' analysis are from coding regions of DNA:

> Q. Do you know which ones they are? Is number one, for example, a pre-
> disposition for diabetes and number two blue eyes or can you tell us just
> that?
> A. These probes do not recognize anything that is understandable in terms of
> those, those kinds of physical traits.

20. See Aronson (2007: chap. 3) on the introduction of DNA analysis in the United
States.
21. *Andrews v. State* (1988).

Q. All right.

A. They are called, you know, anonymous DNA regions.

When attempting to discredit prosecution witnesses with summary arguments, the defense made general ascriptions of vested interest to scientists, such as a prominent molecular biologist from MIT who testified for the prosecution:

> I would suggest by that while Doctor Houseman's credentials are impressive, to say the least, that he is not a totally dispassionate, totally disinterested member of the scientific community and may well have a career interest in having this test determined to be reliable by coincidence, since he also draws his paycheck by virtue of doing five to ten of these a week. And if the test were not found to be reliable, he might well suffer some career damage from that. (*Andrews v. State,* 1988 : 66)

This hypothetical argument, stressing the lack of total disinterest on the part of the witness, was easily rebutted by the prosecutor.

> The court heard from an independent witness from MIT, who I doubt seriously has any true vested interest in the outcome of this case. I don't think that his paychecks or his position on the faculty at MIT since 1975 would be severely damaged if this gets into evidence as Mr. Uhrig suggests. (67)

The defense brief in another early case (*New Jersey v. Williams,* 1991: 5), deployed an even more global use of an interest argument to discredit the prosecution experts' testimony about PCR (the defense called no expert witnesses):

> The prosecution called nine (9) witnesses. All were qualified as experts in various fields of molecular biology, microbiology, genetics, immunology, population statistics, polymerase chain reaction, forensic serology, forensic science, forensic biology, DNA molecular biology etc. Each of these witnesses was infirm either by reason of close association and economic and professional reliance upon Cetus [Corporation] and the test in particular or they were not competent to give an opinion in the forensic context of PCR.

The defense attorney associates the fact that the witnesses were recognized as having specialized knowledge about PCR with a vested interest in promoting a corporate product (which at the time was held under patents assigned to Cetus Corporation). Then, to dismiss the testimony by witnesses with academic credentials, the attorney adds that they were not competent to give an opinion about the forensic context of use. Referring to another notable witness—Henry Erlich of Cetus Corporation—the defense attorney associated the scientist's weighty curriculum vitae with his expensive suit (conveying a distinctive sense of "lawsuit," in an attempt to wed class resentment with suspicion of expert authority): "If a juror cannot quite understand allele drop-out or mixed samples, the issue should not be admitted because Dr. Erlich wears a five hundred dollar suit and has a CV four pounds in weight."[22]

Such arguments, which are commonplace in trials involving expert evidence, are variants of what S&TS scholars have called "interest arguments," and the *Williams* attorney also invokes the word "context" in a familiar way to mark specific organizational differences in the configuration of an innovation. Attributing seemingly "objective" evidence to specific social interests and contexts is a well-known explanatory strategy in the sociology of scientific knowledge (see, for example, Barnes, 1977),[23] but contrary to the general explanatory aims of the sociologist of knowledge, the attorneys in these early cases attempted to discredit the credibility of *particular* claims. Interest arguments used in a particular court case can be effective, depending upon the salience of the interests in question and the jury's receptivity to the attorney's line of attack. In the above instance, however, the attorney's argument did not specify just how the witness's alleged interest in promoting the technique biased the specific evidence he presented. Indeed, the attorney was reduced to complaining that Dr. Erlich's evidence was incomprehensible, and he attempted to transform the witness's CV from being a record of impressive expert credentials to being evidence of vested interests.

22. *New Jersey v. Williams* (1991).

23. Interest explanations are both regularly used and regularly discredited in the sociology of scientific knowledge. The problem is connected with the "regress" problem: that the sociologists trades in a form of attribution that can easily be turned against her own account. This possibility of confusing (or, the impossibility of *separating*) the discursive form of a social explanation from that of a social debunking strategy challenged the sociology of knowledge from the outset. Efforts to "strengthen" the sociology of knowledge by refusing to exempt science and mathematics from its purview remained vulnerable to the argument that "interest explanations" involved an inherent contradiction when familiar argumentative tropes were mobilized in a general, nonevaluative, program of explanation (Woolgar, 1981; Gilbert & Mulkay, 1984; Lynch, 1993).

Not only did defense attorneys expose their ignorance in early cases, judges sometimes did not fare very well either when they waded into dialogue with expert witnesses. During the admissibility hearing for *New Jersey v. Williams* (1991), the judge (the Court in the transcript below) played the part of a befuddled and yet authoritative Simplicio confronted by a savant, Edward Blake, appearing for the prosecution) in a Galilean dialogue.[24]

THE WITNESS. The PCR product is evaluated again with a test gel, to see whether or not this 242 base pair DNA fragment has been produced in the sample.

THE COURT. Now when you say—when you reduce it to its pure form, it is about a drop.

THE WITNESS. Well, one has about a drop of fluid. Now—

THE COURT. Of pure DNA?

THE WITNESS. No, no, no. No, no, no. This is, perhaps, the thing that is confusing the Court. The court apparently has the idea that you can see molecules.

You can't see molecules. But you can test for their consequence. You simply—one has the idea that you have one of these cocktails. You have one of these cocktails and there is a lid on this thing. A little cap and that's probably about one hundred times larger than what we have. We have this fluid here and we stick it in a thermal cycler and after 30 or 40 cycles the stuff comes fuming out and all of a sudden your laboratory is taken over by these DNA molecules. That's not what we are talking about here, Judge. It is not like—it is not like in one of these things you see in science fiction movies. (43ff.)

THE COURT. If I were to look at that test tube at the beginning and then look at it at the end of three hours, would I see anything?

THE WITNESS. No.

THE COURT. Would it change its color?

THE WITNESS. No.

THE COURT. It would be either—I'd see nothing different?

24. This sequence of trial transcript is also featured in Jordan (1997) and Jasanoff (1998). Similar expressions of judicial naïvety are quoted by Aronson (2007: 45) from the transcript of the admissibility hearing in *New York v. Wesley* (1988). In this case, Baird was testifying for the prosecution, and the judge asked him, "The term genes, what is the relationship with the term DNA?" and "What is a chromosome, Doctor?"

THE WITNESS. You'd see nothing different.

THE COURT. But you would see something inside.

THE WITNESS. No. I could show you how to visualize the consequences of what has happened in those three hours. And the way that you visualize the consequence of what has happened in the three hours is by taking a little bit of that fluid out and running a test gel on it. And when you run a test gel on a little bit of the fluid that is taken out, you will either see or not see a fragment of DNA that was not present in the fluid before that is of the size of this gene that is being amplified.

THE COURT. When you say see, you mean with your own eyes?

THE WITNESS. Yes.

THE COURT. Not with the use of microscopes.

THE WITNESS. I don't mean see with your own eyes in the sense you can see a molecule and you can sit there and count one, two, three, four.

THE COURT. This is where I am having my trouble. How can you see something that you don't see?

THE WITNESS. How do you know something is separating if you can't see it? Because, Judge, you are asking how do the tools of all science work in general when you ask a question like that, and the way you see it is with some technical procedure that allows you to see the consequence of the molecule with a particular set of properties. (46)

This sequence could be described as a tutorial on how to use the verb "to see" in a particular technical context.[25] The judge is conspicuously stuck in a conception of "seeing" that is tied to naked-eye visibility, whereas the expert witness, Edward Blake, deploys the verb in reference to analytical techniques that produce visible displays that implicate molecular processes that are well below the threshold of naked-eye perception. In the latter sense, "seeing" is bound up with demonstrating what is implicated by material evidence. Earlier in his tutorial, Blake referred to what was visibly shown by exhibits (which were not included in the transcript). From what he said, it is clear that the exhibits were schematic diagrams, much like those that are commonly presented in summary accounts of DNA profiling (see fig. A.1).[26] The cartoon convention exhibits DNA with the standard icon, showing it being extracted from the initial

25. See Coulter & Parsons (1991) for an illuminating discussion of the varied uses of verbs of visual perception.

26. See Halfon (1998) for an analysis of an instance of this type of schematic diagram.

samples shown at the start of the scheme. The judge evidently mistook the visual demonstration to be a reference to the relative size of the molecular object in question.

In other early cases, courts, and even defense attorneys, took expert witness statements at face value, apparently because they had no access to contrary information. Eric Lander (1989: 505) quotes from a Lifecodes scientist, Kevin McElfresh, who testified in *Caldwell v. State* (1990)—a death penalty rape-murder trial—and stated that declaring a match is a "very simple straightforward operation . . . there are no objective standards about making a visual match. Either it matches or it doesn't. It's like you walk into a parking lot and see two blue Fords parked next to each other." Lander presents this statement as an example of misleading testimony that courts accepted without effective challenge.

Castro and the "DNA Wars"

It was not until 1989, in the New York murder trial of José Castro, that the first successful challenge to the admissibility of DNA evidence occurred. Before then, DNA evidence had been used in hundreds of U.S. trials,[27] and it was entrenched in many other national court systems. *Castro* was soon followed by many other admissibility challenges, some of which were successful, and it touched off a controversy that was dubbed, with a dose of journalistic hyperbole, the "DNA wars."[28] Difficulties also arose in the United Kingdom in connection with a series of Manchester-area rape cases, and also in an Australian case (McLeod, 1991). For the most part, the DNA wars took place in two venues: criminal courts, particularly admissibility hearings in U.S. courts, and the pages of the science press, especially *Science, Nature,* and lesser science magazines such as *The Sciences*. The two venues were deeply linked: news reports and

27. Aronson (2007: 56) reports that estimates of the number of U.S. cases included an indefinite, but large, number of guilty pleas as well as at least eighty trials in which DNA evidence was admitted.

28. Thompson (1993: 23) traces the "DNA war" or "wars" metaphor to a combination of press and participant commentaries. For example, science journalist Leslie Roberts (1991: 736) quoted John Hicks, head of the FBI Laboratories, as saying that the debate about DNA typing was no longer a "search for the truth, it is a war." Although coincident with various other so-called culture and science "wars," this particular struggle did not focus on trends in universities and the "culture industry," but was a dispute among scientists played out in the science media and the criminal courts.

articles in the science press focused on forensic evidence in court cases, and these reports and articles fed back into courtroom dialogues and judicial rulings.

The pretrial admissibility hearing in *Castro* was exceptional, because the court-appointed defense attorney came prepared with an arsenal of expertise. The attorney sought the help of Barry Scheck and Peter Neufeld—two young attorneys who later, through their involvement in the O. J. Simpson case and the Innocence Project, became famous for having expertise with DNA evidence. In the late 1980s Scheck and Neufeld were just beginning to learn about such evidence through informal seminars on the subject for defense lawyers.[29] They, and a group of other defense lawyers, had become suspicious about the rapid acceptance of forensic DNA evidence and its unqualified presentation in the courts, and their suspicions were shared by a few biologists as well. Because of its timing and placement in New York City, the admissibility hearing in *New York v. Castro* turned into a highly visible forum for staging a challenge. Scheck and Neufeld prevailed upon Eric Lander, who had recently become concerned about the way molecular biology was being used in forensic science, to give evidence for the defense.[30] Other molecular biologists also signed on, creating a rare situation in which the defense was able to counter the prosecution's expert firepower.

The *Castro* case arose from the stabbing deaths of twenty-seven-year old Vilma Ponce and her two-year old daughter in 1987. José Castro, described as a local handyman, was arrested on the basis of an eyewitness report, and a small bloodstain was recovered from his wristwatch. Expert witnesses working for Lifecodes (one of the first private companies to perform forensic DNA analysis) testified that they extracted 0.5 μg of DNA from the bloodstain. Using the single-locus probe method, Lifecodes analyzed that sample, and compared it with samples from the two victims. Probes for three RFLP loci were compared along with a probe for the Y chromosome locus. According to Lander (1989: 502), "Lifecodes issued a formal report to the district attorney . . . stating that the

29. Interview with Lynch and Cole, 21 May 2003. For further information about the situation with defense lawyers, and Scheck & Neufeld's involvement in challenging prosecution uses of novel forensic evidence, see Aronson (2007, chaps. 2 and 3). Also see Neufeld & Colman (1990); Neufeld & Scheck (1989); and Parloff (1989).

30. Aronson (2007: 60) describes how Neufeld first met Lander at a meeting in Cold Spring Harbor, New York, and later persuaded him to review the evidence from Lifecodes that the prosecution put forward.

DNA patterns on the watch and the mother matched, and reporting the frequency of the pattern to be about 1 in 100,000,000 in the Hispanic population. The report indicated no difficulties or ambiguities." Lander then went on to describe "fundamental difficulties" with the report. One problem had to do with the tiny amount and degraded quality of the DNA extracted from the bloodstain (or, as Lander describes it, "blood speck") on the watch. This resulted in an extremely faint trace on the autoradiogram, and the DNA in the stain may also have been contaminated with DNA from bacteria. As Lander recounts, far from discounting the evidence, Lifecodes used the faint quality and potential contaminants as an interpretive resource for preserving the claimed match with the evidence from the mother.[31]

Figure 2.2 presents autoradiographic evidence for one RFLP locus marked by a radioactive probe. The three lanes compare results from the analysis of the blood of the two victims (M = mother; D = daughter) with the results from the speck of blood on the defendant's watch (W). According to Lander, the prosecution witness (Michael Baird, Lifecodes' director of paternity and forensics) who presented the evidence "agreed that the watch lane showed two additional non-matching bands, but he asserted that these bands could be discounted as being contaminants 'of a non-human origin that we have not been able to identify'" (Lander, 1989: 502). Lander went on to identify several other problems with the laboratory procedures and statistical analysis. During the lengthy admissibility hearing, expert witnesses on both sides agreed to meet without the lawyers present, because they were appalled by some of the evidential problems that emerged, and no less appalled by the disrespectful way the lawyers handled the questioning (Roberts, 1992; Thompson, 1993: 43). A "consensus about lack of consensus" came out of this ad hoc meeting. According to one prosecution witness, Richard Roberts, "We wanted to be able to settle the scientific issues through reasoned argument, to look at the evidence as scientists, not as adversaries" (quoted in Lewin, 1989: 1033). The two prosecution witnesses who took part in this meeting later retook the stand and recanted their earlier testimony supporting Lifecodes' determination of a DNA match (Thompson, 1993: 43). The meeting also resulted in a recommendation to the National

31. A similar interpretive strategy was used by prosecution witnesses in the *Regina v. Deen* case discussed in chapter 5, and during an interview (18 October 2005) William Thompson also informed us of more recent cases, using very different techniques, in which a variant of it was deployed by prosecution experts.

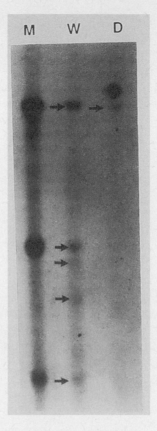

FIGURE 2.2. Single-locus probe, from *New York v. Castro.* Reprinted from Lander (1989: 503).

Research Council—the research arm of the National Academy of Science—to investigate forensic uses of DNA typing.

During the extended pretrial hearing in *Castro,* the defence witnesses presented an array of problems that were not mentioned in the initial Lifecodes report. It seems likely that had Peter Neufeld and Barry Scheck not become involved, and had the defense not called in Lander and others to open up questions about controls, interpretation, sample custody, statistical representation, and so forth, the evidence would have been admitted. After the hearing, the judge excluded the DNA evidence, but in a qualified way. While acknowledging extensive and "fundamental" problems raised by many of the expert witnesses, the judge ruled that forensic DNA profiling "can produce reliable results,"[32] but that in this case the testing laboratory failed to use "the generally accepted scientific techniques and experiments." The credibility of the technique itself was thus preserved, while blame was laid at the door of the way it was implemented.[33] And, as it turned out, Castro pled guilty, even though the DNA evidence was excluded, but the fate of the DNA evidence received far more publicity.

Despite the limited scope of its victory, the defense, aided by Lander and other expert witnesses and consultants, opened up a chamber of methodological horrors to disclose a litany of problems.[34] Where DNA evidence had been presented, and for the most part accepted without effective challenge, it now seemed to unravel. Defense lawyers such as Scheck, Neufeld, and William Thompson held meetings and exchanged transcripts of admissibility hearings in which DNA evidence was effectively challenged.[35]

The Castro hearing provided an object lesson for lawyers on how to attack DNA evidence. Defense lawyers began to learn more about DNA evidence and proponents no longer expected courts to accept their testimony without question. Even in cases in which the defense did not call expert witnesses, it was possible to use the record from Castro and other challenges to rebut the prosecution's witnesses. William Thompson described an amusing example of how an expert witness could be virtually transferred from one case to another:

32. New York v. Castro (1989), quoted in Thompson (1993: 44).

33. Although it was criticized for being too stringent with some of its recommendations, the first NRC (1992) report also affirmed the credibility of what the Castro court called "the generally accepted scientific techniques," while criticizing the way the techniques were administered in forensic laboratories. Such partitioning is characteristic of official reviews of technological controversies (Perrow, 1984): criticism is ascribed to human error, or in this case corporate error, in a particular case, while preserving systematic features of the technology which, of course, would require a more massive and expensive effort to change. See chapter 9 for a striking instance of this sort of error account in connection with latent fingerprint examination.

34. Steve Woolgar (1988: 30ff.) coined the expression "methodological horrors" to describe highly general interpretative problems both in empirical science and empirical science studies. Here, we use the term in a more particularized way to describe a field of contingencies and possible sources of error that adversary attacks on forensic DNA evidence have raised.

35. William Thompson, interview (May 2003). See Aronson (2007: 78–79).

[W]hat I did was I would pull quotes out of the transcripts or the published statements of these critical scientists and I would say to Bruce Weir [a prominent forensic analyst testifying for the prosecution], "Eric Lander says x, do you agree with that or do you not?" Or I would quote something that Lander said that sounded critical—"Do you agree with it or do you not?" And he'd say, that "Well, uh, I, I disagree." So then I'd say that "At least with respect to this client there's disagreement in the scientific community, is there not?"[36]

Publications by Lander (1989) in *Nature,* and Lewontin & Hartl (1991) in *Science* provided lawyers with authoritative sources they could cite to counter DNA evidence.[37]

Lander (1989) and defense attorneys such as Neufeld, Scheck, and Thompson who took the lead in challenging DNA evidence developed two broad areas of criticism: first, elaborating an expansive array of technical problems with the collection, handling, and analysis of samples; and, second, citing a combination of statistical and population-genetic problems associated with quantifying and reporting upon the probative value of DNA matches.[38] To this pair of problems, we can add a third that often came up in court cases and official evaluations: organizational and administrative problems.

(I) TECHNICAL PROBLEMS AND CONTINGENCIES.   In *Castro* and other early cases, radioactive probes were used to mark polymorphic segments of DNA, and these were photographed on x-ray film. These techniques were often glossed with the acronym RFLP (restriction fragment length polymorphism). The autoradiographic images produced through these techniques produced arrays of bands in adjacent lanes, which were visually inspected for matches. The "professional vision" (Goodwin, 1994) of forensic analysts was called into question during *Castro* and later cases, as defense attorneys and their expert witnesses made a point of the "subjectivity" of the work of handling samples, running equipment, and visually assessing the evidence.

36. Ibid.

37. Prosecutors also could turn to the pages of *Science,* as Lewontin and Hartl's arguments were rebutted by Chakraborty & Kidd (1991) in the same issue of *Science,* leading to further complaints about the way the *Science* editors commissioned the rebuttal after Lewontin and Hartl had submitted their article (see Roberts, 1991).

38. Neufeld & Colman (1990) and Thompson & Ford (1988; 1991) also collaborated with scientists in early publications on the problems with DNA evidence.

When cross-examined about their practices, forensic scientists sometimes admitted to using methods to enhance the visibility of bands, both for analytic purposes and for showing evidence in court. Sometimes computer enhanced images were used to produce "cleaner" images than those originally developed on autoradiographs, and computers also were used to analyze bands, which were sometimes said to "create a result where none was apparent before by, for instance, locating very light bands" (Office of Technology Assessment, 1990: 119).

In the *Castro* case, when questioned closely about their procedures, Lifecodes analysts acknowledged that, in order to support their determination that there was a match they found it necessary to discount particular discrepancies and to enhance the visibility and alignment of bands. The *Castro* defense argued that the forensic analysts assumed the existence of the match they set out to test, and treated discrepancies not as evidence of a mismatch, but as evidence of artifacts that partially obscured the match. For example, they could attribute such discrepancies to variations in the composition of a gel, or in the amount of electric current that "drove" samples in different lanes. When their ad hoc practices are considered at a more abstract level, the Lifecodes scientists faced a familiar dilemma in the history of experimental science:[39] either they would have to throw out the evidence, despite an overall conviction that there was a match, or they would have to dismiss particular discrepancies in order to save the evidence. Weighing on both sides of the dilemma were considerations such as, on the one hand, the cost of freeing a murderer because imperfect evidence was discounted, and, on the other, prejudicing the case against the accused person by presuming evidence of guilt in the very analysis of that evidence. In response to criticisms in *Castro* and other early cases of ad hoc accounts of band shifting,[40] forensic

39. Although ad hoc practices are commonly associated with unsound scientific practice, a deep question remains as to whether they are a necessary part of the local, judgmental work of examining data. In a study of sociological efforts to code data, Garfinkel (1967) suggested that ad hoc practices were inescapable resources for making sense of, and classifying, data. Holton's (1978) analysis of Robert Millikan's oil drop experiment is the canonical study of a case that documents an experiment that succeeded *because* the experimenter violated the canons of experimental practice by presuming the correct result in the analysis of the data from the experimental runs that demonstrated that result. Daniel Kevles's (1998) study of the Theresa Imanishi-Kari/David Baltimore affair documents how badly things can go wrong when outsiders appeal to popular conceptions of experimental practice when investigating a charge of fraud.

40. For example, band shifting was a major issue in *Maine v. McLeod* (1989). See Jasanoff (1992: 32–33).

organizations and scientific review panels began to develop standards for visually counting bands as aligned or not. Sometimes, such enhancements were as crude as darkening a faint band with a marking pen, or hand-tracing an autoradiograph onto an acetate slide.

Another common target of technical criticism was contamination, whereby "foreign" DNA from human or nonhuman sources (for example, bacteria) can mix with, or even replace, the "target" DNA in a sample. Several possible sources were mentioned in *Castro* and later cases, and virtually every step of the process from collecting, storing, transporting, and analyzing evidence was cited as a source of possible contamination.[41] Police and other agents with little or no training in laboratory procedures typically collect and handle criminal evidence samples, and defense attorneys often focused on the possibility that they could deliberately or inadvertently cross-contaminate evidence samples.

The exercise of "subjective" (visual) judgment, and the possibilities of cross-contamination were among the most common themes in courtroom challenges, perhaps because they were relatively easy to grasp. However, many other possible sources of technical error at virtually every stage of analysis were aired during *Castro* and other contentious cases.

(2) STATISTICS AND POPULATION GENETICS. The most widely debated issues in *Castro* and other early cases had to do with the appropriateness of reference samples. In order to estimate the probability that a randomly chosen, unrelated individual's DNA profile would match a profile in evidence, it is necessary to specify the reference database. This is because the alleles marked with DNA probes occur at different rates, on average, in different human groups. Closely related people are more likely to share alleles—with identical twins being the limit case. DNA profiling was implemented for criminal investigations before extensive databases had been compiled. Early databases were cobbled together from blood

41. The possibility of cross-contamination became even more acute when PCR-based techniques were widely implemented. A 1990 report by the Office of Technology Assessment (OTA) noted that with techniques using PCR, cross-contamination was especially problematic: "Even flipping the top of a tube containing DNA can create an aerosol with enough sample to contaminate a nearby tube" (OTA, 1990: 69). The report added, "Should a suspect's sample accidentally contaminate a questioned sample containing degraded (or no) DNA, subsequent PCR amplification of the questioned sample would show that it perfectly matched the suspect" (70). (This possibility later provided a basis for argument in the O. J. Simpson trial.)

banks, volunteers from police forces, and other nonrandom samples, and they were divided into very rough "racial" subgroups. In early cases, databases were limited, especially for minority categories such as "Afro-Caribbean" (a category used in the United Kingdom).

Questions about statistical estimation and population genetics were aired during *Castro,* and later became a subject of heated controversy in *Science* (the magazine). Lewontin and Hartl (1991; 1992) raised a thorny population genetic problem, which questioned the procedure of establishing allele frequencies for a general population and broad "racial" groups (or census categories): North American Caucasians, African Americans, Hispanics, Asians, and Native Americans in the United States; Caucasians, Afro-Caribbeans, and (South) Asians in the United Kingdom. Lewontin and Hartl argued that such groups were too crudely defined to accurately measure the probability of finding matching alleles in segregated urban areas, American Indian tribes, or isolated rural villages. This is because persons in such groups are likely to be more closely related to one another than to randomly chosen members of the general population or even of a broadly defined "racial" subgroup. Given patterns of residential segregation, and the fact that for street crimes the most likely suspects other than the defendant tend to live in the same local area and be similar in age, stature, ethnicity, and so forth, persons in the pool of most likely suspects would be more likely to share alleles than would randomly chosen persons from a general population or population subgroup. The problem of tailoring probability estimates to the demographics of the suspect pool in a particular case gets worse when one assumes that a close family member (such as a brother, or a "hidden half-brother" covertly sired by the suspect's father) is a possible suspect. In theory, at least, the relevant frequency estimates will vary with the constitution of the suspect pool, which, in turn, interacts with the substantive details of the case and assumptions about neighborhoods and crime patterns.

Further debates concerned the common practice of generating forensic estimates by multiplying estimated frequencies of alleles in the relevant human population for each matching band (probe). Critics—most notably Lewontin and Hartl (1991; 1992)—questioned the Hardy-Weinberg equilibrium assumption that each matching allele is statistically independent, and that calculations of very low frequency in the population can be made with relatively small samples.

(3) ORGANIZATIONAL CONTINGENCIES.   As noted earlier, evidence samples that forensic laboratories analyze do not originate in the laboratory: they are collected at crime scenes by police employees, many of whom have little or no training in biology or forensic science. The samples commonly are transported in police vehicles (or the vehicles of private services contracted to the police), and stored at police facilities before being housed in a laboratory. If one extends the contingencies of the laboratory to include the sources of the material samples analyzed, then "laboratory error" can arise at any point in the continuum of practices (or "chain of custody") from crime scene to court. Chains of custody are shot through with organizational contingencies, and they are addressed through bureaucratic and administrative work, including work performed at crime scenes by police and pathologists, and in laboratories by administrators and staff scientists.

The summary term "preanalytical error" is sometimes used to describe the array of mistakes that can result in the possible spoilage, mislabeling, and misplacement of evidence samples. One can even extend the preanalytical sources of problems, if not errors, to the condition of evidence samples at the scene of the crime: conditions such as burial in damp earth, freezing and thawing, baking and irradiation under the sun, and contamination by animal deposits, or bacterial or other microbial infestation. They also can include conditions under which samples are stored in a laboratory or police station.[42]

A problem that emerged very early (and which emerged more recently in connection with a scandal about the handling of DNA samples in a Houston crime laboratory) involved mundane errors in the labeling and handling of samples.[43] For example, in a case in Reading, England, the laboratory mixed up two samples and the suspect was convicted of rape on the results of the analysis of another suspect's blood. At the time of the *Castro* case, and to some extent still, forensic organizations employed different local protocols, maintained proprietary secrets, and incompletely implemented recommended proficiency tests and controls (Thompson et al., 2003).

In *Castro*, the defense charged that Lifecodes kept poor records of the probes and controls it used. The company's expert witnesses, when questioned about the identity of the control sample for the XY chromo-

42. See McLeod (1991: 585, note 15).
43. See chapter 8.

some test, first said the control was from HeLa cells (an "immortal cell line" named after Henrietta Lacks, the source of the lethal cancer from which the cell line was first developed in the 1950s), which was odd, since the profile showed no evidence of Y chromosome. Then, it was traced to a male Lifecodes employee who, it was alleged, had an unusually short Y chromosome. Lander questioned this, and Baird (the Lifecodes witness) later testified that the source of the evidence was a woman identified via the resultant RFLP profile (which, Lander noted, was an odd way of treating a control sample). "The confusion had probably resulted from faulty recollections (by Baird and the technician) and faulty inferences (about the male scientist), but it underscored the need for meticulous record-keeping in DNA forensics, which may not originally have been so clear" (Lander, 1989: 503). As this quotation makes clear, mundane administrative practices (record-keeping, in this case) often framed the credibility of DNA data and interpretations of those data. And, as we shall see in chapter 7, recommendations for assuring the quality of administration often stood proxy for technical procedures and the credibility of analytical results.

Conclusion

The problems discussed in this chapter barely scratch the surface of the often highly technical and circumstantial problems that were aired in court and in the science press in the late 1980s and early 1990s . Problems grew more numerous, and more technically complicated, the more that lawyers, expert witnesses, and critical analysts pursued them. Moreover, the pursuit of such problems, especially in the context of the adversary legal system, infused mundane police and laboratory protocols with monumental significance. The "deconstructive" operations of adversary interrogation peeled back the impressive veneer of DNA matches and the impressive statistics that accompanied them, creating a set of problems that criminal justice organizations and science advisory panels set out to repair in the early 1990s.

Despite the effectiveness of "deconstructive" strategies in some cases, DNA profiling continued to be used, for the most part effectively, in hundreds of cases in the United States and United Kingdom. And, within a few years after the *Castro* case, it became common to read that the controversy was over. By the late 1990s, it was widely assumed that the most significant of the problems raised during the DNA wars of the early 1990s had been solved, due to technical and administrative efforts to standardize protocols, reorganize laboratory practices, and train personnel. The *assumption* that the most serious problems had been solved was an important element of the closure of the controversy, but we shall have occasion to revisit at least some of those problems in the aftermath of closure.

Before delving into the closure of the controversy, and into the aftermath of closure, we shall examine specific themes—protocols, chains of custody, and the uses of statistical estimates—which were problematized in specific cases in the United States and United Kingdom. An appreciation of those themes and how they became problematic will enable us to better appreciate how the currently "unassailable" status of DNA evidence was achieved.

Admissibility, Controversy, and Judicial Metascience

In order to understand the scope and configuration of the controversy over DNA profiling, it is necessary delve into some aspects of U.S. law governing expert testimony. The techno-legal controversy about DNA typing was not limited to experts in a specialized research field. The dispute was aired in trials and admissibility hearings, and while experts were central players, their contributions were solicited and interpreted by lawyers, judges, and jurors. This interlude summarizes key U.S. federal legal decisions about the admissibility of expert evidence. Though focused on a specific court system, these decisions (especially the Supreme Court's 1993 ruling in *Daubert v. Merrell Dow Pharmaceuticals*) had broad significance within and beyond the U.S. federal courts.

From the standpoint of Anglo-American jurisprudence, an *expert* possesses specialized knowledge that is relevant to the evidence in the case at hand. Ordinary witnesses are restricted to testifying about "first-hand" evidence that was directly seen or heard. Inferences based on general experience and statements gained through "hearsay" are inadmissible. Expert witnesses are partly exempted from the hearsay exclusion: they are allowed to comment on general knowledge, including abstract knowledge learned from textbooks or teachings as part of specialized training, rather than through "direct" personal acquaintance. This license to give "hearsay" testimony is limited to knowledge based on the expert's specific field of competence. Science and medicine are canonical forms of expertise, but the law has long recognized an open-ended variety of practical arts and technical skills that do not claim to be scientific. An expert is recognized not as a virtuoso with idiosyncratic insight, but as a spokesperson for a specialized body of knowledge that is relevant to the case at hand. In the Anglo-American system, experts are called to testify for the contesting parties (plaintiff and defendant in a civil case; prosecution and defense in a criminal case), and are often paid a negotiable fee by the party for which they testify.[1]

Even though, with rare exceptions, courts have paid scant regard to the epistemological problems associated with distinguishing direct from inferential knowledge or demarcating science from nonscience, they recognize that judges and juries can face thorny problems with evaluating expert evidence.[2] Especially in the past century, with the increasing prevalence of expert evidence in civil and criminal cases, jurists have long worried about the possibility of clashing expert testimony being played out before a bewildered jury:

> The whole object of the expert is to tell the jury, not the facts, as we have seen, but *general truths derived from his general experience.* But how can the jury judge between two statements each founded upon an experience confessedly foreign to their own? It is just because they are incompetent for such a task that the expert is necessary at all . . . when the conflict [between experts] is direct and open, the absurdity of our present system is apparent. (Hand, 1901: 54)[3]

A solution to the potential "absurdity" noted by Justice Learned Hand is the vetting of expert witnesses by the trial judge before the jury is exposed to their testimony.[4] Two procedural institutions are involved: the *voir dire,* in which an individual witness is questioned by the adversary attorneys (with the judge also interjecting questions and comments) to establish his or her relevant credentials and competency; and the admis-

1. Under Federal Rule of Evidence no. 706 it is possible for the judge to summon expert witnesses, but this is rarely exercised. In some circumstances, the state will cover expert witness fees for indigent defendants.

2. Thomas Gieryn uses the term "boundary work" to denote the difference between philosophical demarcation and practical distinction: "Even as sociologists and philosophers argue over the uniqueness of science among intellectual activities, demarcation is routinely accomplished in practical, everyday settings" (Gieryn, 1983: 781).

3. Quoted in Solomon & Hackett (1996: 150).

4. It might seem that avoiding jury trials would be a way to avoid the problem, but judges are largely in the same position as jurors when it comes to evaluating expert evidence (Edmond & Mercer, 1997; Taroni & Aitken, 1998).

sibility hearing, in which the witness is questioned about the field of expert knowledge that he or she is prepared to represent. Although expert witnesses are always quizzed about their competency and credentials before being allowed to testify, admissibility hearings are not held in every case. Such hearings must be requested by one or the other party, and legal grounds must be given as part of the request.

The Frye Rule

The 1923 federal court case *Frye v. United States* established a rule for determining the admissibility of expert testimony. This rule was applied in federal courts, but variants of it were adopted in many state courts, such as New York and California. *Frye* was a murder case in which the defendant attempted to submit evidence from an early form of lie-detector test that measured changes in the subject's systolic blood pressure. The person being tested would be asked questions, and rapid rises in systolic pressure were taken to indicate an emotional response to the immediate question. The testing methodology presumed that an expert could infer when such leaps were associated with the respondent's struggle to hold back an incriminating answer. The brief judicial ruling in *Frye* noted with a hint of skepticism that "the theory seems to be that truth is spontaneous, and comes without conscious effort, while the utterance of a falsehood requires a conscious effort, which is reflected in the blood pressure." The appeal court affirmed the district court ruling that such evidence, though relevant to the case, was inadmissible. In a much-quoted passage, the court gave a general reason for this decision:

> Just when a scientific principle or discovery crosses the line between the experimental and demonstrable stages is difficult to define. Somewhere in this twilight zone the evidential force of the principle must be recognized, and while courts will go a long way in admitting expert testimony deduced from a well-recognized scientific principle or discovery, the thing from which the deduction is made must be sufficiently established to have gained general acceptance in the particular field in which it belongs.

The *Frye* rule, or "general acceptance standard" as it is often called, was a brilliant practical solution to an intractable problem in the social distribution of scientific knowledge. The problem, a variant of what H. M.

Collins (1987) has dubbed "the knowledge gap," applies to the judge who makes an admissibility ruling as well as the jury on whose behalf the ruling is made: How can a person who is not competent in the relevant specialty assess the credibility of (alleged) knowledge put forward by a specialist?[5] The *Frye* court's solution ostensibly relieves the court of the burden of delving into the content of the knowledge in question in favor of deciding whether it is generally accepted in the relevant field. A social-historical inquiry—akin to a social-historian's question of whether there is *consensus* in a particular field on a particular factual or theoretical matter—stands proxy for an evaluation of the substantive content or quality of the evidence in question. Accordingly, a judge who claims no competence in the field can take testimony from representative experts, and read depositions, "friend of the court" (*amicus curiae*) briefs, and other documents in order to determine if the source of the evidence is "generally accepted" in the relevant fields. The *Frye* rule is circumscribed by two considerations: first, it applies only to evidence allegedly based in a *science;* and, second, it applies only to techniques or theories that are *novel* sources of legal evidence.[6]

The *Frye* "general acceptance" standard remained in play in federal courts for seventy years, and it remains in force today in many state courts, but as critics of that standard pointed out over the years, determinations of *relevancy* and *general acceptance* are far from straightforward. For some forms of evidence, more than one field of expertise may be deemed relevant, and determining the outlines of any given field can be uncertain and contentious. Judgments of relevancy involve considerations of evidential quality: a judge would not deem the "knowledge" from, say, astrology relevant for deciding the facts of a case, even if astrologers agreed among themselves.

5. As Collins recognized, this problem has at least two general loci in the sociology of knowledge: first, it is located in the communicative practices that sociologists examine (in this case, the courts, but also in many other occupations and professions), and, second, it applies to the sociologist's own efforts to apprehend the practices studied. In later work, Collins and Evans (2002) put forward a solution, of sorts, by formulating a category "interactional" expertise which falls short of making substantive contributions to a field, but which can be sufficient for some purposes of engaging with, understanding, and even evaluating practitioners' discourse.

6. In fact, *Frye* hearings are sometimes held in cases in which established forms of expert evidence (and not necessarily scientific evidence) are subjected to renewed challenge—as we shall see when we examine recent challenges to fingerprint evidence.

The Daubert *"Factors"*

The *Frye* rule is sometimes said to involve judicial deference to experts (Jasanoff, 1995: 62), because the judge takes agreement among the experts who testify in the case as a proxy for consensus in the relevant fields, and does not independently evaluate the technical evidence in question. The judge must decide from the evidence at hand whether a given witness is, in fact, a bona fide member of the relevant field, as opposed to a crank, charlatan, or incompetent. Judicial assessments of reliability thus are implicit whenever the *Frye* rule is implemented in particular cases, but there often is little basis for such assessments other than the testimony of the witnesses in question. Consequently, the Supreme Court in 1993 used its decision in the case *Daubert v. Merrell Dow Pharmaceuticals* as an occasion to review and clarify admissibility standards.

The *Daubert* ruling arose from a series of class action lawsuits that began in the late 1970s. The lawsuits involved a pharmaceutical drug, Bendectin, which was prescribed for morning sickness (nausea and related symptoms experienced during pregnancy) and was taken by more than 30 million women between 1953 and the mid-1980s, when it was pulled off the market in response to thousands of lawsuits alleging that it was a cause of a specific type of limb defect in children born to mothers who had taken the drug. Different companies were named in the suits, though Merrell Dow Pharmaceuticals Inc. eventually held title to Bendectin after a series of mergers, and was the party named in the lawsuit that reached the Supreme Court. The plaintiffs and defendants put forward systematically different, and contradictory, expert evidence. In most cases, courts ruled that the epidemiological evidence presented by the defendants was "generally accepted" (with publication in peer-reviewed journals being the main proxy for general acceptance), whereas the plaintiffs' evidence (which included results from animal studies, structural chemistry, and critical metaanalyses of the epidemiological studies) was unpublished, speculative, and therefore not admissible. Given the importance of the expert evidence in these cases, decisions not to admit the plaintiffs' experts effectively resolved the trials. In some cases, however, the plaintiffs' evidence was admitted, and in a few notable cases, juries awarded large damages to the plaintiffs. Critics complained that jurors were swayed less by the scientific evidence, and more by the pathetic spectacle of the damaged children and by the savvy lawyers' choice of exemplary mothers who were likely to evoke sympathy. A polemical tract written by Peter Huber—*Galileo's Revenge: Junk Science in the Courtroom* (1991)—chronicled these trials and bemoaned the quality of the "junk science" that was purveyed by plaintiffs to unwitting and credulous juries.

> [Junk science] arises when a witness seeks to present grossly fallacious interpretations of scientific data or opinions that are not supported by scientific evidence. Junk science is a legal problem, not a scientific one. It is cultivated by the adversarial nature of legal proceedings, and it depends on the difficulty many laypeople have in evaluating technical arguments. (Foster & Huber, 1999: 17)

Junk science became a target of a popular campaign, often but not exclusively run by corporate-funded opposition to toxic torts and environmental and safety legislation. The U.S. Supreme Court is well insulated against such popular campaigns, but it is clear that when the Court agreed to review one of the Bendectin lawsuits on behalf of two children born with birth defects—Jason Daubert and Eric Schuler—it shared Huber's concern about the quality of scientific evidence that had been admitted in some of the lower court trials. The most immediate reason for taking up the case was to advise trial judges on whether to follow the common law standard provided by *Frye* or the legislatively enacted Federal Rules of Evidence (which had been reviewed and modified by the U.S. Congress in 1975). In its landmark ruling, the Court stated that the Federal Rules of Evidence should supersede *Frye*. As we shall see, however, the Court did much more than that.

The *Daubert* ruling mentions several of the Federal Rules of Evidence, but the focus was on Rule 702, which stated at the time (it was later modified): "If scientific, technical, or other specialized knowledge will assist the trier of fact to understand the evidence or to determine a fact in issue, a witness qualified as an expert by knowledge, skill, experience, training, or education, may testify thereto in the form of an opinion or otherwise."[7] On the face of it, this rule is far from restrictive:

7. Rule 702 was amended in the year 2000, in a way that brought it more closely in line with the *Daubert* ruling (see Gross & Mnookin, 2003), with the qualification that such a witness might testify "if (1) the testimony is based upon sufficient facts or data, (2) the testimony is the product of reliable principles and methods, and (3) the witness has applied the principles and methods reliably to the facts of the case."

"expert" is given an eclectic definition, and the grounds of expertise are open-ended. And, as the *Daubert* ruling states, the legislative intent expressed in Rule 702 appears to be to encourage judges to take a relatively relaxed view on the admissibility of "scientific" and other expert evidence, so that the "trier of fact" (the jury or, in a case with no jury, the judge) can decide if such evidence is credible after it is tested in the trial court. Indeed, immediately after the Supreme Court sent the case back to the lower courts with instructions to use the Federal Rules of Evidence rather than the *Frye* rule, many press reports interpreted the decision to be a victory for the plaintiffs, on the assumption that Rule 702 was less restrictive than the "general acceptance" standard (Foster & Huber, 1999: 263ff.). It seemed to follow that the plaintiffs' evidence, which had been excluded by the district and appeals courts, would now be admitted in trial. However, such reactions overlooked the way the Supreme Court majority reread and elaborated upon Rule 702.[8]

Justice Harry Blackmun, writing for the majority, noted that the Federal Rules do, in fact, oblige the judge to "screen" evidence prior to trial, to "ensure that any and all scientific testimony or evidence admitted is not only relevant, but reliable." To support this conclusion, he performed an interesting hermeneutic exercise:

> The primary locus of this obligation is Rule 702, which clearly contemplates some degree of regulation of the subjects and theories about which an expert may testify. "*If scientific,* technical, or other specialized *knowledge* will assist the trier of fact to understand the evidence or to determine a fact in issue" an expert "may testify thereto." The subject of an expert's testimony must be "scientific . . . knowledge." [n.8] The adjective "scientific" implies a grounding in the methods and procedures of science. Similarly, the word "knowledge" connotes more than subjective belief or unsupported speculation.

Blackmun's use of the ellipsis ("scientific . . . knowledge") focuses attention on one item from the open-ended list in the original passage that included "other" forms of specialized knowledge.[9] He goes on to quote from an *amicus curiae* brief to make the point that scientific knowledge

8. Also important was Judge Kozinski's appellate court decision on remand, released in 1995, which explicated a restrictive interpretation of Rule 702 (*Daubert v. Merrell Dow Pharmaceuticals, Inc.,* 1995).

9. See Haack (2005: S68) for comments on the confusions wrought by Blackmun's elision of "other" forms of knowledge.

is tentative and fallible, and that it should not be held up to standards of absolute certainty:

> "Instead, it represents a *process* for proposing and refining theoretical explanations about the world that are subject to further testing and refinement." . . . But, in order to qualify as "scientific knowledge," an inference or assertion must be derived by the scientific method. Proposed testimony must be supported by appropriate validation—*i.e.,* "good grounds," based on what is known. In short, the requirement that an expert's testimony pertain to "scientific knowledge" establishes a standard of evidentiary reliability.

Then, he explicated the practical consequence of this reading for trial judges:

> Faced with a proffer of expert scientific testimony, then, the trial judge must determine at the outset, pursuant to Rule 104(a), [n.10] whether the expert is proposing to testify to (1) scientific knowledge that (2) will assist the trier of fact to understand or determine a fact in issue. [n.11] This entails a preliminary assessment of whether the reasoning or methodology underlying the testimony is scientifically valid and of whether that reasoning or methodology properly can be applied to the facts in issue.

Blackmun presented this reading as an exegesis of what was *already present* in its original formulation (although it is notable that a later revision of Rule 702 brought its language into more obvious alignment with the *Daubert* ruling). In a further explication of what is meant by reliable scientific knowledge, Blackmun then articulated a list of four or five (later interpretations differ on the number) "general considerations." Citing philosophers of science Carl Hempel and Karl Popper, he asserted that scientific knowledge is a matter of generating and testing hypotheses so that they can be falsified, and that falsification is what demarcates science from other fields of inquiry.[10] He then went on to cite a number of sources to make the point that "peer review" is another "pertinent consideration," albeit not "dispositive" for consideration of scientific validity. He then cites "known or potential rate of error" and "standards

10. Haack (2005: S67) argues that Blackmun conflated Popper's and Hempel's philosophies of science, resulting in "an unstable amalgam of [. . . their] very different approaches—neither of which, however, is suitable to the task at hand."

controlling the technique's operation," and finally notes that "general acceptance" still can have a bearing on assessments of reliability.

When introducing this list, Blackmun qualifies by warning against using it as a "checklist," and he adds that the "inquiry envisioned by Rule 702 is, we emphasize, a flexible one." Indeed, his interpretation is highly flexible. It is difficult to say what it is a list *of*. The majority opinion and various later commentaries describe it as a list of "considerations," "factors," "standards," "rules," or "criteria." Despite Blackmun's caveats, it is often cited as a discrete list of four or five items, and in later cases such as *United States v. Mitchell* (2004), modified versions of it are treated for all practical purposes as a checklist.[11] At the same time, it grants trial judges discretion to use it "flexibly," so that when deciding on the admissibility of a particular technique (such as fingerprinting in *Mitchell*), the trial judge can overlook the absence of one or another (or perhaps even all) of the listed "factors."

The ruling begged the question of whether it applied to the "technical and other specialized knowledge" that Justice Blackmun elided from his reading of Rule 702. As originally written, the rule surely covered all forms of expert evidence, and yet Blackmun's definition of "knowledge," his emphasis on "reliability" and his list of "factors" explicitly applied to *scientific* knowledge. This ambiguity was taken up by Chief Justice William Rehnquist in his dissenting opinion, joined by Justice John Paul Stevens, which excoriates the majority for having put forward "vague and abstract" guidelines for lower courts. Rehnquist writes incredulously about the Court majority's "parsing" of the Rule's language:

> Questions arise simply from reading this part of the Court's opinion, and countless more questions will surely arise when hundreds of district judges try to apply its teaching to particular offers of expert testimony. Does all of this dicta apply to an expert seeking to testify on the basis of "technical or other specialized knowledge"—the other types of expert knowledge to which Rule 702 applies—or are the "general observations" limited only to "scientific knowledge"? What is the difference between scientific knowledge and technical knowledge; does Rule 702 actually contemplate that the phrase "scientific, technical, or other specialized knowledge" be broken down into

11. *Mitchell* is discussed in chapter 9. It seems not to be an isolated case. There appears to be a general tendency for judges to ignore Blackmun's caveats and to cite the "flexible guidelines" as a checklist (see Denbeaux and Risinger, 2003: 44–45).

numerous subspecies of expertise, or did its authors simply pick general descriptive language covering the sort of expert testimony which courts have customarily received?

Rehnquist also complained that the majority's "definitions of scientific knowledge, scientific method, scientific validity, and peer review," were "far afield from the expertise of judges." He closed his opinion with memorable lines that warned of the hazards of engaging in judicial metascience:

> I defer to no one in my confidence in federal judges; but I am at a loss to know what is meant when it is said that the scientific status of a theory depends on its "falsifiability," and I suspect some of them will be, too.
>
> I do not doubt that Rule 702 confides to the judge some gatekeeping responsibility in deciding questions of the admissibility of proffered expert testimony. But I do not think it imposes on them either the obligation or the authority to become amateur scientists in order to perform that role. I think the Court would be far better advised in this case to decide only the questions presented, and to leave the further development of this important area of the law to future cases.

Rehnquist conflates amateur science with amateur philosophy of science, but his point is that the *Daubert* ruling promises to embroil trial judges both in an exercise of assessing the content of technical evidence and a decision as to whether or not the practices in question live up to abstract standards of scientific methodology. He does not point out that the standards the majority mentions—particularly falsifiability—are contested, and even widely rejected, by philosophers of science.[12]

12. Caudill and LaRue (2006: 5) observe that Blackmun's reading of Rule 702 posed a "daunting task" for judges, that is, to determine whether, in fact, the testimony in question met standards of "scientific knowledge." For a debate about another judicial exercise in philosophy of science, see Ruse (1982; 1986), Laudan (1982), and Quinn (1984). These philosophers debate different positions about Ruse's testimony as an expert witness in *McLean v. Arkansas*. Ruse provided a list of demarcationist criteria that the court cited for justifying the ruling against teaching "creation science" in public schools as an alternative scientific theory to evolution. Laudan and Quinn criticized Ruse's arguments and disputed their acceptability in contemporary philosophy of science. Haack (2005) cites this earlier round of arguments in her criticisms of the *Daubert* court's treatment of philosophy of science.

Kumho Tire v. Carmichael

Six years later, the Supreme Court took the opportunity to clarify the matter of whether and how the *Daubert* ruling applies to nonscientific expert evidence. The case that provided the court with the opportunity for this review was a product liability suit. According to the judicial summary, a van driven by Patrick Carmichael overturned after a tire blowout, resulting in the death of a passenger and injuries to others. The Carmichael family sued the tire manufacture and distributor (collected under the name Kumho Tire Co.), claiming that the tire failed due to manufacturing defects. A key witness for the plaintiffs was an "expert in tire failure analysis" named Dennis Carlson, who gave a deposition based on a visual inspection of the damaged tire after the accident. Carlson acknowledged that the tire was badly worn, with barely any tread left, and that two punctures had been inadequately patched prior to the accident. Nevertheless, he claimed that his inspection of the tire showed that it met two of four criteria for determining defects in manufacture or design, and on that basis he supported the plaintiff's allegation.[13]

The Eleventh Circuit Court of Appeal agreed with the plaintiff and reversed:

> It noted that "the Supreme Court in *Daubert* explicitly limited its holding to cover only the 'scientific context,'" adding that "a *Daubert* analysis" applies only where an expert relies "on the application of scientific principles," rather than "on skill- or experience based observation." *Id.,* at 1435–1436. It concluded that Carlson's testimony, which it viewed as relying on experience, "falls outside the scope of *Daubert,*" that "the district court erred as a matter of law by applying *Daubert* in this case," and that the case must be remanded for further (non-*Daubert*-type) consideration under Rule 702. *Id.,* at 1436. (Quoted in *Kumho Tire Co. v. Carmichael* (1999: 6–7)

According to this line of reasoning, Blackmun's specification in *Daubert* of Rule 702 explicitly referred to science. We can add that, while

---

13. The preponderance of evidence standard in civil law requires the plaintiff to establish that the harmful effect was more likely than not to have been the defendant's responsibility. This standard is less stringent than the "beyond a reasonable doubt" standard in criminal trials, and is less demanding than the explicit and implicit standards of causal demonstration used in many other contexts of argument.

he made clear that the "factors" were guidelines to be used flexibly, he defined "reliability" in accordance with various scholarly writings and popular assumptions about science. Consequently, to impose such a conception of reliability on practices and practical judgments that claim an "experiential" foundation would be like insisting that a master chef must justify claims to expertise by citing criteria appropriate for an experimental chemist (publication in peer-reviewed journals, measured error rates, and so forth). This would involve what philosophers call a category mistake (Ryle, 1954; Winch, 1970) of imposing criteria from one form of activity onto another to which they do not apply. It also was liable to the accusation of "scientism": treating an abstract conception of *scientific* knowledge as the only reliable form of *expert* knowledge. A radically different world from the one in which we live would be required if every species of established expertise—clinical judgment, carpentry skill, artistic judgment, historical scholarship, property appraisal, and so forth— were to be held to the same standards as those the court assumed apply to science. (Indeed, many of the sciences that we currently recognize would fail to meet such standards.)

The Supreme Court's ruling in *Kumho Tire* managed to resolve the ambiguity, first, by asserting that the *Daubert* "factors" (and Rule 702) were meant to apply to all forms of expert evidence and, second, by stressing that district court judges have discretion to apply all, some, or none of those "factors" to a given case. In his majority opinion, Justice Stephen Breyer referred to Rule 702, and asserted,

> This language makes no relevant distinction between "scientific" knowledge and "technical" or "other specialized" knowledge. It makes clear that any such knowledge might become the subject of expert testimony. In *Daubert,* the Court specified that it is the Rule's word "knowledge," not the words (like "scientific") that modify that word, that "establishes a standard of evidentiary reliability."

In other words, the fact that the *Daubert* opinion does not explicitly discuss technical and specialized knowledge does not mean that it applies only to scientific knowledge; it implies that all reliable knowledge is covered by what Blackmun attributed to science. Accordingly, any form of expert knowledge can be admissible, as long as it meets standards associated with scientific knowledge.

Taken at face value, the *Kumho Tire* ruling encourages trial judges to exclude bodies of expert evidence that lack peer-reviewed literature, experimental tests, quantitative error rates, and so forth. However, the ruling also provides judges with an escape clause. The court cites its 1996 judgment in *General Electric Co. v. Joiner* (1997): "As we said before, *supra*, at 14, the question before the trial court was specific, not general. The trial court had to decide whether this particular expert had sufficient specialized knowledge to assist the jurors 'in deciding the particular issues in the case.'" In other words, despite the lengthy discussion in *Daubert* and *Kumho* about general scientific methodology, the question at hand "was not the reliability of Carlson's methodology in general, but rather whether he could reliably determine the cause of failure of *the particular tire at issue.*"

The opinion goes on at length to review the lack of reliability of the particular criteria that Carlson used for making his determination about the tire: there was little evidence that other tire experts used his criteria and there appeared to be no published literature supporting it. The opinion goes so far as to say that not only did Carlson's procedure fail by the *Daubert* standards, it failed by any conceivable standard of reliability. Having noted this, the Court then allows that trial judges may use discretion to assess the admissibility of expert evidence in accordance with *any reasonable standard of reliability.* Consequently, the Court asserts that all expert evidence should be judged under the same standards of reliability (the standards articulated in *Daubert* for "scientific . . . knowledge"), but then says that such evidence can be judged under any reasonable standard of reliability. Anything goes, as long as it is reasonable. And yet, the *Daubert* list remains in place, inviting judges to use it as a set of criteria for excluding expert evidence that does not meet a particular (arguably outmoded, or even incoherent) philosophical definition of science (Edmond, 2002; Jasanoff, 2005).

Implications for Forensic Science

The *Daubert* and *Kumho* decisions provided trial judges with grounds for restricting the admissibility of scientific evidence in civil cases, thus relieving the burdens imposed on civil defendants in "toxic torts," which often are large corporations (Edmond & Mercer, 2000; 2004; Melnick, 2005). Although neither the majority opinion nor dissent addressed forensic evidence, the *Daubert* decision potentially threatened to cause trouble for traditional forensic practices, which have long been accepted under *Frye* but have little to show for themselves in terms of peer-reviewed literature, probabilistic measures, quantitative reliability tests, and so forth (Saks, 2000; Risinger, 2000b).[14]

14. The implications of *Daubert* for forensic science were not necessarily an unintentional side effect. In an interview, Scheck and Neufeld claim that they foresaw *Daubert*'s potential to force forensic science reform, and that they had actually formed a covert alliance with some of the lawyers litigating *Daubert*. (Interview, Michael Lynch & Simon Cole with Peter Neufeld and Barry Scheck, 21 May 2002. See Neufeld [2005] for a discussion of the disappointing impact of *Daubert* for defendants' challenges to prosecution experts in criminal cases.) See Nordberg (2003) for statistical support for Neufeld's complaint, showing that "criminologists & forensic experts" are admitted in 85 percent of cases examined, higher than any other category in the list. Toxicologists, for example, average barely above 30 percent in gaining admission.

CHAPTER THREE

Molecular Biology and the Dispersion of Technique

At about the time of the *New York v. Castro* (1989) case, and in some cases in response to it, several inquiries were undertaken in the United States by the National Research Council (NRC, 1992), the Office of Technology Assessment (OTA, 1990), the National Institute of Standards (NIST), and the FBI's Technical Working Group on DNA Analysis Methods (TWGDAM, 1991).[1] For the most part, these inquiries assumed that the molecular biological principles behind forensic DNA testing were unproblematic, and that the "same" laboratory techniques were used reliably for population genetic research, prenatal testing, paternity testing, and many other applications. The problem was to determine whether or not *forensic* uses of these techniques were acceptable to the larger scientific community, and not just to a narrower community of forensic specialists. This form of comparison—forensic practices held to standards supposedly met by biomedical uses of "the same" techniques—framed much of the public debate as well as many of the official evaluations of forensic DNA typing.

Depending upon the position taken in an adversary dispute, the relation between forensic DNA typing and biological and diagnostic uses of the "same" techniques (gel electrophoresis, Southern blotting, PCR, etc.) could be treated as a source of credibility or a basis for making invidious comparisons. Proponents stressed sameness of technique, and in

---

1. Inquiries also were undertaken in Britain (for example, Royal Commission on Criminal Justice, 1993).

many early cases judges agreed. For example, in a preliminary hearing of *California v. Moffett* (1991), Hon. William D. Mudd commented, after a defense expert had testified about problems with forensic applications of the PCR DQ-alpha technique,

[I]t appears to me that the P.C.R. technique is not an issue. All of the evidence in front of me indicates that the P.C.R. technique at the DQ-alpha loci or the technique itself is not in question. In fact, all of the witnesses agreed that in certain spheres it is acceptable. And I'll comment on some exceptions in a moment. The question really is only its application in the area of forensics, and in that particular vein I'm going to indicate that I find that the particular field of acceptability must include the entire field, research, diagnostics, as well as forensics, because the same system, the same test, the same technique is being used in all of these areas. . . . I somehow find it patently offensive, almost, to have a research biochemist come into court and point a finger at forensics when he's never done any of the work or never even talked to people in the field, and yet that's exactly what's being done.[2]

In contrast to Judge Mudd's remark, doubts were raised in courtrooms, the science press, and science advisory panels about whether the constituent techniques remained *the same* when transferred from biomedical to forensic contexts. This question was prominent in the first of two reports by the National Research Council of the National Academy of Sciences.[3]

The forensic use of DNA typing is an outgrowth of its medical diagnostic use—analysis of disease-causing genes based on comparison of a patient's DNA with that of family members to study inheritance patterns of genes or comparison with reference standards to detect mutations. To understand the

2. *People v. Jessie R. Moffett* (Superior Court of California, San Diego County, May 21, 1991, no. CR 103094, Hon. William D. Mudd, Judge), at 1007.

3. As mentioned in chapter 2, the first National Research Council report (NRC, 2002) was commissioned in the aftermath of the *New York v. Castro* (1989) admissibility hearing. During that hearing, experts from both sides met in a private session, without the lawyers present, and agreed that there were serious problems with forensic DNA analysis, which should be addressed by an appropriate panel. The NRC (the research arm of the National Academy of Science) put together a panel of scientists, lawyers, and forensic scientists to study the problems. The first report was published in 1992, but objections were made to its recommendations both by proponents and critics of forensic DNA testing, and so a second panel was commissioned, which published a report in 1996.

challenges involved in such technology transfer, it is instructive to compare forensic DNA typing with DNA diagnostics. (NRC, 1992: 6)[4]

The report then recited a litany of contrasts: diagnostics involves clean tissue samples from known sources; diagnostic procedures usually can be repeated to resolve ambiguities; diagnostic comparisons are made with discrete alternatives, thus providing "built-in consistency checks against artifacts" (NRC, 1992: 6); and they require no knowledge of the distribution of patterns in the general population.

In contrast, forensic DNA typing "often involves samples that are degraded, contaminated, or from multiple unknown sources" (NRC, 1992: 6). Procedures often cannot be repeated, because of limited amounts of sample material collected at a crime scene. Moreover, in most investigations, the "donor"[5] of a sample collected from a crime scene is an unknown member of a population of possible suspects. To assess the likelihood that a given suspect who matches the crime sample is the "donor," it is necessary to estimate how often the specific DNA profile occurs in the population of possible suspects. Consistent with this view, the NRC recommended a set of administrative quality assurance/quality control procedures to bridge the gap between biomedical and forensic fields of application.[6] Standard protocols were one of the recommended means for assuring the reliability and transparency of technique:

4. DNA typing involves a number of different techniques that were invented at different times and for different purposes. However, contrary to this quotation's conception of "technology transfer," the method of "genetic fingerprinting" developed by Alec Jeffreys and his associates at Leicester University in the mid-1980s was initially developed for paternity testing in connection with British immigration cases, and it was developed for criminal investigations at least as early as for any medical applications.

5. This startling use of the clinical term "donor" represented a linguistic displacement that paralleled the effort to "transfer" techniques from the biomedical world to that of criminal forensics. The term "sample" also extends the language of the clinical laboratory to encompass evidence collected at a crime scene.

6. For the NRC and many critics, the "gap" was temporary and remediable. For William Thompson, it is a basic social fact: "Forensic scientists play a fundamentally different role in society than do academic scientists. The major imperative of the academic scientist is to advance scientific knowledge—to find truth through the use of the scientific method. Forensic scientists also use scientific techniques and seek to find truth in specific contexts. However, forensic scientists' major purpose is to provide a service to a client by answering specific questions about evidence" (Thompson, 1997: 1114–15). Thompson goes on to say that forensic scientists tend to have strong interests in supporting their clients (in most cases, prosecutors). Although prosecutors can make the same argument about defense witnesses, Thompson's argument about forensic science remains pertinent today, despite popular portrayals of criminal forensics as an objective, and even heroic, science.

An essential element of any forensic DNA typing method is a detailed written laboratory protocol. Such a protocol should not only specify steps and reagents, but also provide precise instructions for interpreting results, which is crucial for evaluating the reliability of a method. Moreover, the complete protocol should be made freely available so that it can be subjected to scientific scrutiny. (NRC, 1992: 53)

This chapter investigates protocols for DNA typing, but contrary to the NRC's conception of technology transfer, we draw upon empirical research to question whether the same techniques are, in fact, less problematic in contexts of biomedical research and commercial diagnostics than they are in forensics. We also question the faith invested in protocols as essential elements for reproducing practices and assuring their reliability. We suggest instead that, regardless of their value for assuring practical reliability, protocols are rhetorically significant for claiming credible links between forensic evidence and "scientific" standards. Finally, we briefly examine how the unresolved tension between protocols and practices plays itself out in an adversary dialogue between defense lawyer Peter Neufeld and Dr. Robin Cotton, a forensic scientist testifying for the prosecution, during the 1995 O. J. Simpson trial.

Dispersion of the "Same" Techniques

Although the differences between forensic, diagnostic, and academic applications of the same techniques were, and continue to be, salient for making judgments about the credibility of forensic DNA evidence, our research has convinced us that academic and clinical uses of DNA typing are not inherently more stable and reliable than forensic applications of the same techniques. Researchers in academic laboratories often experience persistent problems with making techniques such as PCR work. It is not as though these techniques were first perfected in academic and diagnostic laboratories, and then simply transferred to forensic laboratories. Instead, in every context of application, researchers face distinctive problems that require them to develop local protocols and skills for making the techniques work (Jordan & Lynch, 1993; 1998).[7]

7. The struggles involved in "making things work" was prominent in early ethnographies of laboratory practices by Latour & Woolgar (1986 [1979]), Knorr (1981), and Lynch (1985a). See Knorr-Cetina (1983) for a summary account.

In line with current thinking in science and technology studies, we question the presumption that the same techniques were transferred or diffused from basic research sites to sites of forensic application.[8] We propose a different picture of the dispersion of technique: an organized movement that is less a transfer or reproduction, and more a localized improvisational struggle that creates variation alongside identity. Moreover, as the term "dispersion" suggests, a technique can lose its initial unity and identity as it moves from one site of production to another.[9]

Biotechnology protocols tend to have a "soft" and malleable design. Although labs often make use of off-the-shelf instruments and reagent kits, and many protocols have been partly or wholly automated, so-called wet-lab techniques remain in play, and they are subject to open-ended variation in different contexts of investigation and production.[10] Consequently, it makes sense to say that protocols are reinvented, sometimes in radical but often in minor ways, in different organizational circumstances.

Protocols

The term "protocol" derives from the Greek, originally referring to an early version of a table of contents: a leaf glued to the front of a manuscript that outlines the contents. Contemporary definitions of the term

8. For an early, and concise, criticism of the idea of "diffusion" of innovation, see Latour (1987: 133). H. M. Collins's (1974) research on the contextual problems of building a laser from plans also indicates that reproducing technology requires an alignment of tacit knowledge and local practice from one site of production to another. At every point, questions of sameness and difference are addressed in material detail. For a historical example, see Simon Schaffer's (1989) account of the uptake of Newtonian optics in seventeenth-century France and Italy. Contrary to the idea that Newton's experiments and principles traveled freely, and with minimal resistance, Schaffer chronicles the way the material qualities of the glass used in Newton's prisms became a site of contestation and a source of skepticism about the reproducibility of experiments demonstrating the refraction of light into monochromatic constituents.

9. See Jordan & Lynch (1992). Also see David Kaiser's (2005) deft treatment of the dissemination of Feynman diagrams for further refinement of the theme of "dispersion." Also see Cambrosio et al. (1990) on how patent disputes draw adversaries into presenting rival, and excruciatingly detailed, specifications of technical similarity and difference (also see Swanson [2007] for a patent dispute about PCR).

10. Karin Knorr-Cetina (1999) attributes to molecular biology a relatively open-ended process of "blind variation" in technique, which distinguishes its epistemic and existential character from the more rigidly closed and controlled environments of particle physics research. For an alternative view of experimental practice in physics that effaces such ideal-typical demarcation from biology, see Doing (2004).

include a set of customs and regulations associated with etiquette or dip-
lomatic affairs; an initial draft of a treaty; and a supplement to an international agreement. A more technical use of the term in analytic philosophy of science—"protocol statement" or "protocol sentence"—refers
to an exact formulation of the sequence of procedures followed in an
observation. Such a statement is supposed to describe the observational
procedure without adding any interpretation.

In biology, protocols are written instructions that specify ingredients,
equipment, and sequences of steps for preparing and analyzing research
materials. Some protocols are published in widely used manuals, such
as *Molecular Cloning: A Laboratory Manual* (Maniatis et al., 1982), informally referred to as "the bible" in the early 1990s (Jordan & Lynch
1992: 80). Other protocols are variants of standard procedures handwritten by, and for, particular scientists and technicians. Whether written for
broader or narrower readerships, all protocols presuppose a degree of
competence on the part of readers who must adapt them to particular
contexts of use. The original sense of the word "protocol," as a written
note or draft that adumbrates a more complicated text, has some relevance to the vernacular conception of protocols in biology, except that
the relationship in question is not between a shorter and longer text but
between a text and a performance.

Analogies with household cooking are often mentioned in connection
with molecular biology, and these analogies work remarkably well. A lab
is like a kitchen, with cabinets stocked with ingredients, drawers full of
utensils, and countertops and shelves fitted with appliances for supplying water, washing glassware, measuring ingredients, heating preparations, and venting the resultant fumes. According to the analogy, a laboratory protocol is like a recipe: it lists the ingredients, specifies measured
amounts of each ingredient, and gives step-by-step instructions for the
preparation. Trends underway in biology also seem to follow closely the
technology of the kitchen and the labor of cooking: commercial firms
supply packaged "kits" analogous to cake mixes purchased from a supermarket, and external services now supply the equivalent of fast food
(complete with similar promises in trade advertisements for speedy delivery of "oligos" [oligonucleotides] and a high standard of quality). Members of the laboratory "family" no longer perform the labor; instead,
anonymous machines and employees of service franchises perform the
manual tasks and deliver the product.

Like a recipe, a protocol is an ideal-typical set of instructions formulated independently of any particular attempt to do what it says. Unlike the philosophical concept of protocol statement, a molecular biological protocol does not aim to describe the exact sequence of a particular performance. A protocol can be simple or elaborate, depending on the circumstances. When given to a skilled and trusted practitioner, instructions are often minimized. For example, in an interview, a lab director described an assistant with "golden hands" who needed only the barest of instructions:

> [She] is the kind of person that I don't have to sit down [with] on a daily basis and say "today I want you to do this and this." What I do with . . . [her] is informally every couple of weeks, I'll say where we are going, "Let's look at this and this." And all of the details of the experiment she does herself. I can walk into the lab and look at the notebook, look at what's running . . . she doesn't even need to be there. What bands are coming up in the tray. And I know exactly what she's doing and why she's doing it, and it's a very comforting position to be in when you don't have to do daily supervision.[11]

In this account, the lab administrator minimizes instructions and assumes a *transparent* relation to the research assistant's work, a transparency facilitated by conventional hierarchical and gender relations in the workplace.[12]

Initial instructions given to novices and low-grade technicians take a more elaborate form, as less background knowledge is presumed, or in some cases such knowledge is actively discouraged. Explanations are given to varying degrees of depth on why one step follows another, or why particular ingredients, temperature settings, and latency periods are used. For example, in a teaching laboratory such explanations instruct students on the underlying biology and chemistry. Explanations

11. Interview of manager from a biotech corporation, conducted by Kathleen Jordan (24 December 1993).

12. This assistant is an "invisible technician" (Shapin, 1989) because her competence at implementing the experiment is taken for granted and trusted without need for elaborate specification. Though the boss trusts her to perform complicated tasks with the barest instructions, he also takes for granted her subordination to the task at hand. This is not to say that she was unappreciated, or taken lightly. He emphasizes that he very much appreciated her work, and she was not a "technician." When asked if his assistant was a "technician" or a "scientist," he responded, "No, she's a master's degree chemist." When asked further if this implied that she was a "scientist," his response was, "She's a scientist." Sometimes a PhD degree marks the boundary between "technician" and "scientist," though in this case the master's degree was treated (with some hesitancy) as the relevant criterion.

of a different sort are used to prepare technicians to be able to adapt or "tweak" the protocol under different circumstances, but in highly routinized work, the lowest-grade technicians are required to act with machine-like regularity, keeping thought and creative variation to a minimum.

Remaking PCR

Protocols for widely used procedures such as the polymerase chain reaction (PCR) are subject to endless variation, depending upon the specific aims of a preparation, equipment and ingredients at hand, degrees of precaution, and many other circumstantial considerations. PCR is a ubiquitous technique for "amplifying" DNA samples. "Amplifying" in this context means greatly increasing the amount, and enabling the analysis of, an initial sample of DNA. The invention of PCR at Cetus Corporation in the early 1980s is a very interesting and contentious story (Rabinow, 1996a), but the technique itself rapidly became a humdrum feature of laboratory work in a broad array of fields. Within a few years of its invention, PCR was used in archaeology, population genetics, and many other fields. In forensic science it became an important part of two DNA profiling techniques: the DQ-alpha system, denoting a single hypervariable locus of the human leukocyte antigen, and the STR (short tandem repeat) system, which uses a combination of markers. By the late 1990s, the STR system was widely adopted internationally as the technique of choice. The principal advantage of PCR for purposes of criminal identification is that it requires a miniscule amount of DNA, and it can also work with degraded samples.

After a few years of controversy about credit for its invention and rights to its commercial exploitation, PCR became a named and patented technique. Even though the controversy was settled years ago, with Kary Mullis being awarded the Nobel Prize in chemistry, and Cetus Corporation awarded rights to the patent, making PCR is a continual project. Just how it is "made" in any given instance depends upon which genetic patterns are of interest, how much sample is needed for available applications, and what consequences (including legal consequences) may arise from false-positive or false-negative results. In the early 1990s, in many areas of laboratory work and teaching, PCR involved a cottage industry style of work with small batches of material, hand-operated tools, and table-top appliances.[13] In such circumstances, material ingredients of PCR typically included the following: small plastic test tubes with labels for types and amounts of primers, buffers, and enzymes; pipettes with disposable tips for transferring measured amounts of ingredients from one container to another; a thermal cycler for repeatedly heating and cooling batches of preparations; and an electrophoresis gel apparatus for analyzing and visualizing the product. For example, during a session in which a staff scientist at a biotech firm showed a novice how to perform PCR, a handwritten protocol used for that occasion stated:

1. Label the tubes.
2. Add appropriate volume of water.
3. Make reaction mix for 10 reactions.
4. Add 30 microliters of reaction mix to each tube.
5. Add appropriate volume of appropriate DNA sample to each tube.
6. Overlay with two drops, 50 microliters of sterile mineral oil.
7. Place tubes in thermal cyclers set for 90 degrees C, 50 degrees C, and 72 degrees C.
8. Run 40 cycles.

This version of PCR exemplifies what social-phenomenologist Alfred Schutz (1964b: 71) once called "cook book knowledge," practical know-how that requires little understanding of underlying principles or mechanisms. This recipe is incomplete in many respects, and the performance is accompanied by further instructions about what is in the various test tubes, what happens when the materials are heated in sequence and at different temperatures, and so forth. In this particular instance, when it was used in order to instruct a novice, the recipe was accompanied by more detailed oral instructions referenced to unfolding actions and particular items of equipment in the laboratory setting:

Let's go to the freezer.

First thing, is I take a bunch of tubes—we're going to need and label them. I usually label them with the name and then the primers that I'm using. These are just "1" plus "2" and the date. . . .

13. The "cottage industry" analogy was suggested to us by Richard Lewontin during an interview with Kathleen Jordan (27 June 1990).

So, you want water or the 10x buffer which we've taken out of the freezer
and these are the primers, and these are the volumes. The way this is figured
out is I add water and DNA to a total of 20 microliters, so that leaves a remainder of 30 microliters of volume for the total of 50 microliter reaction . . .

Unlike a "cookbook" protocol, this running commentary includes references to the immediate scene, as well as "steps" that are not mentioned
in any recipe—"Let's go to the freezer" formulates a step of an especially
trivial order. These mundane instructions embed the protocol within the
immediate field of action. Even the "DNA" has a tangible point of reference to a visible, palpable container whose contents can be "picked up"
in a pipette.

Some of the competencies implied in this running commentary are
technical, in the sense that they require training and experience with
similar laboratory procedures, whereas others make use of ordinary language and require commonplace "skills" like locating the freezer, opening a container, and pouring water from a faucet. There is considerable
interplay between understanding ordinary expressions (for example,
"add water" and "appropriate volume"), and developing local competencies that establish just what those expressions imply for the technical task
at hand (for example, knowing if tap water or distilled water is required,
and understanding what "appropriate" means).[14] While the referential
terms have no specific identification with molecular biology, the competencies they gloss are mastered in specific laboratory circumstances.

The distinction between formal instructions and situated actions resonates with familiar experiences such as using a published recipe, con-
sulting a road map, or following a laboratory manual (Suchman, 1987).[15]
When engaged in the "here-and-now" tasks of following the recipe,
reading the map in the course of a journey, or enacting a step in a lab
manual, one is frequently reminded of the difference between the instructions and their singular applications. The recipe "asks for" lemon
juice, but you only have a lime; you find that you happen to be driving

14. Some of the procedural requirements recall Harold Garfinkel's (1967: 20–22) list of
ad hoc practices (viz., "the etcetera clause," and "enough's enough").

15. See Garfinkel (2002: 197ff.) for a published account of a long-standing theme in his
ethnomethodological studies. For a discussion of its relevance to molecular biology, see
Lynch & Jordan (1995). Lewontin's (1992) critical account of the relationship between genetic "instructions" and their "expression" deploys a conception of language remarkably
like Garfinkel's account of the uses of instructions (of the more ordinary kind), as well as
Wittgenstein's (1958) picture of language use.

on a street whose name is not on your map; you wonder what to do now that you have spilled slightly more than the prescribed amount of an ingredient into a preparation. Whether or not an improvisation performed on the spot will work successfully might not be apparent until much later. Descriptions of such improvisations in specific cases can be revealing, as they provide rich elaboration on such themes as tacit knowledge and the role of common sense in the performance or repair of formal procedures.

Situated Enactments

The distinction between formal instructions and situated actions is far from simple in the case of protocols. Even for some highly specific laboratory techniques, there is no single protocol. Nor is there a linear relation between the instructions in a protocol and the actions of following them. Trouble with performing PCR and other molecular biological techniques is far from unique to forensics and has also beset many academic researchers and diagnostic technicians. Consequently, invidious comparisons between scientific and forensic uses of "the same" techniques is fraught with empirical and conceptual difficulty. A set of interviews and observations we made in the late 1980s and early 1990s illustrates this point.[16]

Initially, we did not set out to compare forensic DNA profiling with other uses of the "same" techniques. Instead, we were interested in the more general question of how commonplace laboratory protocols vary among laboratories in different fields, and even among different practitioners within the same laboratory. Our method was to "follow techniques around" from one organizational site to another.[17] The guiding idea was that common laboratory protocols are templates upon which

16. This phase of the research was a collaboration between Kathleen Jordan and Michael Lynch. The research was partly funded by a grant from the National Science Foundation, Studies in Science, Technology & Society Program (Award no. 9122375). Title: "The polymerase chain reaction: The mainstreaming of a molecular biological tool."

17. This research strategy was a variant of Bruno Latour's (1987) injunction to "follow scientists around." Latour suggested that if one were to follow a (prominent) scientist around, the path traced would connect the laboratory to a large variety of other places: government offices; corporate boardrooms; public speaking engagements, and so on. Our aim to follow a technique around was not part of an attempt to outline a network; rather, it was a way to develop detailed comparisons between the way the nominal (and in the case of PCR, patented) technique was composed on different occasions of use.

practical, circumstantially specific variations are inscribed. Accordingly, protocols for "the same" technique would contain specific variations related to the purposes of a project, the scale and standardization of production, and the costs of failure and error.

We visited academic laboratories in different university departments, as well as diagnostic and forensic laboratories. At each site we spoke with administrators, staff scientists, and technicians. Our interviews focused on two basic techniques that were in widespread use at the time. These were plasmid purification and isolation ("the plasmid prep"), a recombinant DNA technique, and PCR (the polymerase chain reaction). At the time (the late 1980s and early 1990s), the FBI and other forensic organizations were involved in developing and testing forensic techniques that used PCR as a key phase.

Our research on these common laboratory protocols supported the idea that PCR and other constituents of DNA testing were generally accepted in many fields. However, although they were widely relied upon, they also were widely cursed. Scientists and technicians expressed frustration, and confessed to uncertainty about how to make the techniques work reliably. Difficulties, uncertainties, and remedies for overcoming problems in one field did not necessarily translate into difficulties, uncertainties, and remedies in another. Nor was it the case that "applied" work with, for example, PCR was less certain, and more problematic, than "basic" research. Indeed, the opposite often seemed to be the case. In addition to variations in organizational purposes and disciplinary practices, there were institutional concerns about costs and liabilities. Legal issues, though increasingly salient in many areas of biomedical science, were especially prominent in forensics. The circumstances of criminal investigation, and the possibility of challenge in an adversary system, framed the significance of error, mistake, and fraud, and far more stringent efforts were made to demonstrate that proper controls had been followed and checks against error and contamination had been implemented. Though reputed to be a domain of "dirty work," criminal forensics was more preoccupied with purity, in contrast to contamination, than the small university laboratories we visited.[18]

Partly because of the open-ended range of variation, it can be difficult

18. The contrast between purity and contamination calls to mind Mary Douglas's (1966) classic work on the subject. For an illuminating effort to develop Douglas's conception of the significance of "dirt" in laboratory science, see Mody (2001).

and frustrating to follow PCR protocols. Standard tools, packages, and instructions are never completely suited for singular applications. Even when a protocol is developed within the confines of a single laboratory, different practitioners work out their own sequences of steps (Jordan & Lynch 1992: 93). Some practitioners prefer to leave out specific precautions, while others prefer more elaborate checks against error and contamination. Practitioners freely acknowledge that their preferences often lack biological rationality, and are developed through local, and even personal, regimes of trial and error. In a playful way, they often speak of the "black magic" involved in getting the techniques to work (Cambrosio & Keating, 1988), or when faced with continued failure they speak of "PCR Hell" (Jordan & Lynch, 1993: 170): "I don't think you know how I hate to talk about PCR. It has given me nothing but grief. I wish it never was invented. It just leads you down the rosy path to hell."[19] Although at that time PCR was widely used in many fields, it was common to get accounts such as the following: "The problems come and go. It's a mystery. Sometimes it works, and sometimes it doesn't. Sometimes the DNA is very sequence-able and sometimes it isn't. I mean, at the moment it's pure art. And, ahh, it's not even art, it's magic!"[20] No single formulation of the PCR protocol gave a precise description of the range of techniques performed under its name. Nor did any protocol exhaustively describe the actions involved in a singular attempt to reproduce the technique. Practitioners also acknowledged that they had limited control over how the procedure worked. They mentioned that efforts to contend with problems arising from one performance to another of "the same" procedure inevitably resulted in further variations in performance. At times uncontrolled variations led to discoveries of new, and arguably better, ways to perform the procedure. Consequently, the dissemination of protocols within and between laboratories involved local rewriting, adaptation, and tweaking. Even in the case of a named, patented, and commercially available protocol, the transfer of the technique from one application to another was as much a matter of dispersion and reinvention as it was of the diffusion of a stable technical object from one site of practice to another.

Complaints about "PCR Hell" were commonly heard in the late 1980s and early 1990s when many laboratories first adopted the technique. Since then, the technique has become ever more widespread and increas-

19. University biotech laboratory director, interviewed by Kathleen Jordan (1997: 51).
20. University laboratory director, ibid., 116–17.

ingly automated. Biotech advertisements hawk packaged reagent kits and numerous other products that promise speedy and trouble-free performance.[21] Nevertheless, one can still find evidence of personal struggle with getting the techniques to work. Electronic exchanges among practitioners in a biotechnology newsgroup provide evidence of such struggle. Take, for example, the following string addressed to a problem with using a DNA extraction method:[22]

[A:] Hello, I was wondering if any one else has experienced this problem and if so, what they did to solve it.

I am using a PCR-restriction digestion assay to genotype some blood samples, and I find that one of the polymorphisms is completely undetectable when I use the "Nucleon" kit from Scotlab to extract the DNA, but when I take the same blood sample, boil for 15 mins, spin out the crud and use the supernatant, there is no problem in detecting this polymorphism.

I've done extensive experiments to ensure that it is not a variability in the PCR or digestion steps. And IT IS NOT.

Please can anyone help me??

Thanks in advance.

[B:] Hello

I had the same problem of PCR band disappearance after extraction of DNA from blood using Quiagen extration kit. In fact latter I read that Hemoglobine which some time contaminates samples is an inhibotor of DNA polymerase. And it's necessary to heat PCR mixture-samples at 94 C for 10 to 15 min prior addition of nucleotides and enzyme. When I did this, suddenly the bands appeared. So when you boiled your samples I suppose you distroy the hemoglobine and then your PCR reaction is good. In fact recently I shift from quiagen and I use a kit from DYNAL: Dynabeads DNA direct kit. It's very easy to use and I never had problem of PCR inhibition after DNA preparation. I even never need to boild my samples anymore.

Hope this can help you.

---

21. See Haraway (1997) for well-chosen examples of such biotech advertisements.

22. We are grateful to Christine Hine for locating these materials for us. See Hine (2000) for the significance of such materials for "virtual ethnography." Note that misspellings, typos, and grammatical errors are in the original messages. We deleted signature lines that give the writers' names and affiliations, and substituted the letters A, B, C, and D for these identifying details. For convenience, we have combined four separate messages, but in the original strings, B and C respond to A, and D responds to B.

[C:] Hi:

I had a similar problem when I was doing microsatellite typing. Samples extracted with kits did not work so well as samples extracted with phenol. Even when I tested samples sent to me, some extracted with a different kit then mine, and some with phenol, the results were the same.

[D:] We have also had problems doing human microsatellite typing when preparing the DNA using the QIAGEN blood kit. Sometimes works, sometimes doesn't. After talking to QIAGEN, and trying many variations (at their suggestion), we still have problems with reproducibly preparing DNA that will amplify. We basically solved the problem by simply doing an ethanol precipitation of the QIAGEN prepared material. Add an hour to the procedure, but the DNA always amplifies now.

A's problem involves the use of a PCR kit to "extract" (isolate and detect) distinct segments of DNA from the cellular chromosomes in blood samples. The word "polymorphism" refers to a stretch of DNA at a particular locus on a chromosome, which is unusually variable in a human (or other species) population. Polymorphic alleles (alternate forms of a genetic sequence) are inherited, and so extracting and marking them is an extremely valuable tool for population genetics as well as many other scientific, diagnostic, and forensic applications of DNA typing. When a particular polymorphism has been successfully isolated and marked, it shows up as a discrete "band" in an autoradiograph (a display of the product of an electrophoresis procedure). The problem, as A describes it, is that a particular polymorphism fails to show up when the "Nucleon kit" is used, but it does show up when an alternative, more laborious, procedure is used. To pursue our kitchen analogy, the problem is that the "cake mix" fails in certain respects, necessitating an alternative method for concocting ingredients "from scratch."

B offers two solutions to the band disappearance problem. First, B offers a possible diagnosis: residual hemoglobin in a blood sample from which DNA is extracted may inhibit the action of polymerase (an agent of DNA duplication that is a key constituent of PCR). B then offers a possible remedy for the problem, which consists in a preliminary step of heating the PCR mixture (this is similar to the solution A had already mentioned). However, B then goes on to give a simpler alternative, which is to use a different brand of reagent kit. The citation of the brand name serves to index an unspecified difference between the two kits, and the instruction simply endorses the effect of using the second kit (the makers of the

"Dynabeads" kit would be happy to read this spontaneous endorsement of their product). D offers a different remedy for the problem, but like B's first remedy, it involves adding a further step to the procedure.

The exchanges between A, B, C, and D address the difference between explicit and tacit knowledge. They formulate, and thereby make explicit, problems and solutions that arise in situ when using prepared kits according to standard instructions. Contrary to Polanyi's (1964) original conception of tacit knowledge, in this case there is no clear division between explicit and tacit dimensions of the practice. What is, or was, "tacit" is not essentially mysterious, ineffable, private, or unconscious. Instead, it is a matter of what is mentioned or unmentioned in a particular written recipe or other formulation, and, correlatively, what is or is not "grasped" or even "discovered" by someone using the instructions. What is mentioned, and what goes unmentioned, is highly circumstantial and occasionally problematic. Copies of a standard protocol take on a stable form when published in a laboratory manual, and even local recipes circulated informally are copied more or less exactly. However, the form of a protocol is not that of an "immutable mobile" (Latour, 1990): an inscription fixed in a medium that is passed from one site of local production to another, thus facilitating its reproduction in remote places and giving it an archival permanence. Instead, local practices and written protocols intertwine and supplement each other. It is common, for example, to amend and supplement the prior text of a protocol. What had been, or might have been, "tacit" is then incorporated into the recipe, though evident differences may remain between typed steps and handwritten supplements, or between written instructions and verbally conveyed advice. Note that none of the suggested remedies for A's problem radically breaks with the format of an explicit protocol, and all of the advice employs concise recipe-like instructions that gloss over the laborious work involved in implementing the trademarked kits and patented techniques. And, while it seems safe to say that an imaginable reservoir of tacit knowledge is never exhausted, there is no separate "tacit" register. Instead, one finds a series of amendments, substitutions, and other modifications to the protocol in question, and all of the remedies implicate unformulated understandings of what might be required to "work out" the procedures.[23]

23. See Bobrow and Whalen (2002) for a study of a knowledge-sharing system, "Eureka," that is designed to incorporate practical know-how and personal "tips" into an evolving database that provides on-site instructions for photocopy technicians.

Credibility Transference

Despite what we have said about the dispersion of technique, most commentators speak of a material transfer of the same techniques from biomedical to forensic contexts of investigation. However, a *rhetorical* transfer of credibility from biomedical science to forensic practice was at least as important for establishing forensic DNA profiling in the U.S. legal system. In trials and admissibility hearings, proponents of DNA evidence boosted its credibility by stressing identities with genetics and medicine. For example, in an early U.S. case, the prosecutor's summary stressed such identity with a long list of fields:

> These procedures have been used for years for reliable identification of genes in the broad areas of general research, genetics, infectious disease and cancer research, immunology, evolution, ecology, prenatal diagnosis, and transplants, and in the specific areas of detecting diabetes, sickle cell anemia, hemophilia, rheumatoid arthritis, multiple sclerosis, cystic fibrosis, muscular dystrophy, AIDS and AIDS research, thassalemia, Huntington's disease, Lyme disease, bone marrow grafts, paternity testing, missing persons identification and tissue typing. (*New Jersey v. Williams:* 11)

This link to biomedical applications sometimes was stressed to the point of strain; expressing a defensiveness that acknowledged that the "transference" was fragile and forceful persuasion was required to secure it. For example, an FBI unit chief, who described himself as an administrator whose job it was to "keep up the protocol and keep up the QA/QC [quality assurance/quality control]," objected to the interviewer's characterization of the "controversy":

> I don't think there is much of a controversy. But let me explain to you why I don't think that's the case. Number one, for the RFLP technology I don't think anyone disagrees that it's a solid technology. There is absolutely no disagreement on that, in the field of molecular biology. . . . It has been used in research laboratories throughout the world and it continues to be. So, as far as the technology itself . . . there is not controversy about its scientific validity, period.[24]

24. Interviewed by Kathleen Jordan in December 1991, just when it was popular to speak of the "DNA wars." Thompson (1993: 24) characterizes a view similar to the one

The FBI administrator went on to complain about "some high named individuals" who testified for the defense in cases such as *New York v. Castro,* some of whom, such as Eric Lander, were serving on the first NRC panel at the time, and he added that "to my knowledge none of the people who are in the fray have ever done this technology, actually done the RFLP technology, sat down and in fact done it. . . . Ask the people on the NRC panel . . . how many of them in fact perform RFLP technology, and have they ever done it on any forensic cases." He referred to "an enormous body of evidence out there" which supported the "conservative" probability estimates used by the FBI, and reiterated that "there is data throughout the world which supports the use of frequency counting calculations that we in fact have been doing at the FBI. That's the whole point, nobody wants to sit down and digest all this. . . . Now I would ask you to go to anybody, Lewontin, Hartl, Eric Lander, anybody else and ask them where their data is to support their alleged controversial stance." He then mentioned PCR, offering to erase "any doubts as to whether this has any acceptance or validity in the scientific community"—assuring the interviewer that "I will point out to you that Kary Mullis this year won the Nobel Prize for inventing PCR technology," adding that it had "gained universal, global use throughout the world in terms of molecular biology, clinical applications. . . ."

This administrator's defense of the scientific status and legal credibility of DNA profiling exhibits a tension that ran through much of the discourse about the techniques. On the one hand, the proponent stresses the "global" use and the connections with "molecular biology" and "clinical applications," asserts that there is "absolutely no disagreement on that, in the field of molecular biology," and cites the Nobel Prize won by Kary Mullis for inventing PCR. This interview occurred prior to the O. J. Simpson case, and in retrospect it is ironic that Mullis offered to ap-

---

expressed by this FBI administrator, noting that many proponents of DNA typing complained that "the major lesson of the controversy is that the legal system offers powerful incentives for ersatz scientific dissent and that the courts take such dissent too seriously when evaluating the admissibility of new scientific techniques." Such a dismissal of dissent is a counterpart of Peter Huber's (1991) influential treatment of "junk science" as pseudo-scientific testimony encouraged by litigation: in this case, the proponent of a criticized science returns a "junk controversy" charge to the skeptic. This is but one of many illustrations of a fact that should be obvious to anyone who has followed the careers of the junk science/sound science pairing: it is a game that both sides have learned to play.

pear as a witness for the defense (see Mullis, 1998: 44ff.).[25] On the other hand, after acknowledging that "some high named individuals" (prominent molecular biologists and population geneticists) had criticized forensic DNA analysis, the FBI administrator objects to their lack of practical, hands-on engagement with the techniques in question. The administrator thus works both sides of the boundary between molecular biology and forensic practice,[26] emphasizing continuity with global science while also distinguishing the local knowledge of the forensic practitioner from ideological criticisms launched by nominal experts. It is rare to see both arguments stressed by one voice and in such brief compass, but, as we shall see, the continuities and discontinuities between forensic practice and science animated many of the courtroom dialogues and official evaluations about the reliability of forensic DNA profiling.

The first NRC report expressed a "high science" attitude consistent with the credentials of many of the committee's members, and its evaluation presumed that forensic DNA analysis was derived from established scientific principles and practices. The validity and reliability of those principles and practices in molecular biology and clinical diagnostics were not in question; instead, the question for the committee was whether or not forensic practice could be normalized sufficiently to meet "scientific" standards. As indicated in the above quotation from the FBI administrator, however, proponents of the techniques would occasionally invert the relation of credibility between high science and street-level forensic practice. For example, during an interview an FBI casework examiner referred to her postdoctoral experience in a biology laboratory, and noted that her current forensic work was far more stringently controlled:

Having been in research, believe me, my work is judged here and reviewed more critically than it was in research *ever.* You know, no one other than my

25. One of Simpson's defense attorneys, the flamboyant Johnnie Cochran, invoked Mullis's name in the 1995 trial with endorsement similar to that of the FBI administrator but for the opposite purpose: "[Mullis] is the man who invented PCR . . . He is the man who received the Nobel Peace [*sic*] Prize for this invention. And he will come in here and tell you about this evidence, how sensitive it is, and how these police departments are not trained in the collection [and] use of it, that this is by all accounts, twenty-first century cyberspace technology that is used by these police departments with covered wagon technology" (Cochran, quoted in Jordan [1997: 137]).

26. This is a specific variation on the theme of "boundary work" (Gieryn, 1983; 1995; 1999).

supervisor ever looked at my notebook. I could publish a paper and no one out there ever came and said: "I want to see [how you got the results]." You know, it's a whole different level of scrutiny when you have to go in front of lay persons such as juries and judges. You have to open your books to defense experts that are out there. . . . (quoted in Jordan, 1997: 164)

This and related accounts of forensic practice emphasized the transparency and public accountability of forensic practice, as compared with the esoteric, idiosyncratic style of university biology.[27] Another FBI casework examiner also inverted the common association between pure science and forensic dirty work by explicating two distinct senses of the word "contamination":

> For me contamination is . . . I mean, we look at dirty things all the time, we don't work with pristine blood samples. We work with items that have been sitting in . . . if a person has been killed their clothing is going to be filthy. If we find the body out by the side of the road, you know. To a scientist that is contaminated, to me that is forensic evidence. Contamination for me is . . . something that has been placed there that didn't belong there as a result of the crime. Like a technician would, you know, sneeze onto something. It applies in all areas, not just DNA. I mean contamination of latent evidence. If I pick up something and I am not wearing gloves, I have left my fingerprints on that and it is not related at all. That would be contamination in my book. (quoted in Jordan, 1997: 140)

As noted earlier, the first NRC report drew a series of contrasts between biomedical science and forensic investigation that placed the latter in a relatively low, derivative, and uncertain status. In this instance, however, the caseworker treats the ideal-typical "scientist" account of contamination as a superficial association between samples and filth, whereas the caseworker lays out a more technical definition of the word that is akin to Douglas's (1966: 36) conception of dirt as "matter out of place"; a disruption of the protocols of orderly society rather than an intrinsic material quality. Contamination in this sense is akin to laboratory artifacts

27. The "transparency" of the FBI's procedures was also subject to dispute in the early 1990s, when the FBI refused to release data from an assessment of error rates with DNA typing.

(Lynch, 1985a; Mody, 2001) that disrupt the integrity of an observational field, potentially leading observers to mistake artifacts for substantive objects of interest. This inversion of the presumptive relationship between science and forensic practice was specifically directed against the members of the (first) NRC panel by the FBI administrator we quoted earlier:

> [Of] the people on the NRC panel . . . how many . . . in fact perform RFLP technology? Have they ever done it on any forensic cases? . . . I think that their experience was extremely limited and the understanding or appreciation of how it's used in the forensic area was *limited*. (quoted in Jordan, 1997: 135)

In this account, the NRC panel, whose 1992 report presumed a normative stance identified with the authority of science, is accused of holding a naïve, and even vulgar, conception of forensic practice; of making naïve associations that obscure the technical integrity of the practice.

Although it is clearly the case that these FBI caseworkers were speaking defensively and resentfully, at a time when their practices were under severe scrutiny, our own concurrent research on the uses of routine laboratory techniques in biology, diagnostics, and forensics also indicated that practitioners in "basic" or "discovery" research experienced the most protracted, baffling, and frustrating difficulties with making those techniques work reliably. The difficulties had to do with the small-scale, highly particularized crafting of technique associated with innovative research. In contrast, in corporate and government laboratories reagents, markers, tools, and protocols had already been worked out for routine modes of sample processing. More elaborate regimes were in place for designing and administering protocols, record keeping, and surveillance. Although the first NRC report's quality assurance/quality control recommendations were presented as though they were designed to raise forensic practice to the level of a science, those recommendations were displayed and approximated in a far more explicit way in forensic laboratories than in any of the "basic research" laboratories we examined.

Protocols as Administrative Devices
and Interrogator's Resources

We can summarize what we have said thus far with two proposals about
the relationship between formal (often written) protocols and situated
practices:

(1) Protocols do not, and cannot, fully explicate the work of reproducing them under singular circumstances.[28] A particular practitioner or
laboratory team may treat a protocol as adequate for their purposes, but
others with different backgrounds, who face different contexts of application, may find it incomplete or even unintelligible.

(2) There is no discrete "boundary" between protocols and practices;
instead, protocols can be appended and elaborated with commentaries
and instructions that include different literary voices, marginalia, and
targeted advice that explicate previously "tacit" properties of knowledge
and competence. Although never exhausting the "tacit" field, such expli-
cations reach into and selectively formulate it.

A third proposal can be added to these two:

(3) Proposals (1) and (2) are well known to practicing scientists and
others who use protocols routinely in their work. For example, the fact
that canonical accounts of scientific method do not describe concrete
laboratory bench practices is well known among scientists. Although
ethnographies of laboratory practices provide ample empirical documentation of this social fact, for practitioners it requires no special insight or acquaintance with the science studies literature to know it as a
fact of laboratory life.

Evidence cited in support of (3) includes the frequent acknowledgment by practicing scientists, including many who vociferously dispute
the claims of social studies of science, that rules of method provide limited and even misleading accounts of actual practices.[29] Further sup-

28. What we have said about the relationship between protocols and situated practices
is an instance of the maxim that rules cannot furnish complete rules for their own application (Wittgenstein, 1958). But rather than expressing skepticism about the adequacy of
rules, plans, and other formulations as accounts of social actions, this maxim can motivate
investigations of how formal structures are deployed within actions, as instructions, explanations, justifications, excuses, and so forth (Garfinkel and Sacks, 1970; Suchman, 1987).

29. The most prominent testimony to this is Medawar (1964). More recently, defenders
of scientific "reason" who voiced objection to constructivist sciences studies (Gross & Levitt, 1994) assert agreement that scientific methods do not provide accurate descriptions of
practices.

port can be found in scientists' and technicians' frequent remarks, jokes, and other informal expressions testifying to the difficulties of making procedures work (Gilbert & Mulkay, 1984). These can take the proverbial form of variants of Murphy's (or Sod's) laws, tailored to laboratory situations.

Despite the familiarity of proposals (1), (2), and (3) in communities of laboratory practice, protocols sometimes acquire a normative status for evaluating practices and enforcing standards. That this is so is no more or less mysterious than the fact that police officers can acknowledge that laws are routinely violated and enforced selectively, and yet this does not prevent them from using formal law and powers of arrest as resources for keeping order (Bittner, 1967). In the case of protocols and practices, we are not necessarily dealing with a simple matter of compliance or violation, but instead with a whole range of activities that implement protocols successfully or unsuccessfully. However, when protocols are administrated as normative standards, the difference between specific protocol formulations and actual practices can be made thematic in a way that implicates the adequacy of the practices in question.

Two interrelated uses of protocols were significant for the discursive work of establishing and undermining the credibility of DNA profiling. The first was the authoritative formulation and citation of standard protocols as a means for demonstrating "sound science" or "good practice." The two NRC reports are prime exhibits in this regard: the reports recommended that protocols should be developed and standardized, and also provided substantive instances of what such protocols should include. The second discursive practice is an interrogative tactic that cites authoritative protocols (including the NRC reports themselves) as a means for undermining the credibility of an expert witness, by soliciting acknowledgment of discrepancies from "sound science" or "good practice." In order to exhibit this pair of discursive practices, we shall briefly review a particular recommendation from the first NRC report, which was cited in the cross-examination of prosecution witness Dr. Robin Cotton, director of research for Cellmark, during the O. J. Simpson trial.

During the cross-examination of Dr. Cotton, defense attorney Peter Neufeld endeavored (in his words) to "impeach the witness with a learned treatise." This refers to a tactic for interrogating expert witnesses by using an authoritative text as a standard for casting doubt upon ("impeaching") the witness's competency (or, in this case, the competency of the witness's organization and the validity of the evidence produced through that organization's practices). The first NRC report was widely regarded as an authoritative treatise, even though some of its recommendations were severely criticized by proponents and critics of forensic DNA typing.

When attempting to impeach a witness with a learned treatise, an interrogator's task is, first, to get the witness to acknowledge that the version of practice described in the text is normatively relevant to her expertise, and second, having so established the authority and relevance of the text, to get the witness to acknowledge departures from the prescriptions. Then, having solicited such acknowledgment, the interrogator suggests that the acknowledged departures are nontrivial lapses and evidences of sloppy work and incomplete testing which cast doubt upon the substantive evidence the witness presents.

Over a three-day period in mid-May 1995, Neufeld laboriously sought to use the first NRC report in this way. With a copy of the report in hand, he solicited from Cotton that she knew the text very well, so well in fact that she carried it with her in her briefcase; and, moreover, that she knew, and respected the reputation, of the long list of distinguished scientists who were members of the NRC Committee on DNA Technology in Forensic Science. In a laborious sequence, he went through the roster one by one, naming a member (Victor McKusick, Paul Ferrara, et al.) and asking Cotton if she was familiar with, and respected the advice of, the particular figure. Cotton more or less went along, though she sometimes added qualifiers about differences in specific fields of expertise, and she seemed to evince a slightly ironic stance toward the interrogative game that Neufeld was playing. For example, in the following sequence Cotton qualifies her acknowledgment of Ferrara's and McKusick's relevant qualifications (*Simpson* trial, May 12, 1995):

NEUFELD. And you know that Paul Ferrara is the director of the Virginia Division of Forensic Sciences.

COTTON. Yes, and he's also the director of the ASCLAD Lab Accreditation Group.

NEUFELD. And I take it that he is someone whose opinions you respect.

COTTON. That's correct.

NEUFELD. And would the same also apply for uh— Doctor McKusick who is the chairman of the committee. Are his opinions those that you would respect.

COTTON. I guess it would depend on his opinion. About any particular issue.

NEUFELD. His opinions about molecular biology and genetics.

CLARK [Prosecutor George Clark]. That's irrelevant.

JUDGE LANCE ITO. Overruled.

(Two-second pause)

COTTON. Despite the fact that he's a very well known scientist, he may have an opinion on a single issue that I might not agree with, and an opinion on another issue that I might. So, I can't make a blanket statement that I would agree with every opinion that Doctor McKusick would have.

NEUFELD. Would you regard Paul Ferrara as an expert in the area of forensic science?

COTTON. Yes.

NEUFELD. And would you regard him as an expert in the application of DNA profiling in forensic science?

COTTON. No, I don't think so.

NEUFELD. Would you regard him as a— (three-second pause) as an expert in establishing the proper controls and quality assurance for handling items of biological evidence that will subsequently be tested on the DNA level?

COTTON. Yes.

Cotton's refusal to give blanket endorsement to the named scientists ("I guess it would depend on his opinion. About any particular issue.") adumbrates a line of defense she held throughout the cross-examination. When Neufeld went into detail, again and again he encountered difficulty getting Cotton's acknowledgment of the practical relevance of specific procedural recommendations. In a rather stark example of this, Cotton simply denies the relevance of a cited recommendation:

NEUFELD. Doctor Cotton, let me ask you this.

(Two-second pause) Even once they go on line would you agree that it. . . that there's no substitute, in assessing the quality of the work generated by the laboratory, to engage in rigorous external proficiency testing via blind trials on a regular basis.

CLARK. Objection, facts not in evidence.

ITO. That's argumentative. Sustained.

NEUFELD. Doctor Cotton, in arriving at your opinions on this particular matter, did you read the section of the NRC Report entitled Experimental Foundation?

(Two-second pause) COTTON. Ahm, if I could just look quickly at that.

NEUFELD. Page fifty-five in your— in your book.

(Long pause as Cotton consults a copy of the book)

NEUFELD. In arriving at your opinions as to what a laboratory should do as a precondition before using new DNA typing did you at all rely on that section of the National Academy of Sciences book *DNA Technology in Forensic Science*?

COTTON. No.

NEUFELD. 'kay.

The section in question (headed "Experiential Foundation," not "Experimental Foundation" as in Neufeld's reference) reads as follows:

> Before a new DNA typing method can be used, it requires not only a solid scientific foundation, but also a solid base of experience in forensic application. Traditionally, forensic scientists have applied five steps to the implementation of genetic marker systems:
>
> 1. Gain familiarity with a system by using fresh samples.
> 2. Test marker survivability in dried stains (e.g., blood stains).
> 3. Test the system on simulated evidence samples that have been exposed to a variety of environmental conditions.
> 4. Establish basic competence in using the system through blind trials.
> 5. Test the system on nonprobative evidence samples whose origin is known, as a check on reliability.

After emphasizing that the five steps should be carefully followed, the report goes on to say, "Most important, there is no substitute for rigorous external proficiency testing via blind trials. Such proficiency testing constitutes scientific confirmation that a laboratory's implementation of a method is valid not only in theory, but also in practice. No laboratory should let its results with a new DNA typing method be used in court, unless it has undergone such proficiency testing via blind trials" (NRC, 1992: 55).

Given the acknowledged authority of the NRC report and the unequivocal language of its recommendation ("No laboratory should . . ."), it might seem hazardous for a forensic laboratory research director to deny following it, but in a characteristic move, Cotton acknowledges that her laboratory did not use the specific text to guide its practice, but that it had developed its own set of guidelines and standards, which she accepted as adequate.

This interrogative struggle went on for hours, as Neufeld attempted to pin Cotton to the normative terms of an authoritative text (using the TWGDAM report as well as the NRC report) and Cotton insisted that her laboratory had developed its own procedures and standards independently.[30] At the start of the next day's session, before the jury was admitted into court, Neufeld voiced a lengthy complaint about Cotton's selective agreement and disagreement with the terms of the "learned treatise" with which he sought to "impeach" the witness. He complained that her unwillingness fully to endorse the terms of the authoritative text (or texts) was an illegitimate maneuver that negated the logic of his interrogative strategy. The judge gave him little satisfaction in his complaint, but by voicing it (and by expressing similar frustrations when the jurors were present) he suggested that the witness was being less than forthright. However, in this and in many other cross-examinations in which interrogators used the terms of a protocol to solicit admissions of lapses, sloppy work, and the like, the witness claimed an independent ground of competency that could not so easily be dismissed. Consider, for example, the following sequence in which Cotton mentions "going to the freezer"—a reference with an uncanny resemblance to the example of the vocal instruction ("Let's go to the freezer.") for performing PCR discussed earlier (p. 91).

NEUFELD. Would you *agree* that the ability to perform successful DNA analysis on biological evidence recovered from a crime scene depends very much on what kinds of specimens were collected and how they were preserved.

(*Two-second pause*)

COTTON. Yes.

NEUFELD. And would you agree, Doctor Cotton, that if the DNA evidence is not properly documented prior to its collection, its origin could be questioned.

CLARK. Objection, relevance. Beyond the scope.

ITO. Sustained.

NEUFELD. Doctor Cotton, you testified on direct examination as to the types of controls utilized by Cellmark in documenting each item of evidence that you process, is that correct?

CLARK. Objection. Misstates the evidence.

30. For a publication based on the TWGDAM report, see TWGDAM (1991).

ITO. Overruled

(*Five-second pause*)

COTTON. You have controls to assess how your scientific procedure worked, if that's what you're referring to, then.

NEUFELD. Well, didn't you also say that there are certain rigorous documentation requirements, such as, um, documenting every transfer of a biological specimen within the laboratory.

(*Three-second pause*)

COTTON. (Yih—) There are rigorous documentation requirements. Um, I didn't say that every transfer ih— eh— and I'm using that term broadly. Like, we don't write down, "It's now going to the freezer." You know, stuff like that.

Neufeld exploits the terms of the NRC report, with its call for "complete" protocols and "precise" step-by-step instructions.[31] Although one can imagine that the authors of the report did not intend to be taken so literally, Neufeld treats their global recommendations at face value, as though they implied that anything short of "documenting every transfer of a biological specimen within the laboratory" would warrant suspicion. Cotton counters this stress on complete documentation with a commonsensical rebuttal that recalls a deep epistemic issue that ethnomethodologists have called the *essential incompleteness* of instructions, which necessitates ad hoc practices for enacting them on particular occasions (Garfinkel, 1967: 18ff.; Garfinkel, 2002: 197ff.). In commonplace terms: you can't possibly document everything you do. "Going to the freezer" is Cotton's example of an evidently trivial action that would not be documented. But, then, this very defense opens up potentially endless questions about whether or not what was not documented, or what departed from the terms of a protocol, was indeed trivial, inconsequential, of no evidential significance. Cotton asserts that she and her staff know when to be literal and when to exercise independent judgment, and her claim simultaneously invites the audience (especially the jury) to exercise judgment about whether the laboratory's procedures were rigorous enough, despite any acknowledged departures from the letter of one or another authoritative protocol.

31. See Oteri et al. (1982: 252) for an example of a strategy for attacking the continuity of evidence transfers in cross-examinations of forensic chemists in drug trials. We shall return to this issue in our discussion of "chain of custody" in chapter 4.

Conclusion

In this chapter we questioned the assumption that techniques that were established in biomedical research and testing were "transferred" to forensics, and that the conditions of forensic investigation were a source of problems that did not arise in other contexts of use. Our research on the uses of PCR and other laboratory techniques in different institutional contexts suggests a picture of the dispersion of technique that conflicts with that assumption. It is important to recognize that we did *not* suggest that there were no distinctive problems associated with forensic investigation. Instead, we suggested that there were distinctive problems with *all* of the uses we investigated, regardless of their scientific repute. Indeed, the most extreme frustration with making them work occurred in small-scale situations of "basic" scientific research.

The 1992 NRC report stressed the use of protocols to repair a possible breach in the transfer between biomedical science and forensic practice. As we saw, emphasis on (or, perhaps, exaggeration of) the possibility of strict adherence to protocols provided proponents of DNA profiling with the hope that quality assurance/quality control could be leveraged by formal administrative rules and structures. At the same time, the same emphasis (or exaggeration) provided cross-examiners with an interrogative device for soliciting acknowledgment from forensic practitioners that their practices were less than rigorous.[32] However, another way of viewing protocols—as specifically vague resources for action—provides a counterpoint to an emphasis on strict adherence.[33] Practitioners such as Robin Cotton were able to claim entitlement to localized judgments that were not specifically mentioned in any given protocol. Such claims required allowances on the part of the audience about the relevance and significance of acknowledged departures from protocols, and the competency and trustworthiness of the practitioners involved.

32. In his study of the extended series of inquiries over possible falsified data in the Imanishi-Kari/ Baltimore affair, Kevles (1998) points out an ambiguity about the condition of Imanishi-Kari's laboratory notebooks, which contained incomplete and messy records of the experiments. The ambiguity had to do with the fact that notebooks which might otherwise have seemed unremarkably messy became suspicious when treated as evidence for official inquiries conducted by outsiders.

33. The near-oxymoron "specifically vague" is an expression Harold Garfinkel occasionally used to indicate, first, that rules are vague; second, that such vagueness is irremediable; and, third, that they are often written, understood, and used in a way that is attentive to that vagueness and its irremediability (see, for example, Garfinkel, 1967: 40).

At the time (mid-1990s), it was widely acknowledged that forensic laboratories and the various administrative organizations were in the process of developing administrative standards. The techniques themselves were changing in a way that sometimes required new standards and sometimes obviated the need for standards that applied to earlier techniques. Nevertheless "standards" (with protocols as important articulations of them) remained the coin of the realm. Rhetorically, this emphasis provided the adversary courts with a useful rhetorical device. When the practices seemed unproblematic, standards could be cited as a reason for why they were unproblematic; and when systematic problems were made accountable, discrepancies from the standards could be cited to expose exceptionally bad performance at the particular site. What became obscure was that such standards did not, and could not, define or determine "sound science." Nevertheless, "standards" came in handy for defending the integrity of forensic practices and attacking particular departures from that integrity.

So, what does this imply about the reliability of forensic DNA typing? First, it suggests that the NRC's effort to ascertain whether forensic DNA typing lived up to "scientific" standards made use of an ideal rather than an empirical conception of scientific practices, and that many unquestionably scientific practices would fall short of that ideal. Second, it suggests that the adequacy of DNA typing, whether in population genetic research, paternity testing, or criminal investigation, was less a matter of adhering to general standards than of managing a local environment of contingencies: material properties of samples; organization of specific work environments; available staff, their training, and credentials; temporal constraints on delivering results; demands for precision; costs of different kinds of error; and so on. The problem was not that the NRC committee held DNA typing up to too high a standard, but that it held it up to an illusory standard. Third, just what made, or makes, any instance of DNA typing reliable—whether in forensics, diagnostics, or population genetics—is a local determination that is contingent, subject to local dispute, and open to change when the circumstances of investigation change. Such a picture of reliability offers little comfort to administrative bodies and courts of law that want (indeed, need) to make simple either/or declarations of reliability. Nevertheless, as we shall see in the following chapters, law courts and administrative agencies were not deterred in their efforts to determine and manage the reliability of DNA typing in terms of general procedures, generic standards, and broad notions of scientific reliability.

CHAPTER FOUR

Chains of Custody
and Administrative Objectivity

During the "DNA Wars" in the early 1990s criticisms focused on two major problems with the techno-legal accountability of forensic DNA evidence: laboratory error, and a combination of population genetic and statistical issues. The two NRC reports recommended administrative solutions for the first set of problems, and standard estimation procedures such as the "ceiling principle" for the second. The statistical and population genetic problems seemed more challenging at the time, and commentaries often stated that problems with evidence handling and laboratory error would become less troubling as DNA typing techniques became established and forensic laboratories and police departments developed appropriate training regimes and standard protocols. However, the practical problems proved less tractable than initially imagined, and they threatened to obscure the meaning and undermine the probative value of DNA evidence. Moreover, a persistent criticism that continues to be relevant, but which has been bypassed (and to an extent obscured) by administrative recommendations and judicial stipulations, is that the unmeasured *possibility* of practical error (as well as fraud) in the collection, handling, and analysis of samples limits the meaning and value of random match *probability* (RMP) estimates. RMP estimates constitute measures of error, in the technical, statistical sense of estimating the chance that a given match would occur by coincidence in a given population (Koehler, 1996). We shall return to questions about such estimates in later chapters.

In the present chapter we focus on arguments about practical errors, especially in relation to what is called "chain of custody" or "continuity of evidence." These terms refer to the entire sequence of agents and practices involved in the identification, collection, transportation, storage, and handling of evidence. Our interest in this phenomenon is both abstract and particular: chains of custody are a key issue in the practical construction and deconstruction of forensic evidence, and their analysis can shed light on more general epistemological themes of reference and representation.

"Chain of custody" and "continuity of evidence" are vernacular terms. The former is more often used in the United States, while the latter is common in Britain. Both terms refer to procedures for collecting, transporting, and handling legally significant material. The material in question can be a person (a suspect or prisoner held in custody), an object or artifact (for example, a weapon used as material evidence), or a sample of bodily material (a urine sample, a blood sample, etc.). Paper trails are important elements of chains of custody, as they stand proxy for other materials and actions: they track the movements of samples, certify that required protocols were followed, and identify responsible agents and agencies at each step in the processing of evidence. Vernacular accounts often refer to the paper trail of records as the chain of custody.

Chains of custody deliver and certify evidential end products; they set up and underwrite forensic determinations of whether or not a given hair, fingerprint, or blood sample taken from a suspect or victim matches one taken from a crime scene. When focusing on matching (or nonmatching) evidence, it is easy to overlook the local history of practices and material transformations that preceded the presentation and interpretation of the result. Savvy lawyers do not overlook such practices and transformations, however, when they cross-examine expert witnesses, and such interrogations provide richly detailed displays on the construction and deconstruction of evidence.

High- and Low-end Agents

As noted in chapter 2, the theme of laboratory error covers much more than mistakes made by technicians and staff scientists working within a forensic laboratory. As the lengthy *Simpson* trial made excruciatingly clear, an effective line of attack against DNA evidence was focused on how the evidence was collected and handled by the police prior to entering the laboratory. Errors committed at an early stage in the evidence-collection process are sometimes called "preanalytical errors," though such errors can also occur within a laboratory when samples are mislabeled, switched, or cross-contaminated while being prepared for analysis. A strategy similar to one the *Simpson* defense used for attacking chains of custody is concisely described in an article written by a team of lawyers to instruct defense attorneys in drug possession cases (Oteri et al., 1982). The lawyers outline one method for dismantling evidence produced by forensic chemists in such cases:

> Each and every link in the chain of possession should be examined with care. . . .
> 1. Was there a time lag between the date the substance was seized and the date it was analyzed?
> 2. If the answer to (1) is yes, is the drug stable? Is it subject to chemical change? Could the analyzed drug be the decomposed result of a time lag?
> 3. Was the officer who seized the drug the one who delivered it to the laboratory? In the interim, where was it kept? Who had access to it? Was it sealed when delivered to the chemist? Was it sealed at all times before that? If not, there is the possibility that other substances were added to it.
> 4. How many other samples did the delivering officer possess? How fastidious was he in keeping the drugs separate?
> . . . [H]e cannot testify to what happened before he received the samples. Was there a mix-up? His only response is "I don't know." Unless the history of the sample from seizure to analysis is carefully authenticated, the validity of the analysis is irrelevant. (Oteri et al., 1982: 252)

As though he had read such advice (and perhaps he had), Peter Neufeld pursued a similar line of questioning in one phase of his cross-examination of Cellmark laboratory director Robin Cotton during the *Simpson* trial:

NEUFELD. Would you agree Doctor Cotton, that moisture promotes bacterial growth.

COTTON. Yes.

NEUFELD. And would you agree Doctor Cotton that the bacteria starts eating up the DNA, and then the DNA deteriorates and degrades.

COTTON. Over time, yes.

NEUFELD. And would you agree, Doctor Cotton that degradation occurs more quickly under the combined effects of moisture and heat.

COTTON. Uh yes, I would.

NEUFELD. In your laboratory, Doctor Cotton, where you are the laboratory director, would it be scientifically acceptable to let wet plastic stains remain in sealed plastic bags in—

JUDGE ITO. . . . You said wet plastic stains.

NEUFELD. I'm sorry. (Pause.) . . . that's what I have written here.

(Cotton and audience laugh)

NEUFELD. At least I can read correctly your honor. I just can't write correctly, I'm sorry.

JUDGE ITO. Here. Let me do it.

NEUFELD. Sir— ahm, Doctor Cotton. (Pause.) In your laboratory, would it be scientifically acceptable to let swatches of wet blood stains remain in sealed plastic bags in the rear of a parked truck unrefrigerated and unairconditioned in the middle of June for up to seven hours.

CLARK. . . . argumentative.

JUDGE ITO. Overruled.

(Pause)

COTTON. I don't think that would be my *first* choice, but please keep in mind that my laboratory doesn't collect evidence. And so, we don't collect it, we don't have a truck, we— we just *receive* it from someone else who has already collected it.

NEUFELD. And that's becau— . . . as a result of that, Doctor Cotton, there's no way that you can *control* for the extent to which the— um, offering agency either degraded those samples or cross contaminated those samples.

COTTON. Of course.

(*People v. Simpson,* May 12, 1995)

This sequence is notable for at least two reasons. First, the humorous interlude touched off by Neufeld's malapropism reveals that he was reading from a prepared text when he interrogated Cotton. In other words, he was working through steps in an argument prepared in advance, while contending with the contingencies of an actual dialogue.[1] His argument follows a line of an attack on chain of custody similar to the one Oteri et al.

1. The repeated preface "would you agree that" for successive questions in the early part of this excerpt provides a way to comply with the institutional demand to ask questions while putting forward statements for the witness to confirm (Lynch & Bogen, 1996: 130ff.). The assertions are designed for agreement—that is, they are formed as unproblematic

outline for drug cases. Second, Neufeld attempts to enroll the prosecution witness (Cotton) in a criticism that threatens to undermine the value of the DNA evidence that Cotton and other prosecution witnesses had presented to the court.

At the start of the excerpt, Neufeld sets out a microbiological rationale for casting doubt upon biological samples handled in a particular way (namely, left in sealed plastic bags for seven hours in a very hot place). Neufeld is not impugning the integrity of Cellmark's practices, but is instead suggesting that contaminated samples were delivered to Cellmark by the Los Angeles Police Department (LAPD). After the humorous interlude about the malapropism, Neufeld resumes with a rhetorical question that elaborately forecasts the answer, and which draws an objection for being "argumentative." The objection is overruled, but Cotton does not give the straightforward answer anticipated by the question—instead, she qualifies her agreement ("I don't think that would be my first choice"), allowing for the possibility that the evidence would be salvageable, and she pointedly excuses her lab (Cellmark) from responsibility for collecting evidence. Neufeld follows her utterance with a collaborative completion that turns her lab's lack of *responsibility* for bad practice into a lack of *control* over the evidentiary import of the laboratory's analysis.[2] Neufeld thus appropriates Cotton's confirmation of an argument that excuses her domain of expertise from responsibility, while casting doubt upon the expert evidence she presented. Neufeld then builds from her response, treating it as testimony against the adequacy of the LAPD's practice.

Rhetorically and interactionally, Neufeld attempts to cast doubt upon the credibility of the evidence that Cotton had presented for the prosecution. He does so by enlisting Cotton's credibility as a scientist (the phrase "would it be scientifically acceptable" explicitly invites a "scientific" evaluation of the practice in question). The hypothetical instance that Neufeld presents—leaving wet blood specimens in a hot vehicle—is a transparent reference to testimony presented earlier in the trial that describes how blood evidence collected from the crime scene had been left in a police van for several hours. It also happens to be a well-chosen example of a "technical" matter (bacterial contamination of DNA in

---

facts that a reasonable witness *should* confirm. They also link each assertion to the assertion just prior, thereby building an argument with the witness's complicity.

2. A collaborative completion is when one speaker in a conversation extends and completes a grammatical sentence begun by another speaker (Lerner, 1989).

a sample) that is transparently (if misleadingly) intelligible by analogy with everyday examples of organic materials that spoil quickly when left in a warm place.

Cotton is at the end of the chain of agents responsible for collecting, analyzing, and presenting the evidence. As director of research in the firm hired by the prosecution, she is at the "high" end of the hierarchy of scientific authority and credentials. The "low" end is occupied by ground-level police employees (such as hapless criminalists Dennis Fung and Andrea Mazzola) who collected crime scene evidence with "swatches" (cotton-tipped swabs), placed them in containers, and transported them to LAPD facilities, from which the samples were later moved to Cellmark's laboratories. As Simpson's lawyers made clear, the low-end employees lacked scientific credentials, and when cross-examined they typically professed little understanding of the invisible constituents of the evidence they handled.[3] The defense held that their actions wittingly or unwittingly degraded, or otherwise ruined, the scientific (and legal) value of the evidence. Forensic laboratories also include many low-end agents with limited credentials. However, the stark difference between the grubby work of collecting samples and the high-end delivery of expert evidence in the form of graphic displays and statistical probabilities provides an especially ripe resource for adversary attack.[4] Following

3. One of the more humorous episodes during the *Simpson* trial occurred when a police officer was interrogated about footprints left at the crime scene. He was asked if he had taken evidence of all of the footprints at the scene, and after he answered affirmatively, he was caught flatfooted, so to speak, by a question about whether he had collected evidence of "invisible" footprints. When the policeman professed not to comprehend the question, the interrogator informed him that it was possible to dust the scene to reveal footprints that were not otherwise visible. The display of a difference between surface visibility and visibility revealed through technical mediation is one of the prime ways of exposing the difference between experts and nonexperts.

4. Although high-end scientists sometimes complain of the rough treatment they get in adversary hearings, they also are addressed in more deferential ways (see Roberts, 1992). There is a possible echo in the contemporary courts of an early-modern theme that Steven Shapin describes: truthfulness was strongly identified with the culture of the seventeenth-century gentleman (with Robert Boyle as the iconic example), and the credibility of the gentleman was associated with his elevated status as someone with little to gain from deception and much to lose from being caught in a lie. Women, servants, and other low-born categories of folk were deemed untrustworthy because of the favors they could gain through deception and ingratiation (Shapin, 1994; 1995). A related logic, though distributed across the biography of the individual rather than across social categories, is discussed by Goffman's (1961) account of the "mortification" and "demoralization" that occur

the jury's not-guilty verdict in the *Simpson* case, one common explanation of how the jury could have discounted the seemingly powerful DNA evidence (which, according to the prosecution's results showed matches between both victims' DNA profiles and blood evidence collected from Simpson's Ford Bronco and from a bloody glove found outside his home, and between Simpson's profile and blood droplets extracted from the sidewalk at the crime scene and from the Bronco), was that the jurors agreed with the defense's "junk in–junk out" argument that the DNA matches had doubtful significance due to incompetent, and possibly fraudulent, police handling of evidential items prior to their analysis (Jasanoff, 1998: 718; Thompson, 1996). Without discounting the charge of incompetence, we should keep in mind that it was an attribution that was derived and dramatized through interrogative strategies and lawyerly arguments. The defense was able to take advantage of a conjunction of factors that we can liken to a perfect storm: the location of the courtroom,[5] the predominantly African American makeup of the jury, and the collective memory (especially among Los Angeles African Americans) of the Rodney King police trial.[6] Police employees are "low" in the chain of custody and thus more readily attacked than credentialed experts, because they lack scientific training and make no claim to understanding relevant molecular biological principles, and yet are implicated in the

---

when the privileges of an inmate of a carceral institution are reduced to the point that he has nothing to left to lose from disruptive behavior.

5. Former district attorney Vince Bugliosi (1996) observed that the trial was originally scheduled to take place in Santa Monica, a predominantly white, well-to-do, community in West L.A., but was moved to a court near the city center (insofar as L.A. has a center) in order to accommodate the television crews and cameras. The makeup of the local neighborhood from which the jurors were drawn was predominantly African American.

6. The 1991–92 criminal trial of several white police officers who were shown on videotape delivering a series of blows with "batons" (police clubs) to Rodney King (an African American man), who was arrested after a lengthy car chase, resulted in a not-guilty verdict that triggered a massive riot in South-Central Los Angeles. The trial was held in Simi Valley, a predominantly white and conservative district. See Goodwin (1994) for an interesting account of the use of "expert" evidence by the defense in the case that resulted in respecifying the details on the video from being (apparent) excessive force meted out to the prone suspect to becoming technically appropriate responses to incipient aggressive actions by a potentially dangerous suspect. Although the makeup of the jury was often mentioned in popular accounts, Goodwin does not consider the contingent relation between the terms and credibility of the rival "analyses" of the video and the prejudicial inclinations of the jury members.

technical results delivered by the laboratory.[7] Moreover, in the *Simpson* case they were vulnerable to suspicions of racist motivation. Although Neufeld and the other lawyers also attacked the "objectivity" of Cotton and others who represented the high-scientific end of the chain, such efforts to undermine the probative value of the DNA evidence proved far more effective when directed against the ground-level agents who collected and handled the evidence.

The *Simpson* case produced a vast popular archive of videotaped testimony that was a valuable resource for this study.[8] However, to elucidate features of "chains of custody" we shall turn to another murder trial in the mid-1990s, which took place in England. We focus on such concrete matters as a muddy (and bloody) shoe and a photocopied signature that became significant in the investigation and prosecution of a particular murder case. We believe that these details have critical bearing on the conceptions of DNA matches that we also shall pick up later in our discussion of fingerprinting.

The Career of a Sample

To examine the issue of chain of custody, we attempted to retrace the movements of samples used in the prosecution of a particular case in the

7. It is commonplace in the social sciences to impute "unconscious" knowledge to agents; that is, knowledge relevant to the agents' activities but about which the agent has nothing to say. Using a similar logic, interrogations of "low-end" agents use contrasts with "high-end" expert accounts to drive a wedge between the agent's account of the relevant actions and a "scientific" account of factors that bear upon the rationality and effectiveness of the action, but about which the agent seems unaware. In such cases, what is "unconscious" is not located in the hidden recesses of the mind or body (motives that are not acknowledged; knowledge that is tacit, and so forth); instead, what might be called the biological unconscious becomes the source of agency. For example, the "low end" agent is shown to be unaware of molecular processes that can spoil or contaminate a sample.

8. The notion of "popular archive" is developed in Lynch & Bogen (1996: 51–52; Lynch, 1999: 75ff.). Briefly, it refers to materials placed in the public domain during a nationally televised tribunal. Not only does such an "archive" include an immense amount of material, but the very fact that it is shown publicly and contemporaneously to a large national audience compounds the documentary materials, and the "original events" they document, with incessant commentaries, public reactions, and reflexive anticipations of those reactions and commentaries. A popular archive is, of course, incomplete, and its incompleteness is a significant feature that is evident and strategically used by participants who deploy secrecy as well as by those who attempt to disclose secrets and secret agendas.

United Kingdom, *Regina v. Smith* (1995).[9] The case resulted in a murder conviction in December 1995. While this case was underway, we selected it for an intensive study of "the careers of samples."[10] We borrowed the metaphor of "career" from Erving Goffman but used it for a rather different purpose. Goffman (1961: 154) uses the concept of "moral career" to describe a regular series of stages, many of which involve bureaucracies and institutional processes through which individuals pass on their way to attaining a recognizable moral status. Goffman was interested in the ideal-typical "career" pathways that establish and certify the moral status of a person as a mental patient. Central to his conception of moral career is that the path to the mental hospital is not an inevitable consequence of a disease internal to the patient; instead "contingencies" arising from key relationships and fateful interactions have a crucial role in deciding the person's status.

The metaphor of career also can describe the pathways that lead a person to imprisonment (see, for example, Sykes [1958]). In the present case, however, we describe the career pathways of a series of inanimate objects which have a material role in convicting a person: samples of bodily material and stained clothing associated with a crime, a suspect, and a victim. Much in the way a suspect passes through a series of stages before becoming a criminal, crime scene and suspect samples pass through a series of stages on the way to becoming criminal evidence. Like the transformations of identity that persons undergo when they become inmates, samples undergo changes in moral and epistemic status when they become evidence. Samples may become "good" evidence of identity: they are identified with guilt and innocence, but in a contested trial they may be judged to be uncertain. And, as was claimed by the defense during the O. J. Simpson trial, they can be alleged to be products of racially motivated police fraud. In our analysis of *R. v. Smith,* we endeavored to reconstruct the organizational pathways through which selected samples traveled as they underwent material and epistemic transformations. These pathways traced movements in time and space

9. *R. v. Smith* was unrecorded. The trial took place in London Central Criminal Court in December 1995. Research on this project was supported by an Economic and Social Research Council grant, "Science in a Legal Context: DNA Profiling, Forensic Practice and the Courts" (R000235853, 1995–98).

10. Materials for this study were initially assembled for a poster exhibition in a department conference (Department of Human Sciences, Brunel University, West London) in 1996, and the case is discussed in Lynch and McNally (2005).

as evidence samples were collected at the crime scene, packaged, transported, stored, unpackaged, analyzed, and presented in court.

The particular case we are examining, though horrifying in its details, is a normal case in many respects. Unlike some of the notable cases examined in other chapters, the *Smith* case was resolved without any major problems with the (real or alleged) handling and analysis of DNA evidence. Although the case involved some minor hitches, the trial did not expose egregious mistakes, possible planting of evidence, slipshod laboratory techniques, or faulty statistical inferences. Instead, the *Smith* trial resulted in a conviction that was not appealed. The case is unlikely to show up in reviews of remarkable cases in which DNA evidence was featured. We chose the case because we were able to get access to some of the key police officers and forensic examiners who took part in the investigation. The duration of the investigation and trial also happened to be convenient to schedule within the frame of our study. However, even though the trial ran concurrently with our study, our reconstruction of the case for the most part was retrospective. For various practical and legal reasons, it was impossible to follow evidence samples from the point of their collection to their eventual presentation in court. Instead, we reconstructed the case through interviews with available participants and by reading evidentiary documents (including written summaries provided by our informants). Our purpose of eliciting recollections was aided by the fact that the case was recent and the disturbing details of the crime were memorable.[11]

The Case

On Sunday morning, 8 January 1995, local police received a report that a dead, partly clad young woman's body had been found lying near a football pitch in Dunstable, a town near Luton, Bedfordshire, about twenty miles north of London (fig. 4.1). The police responded to the report, and throughout that day a series of constables, detectives, scenes-of-crime officers, forensic scientists, a police surgeon, a police photographer, a pathologist, and other police agents visited the site. They erected an evidence tent around the body, cordoned off the crime scene, examined

11. Information about this case was collected by Ruth McNally, during a series of discussions with Bedfordshire police officers in 1996. Quotations in this section are from Bedfordshire Police, Summary of Evidence.

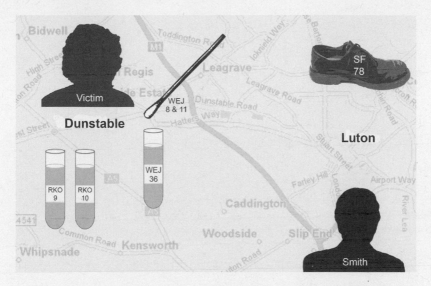

FIGURE 4.1. Origin of Samples. The body of the victim was found in a football pitch in Dunstable at 09:18 on 8 January 1995. WEJ 36 is the victim's blood sample, collected at 17:00 on 8 January 1995 at Luton and Dunstable Hospital Mortuary. WEJ 8&11 are "swabs" that were used to collect translucent stains (presumed semen) found on the victim's body, during examination in an "evidence tent" at the crime scene at 15:00 on 8 January. Smith, the defendant, was arrested in Luton at 08:32 on 11 January 1995, and RKO 9&10 are blood samples from Smith taken at the Dunstable police station at 12:00 on the same day. SF 78 is a shoe recovered early that morning from the residence of Smith's friend, who had told the police that Smith had borrowed the shoes from him on the night of the murder. The shoe had possible blood staining on it. Image prepared by Brian Lynch.

the victim's body and collected evidence samples, and investigated the area immediately surrounding the scene.[12] According to police reports, the victim was a twenty-year-old South Asian woman who had grown up with an adoptive family in Switzerland and recently moved to England, where she had been working as an au pair with a local family. The mother of the family reported her missing early that morning, and later identified her body.

The police collected a number of items of the victim's clothing, and the pathologist took evidence swabs of blood, and also collected swabs of

12. We refer to the woman as "the victim," because the police we interviewed requested that we not use her name. The term "victim" presumes a particular, potentially contestable, relationship to a criminal act, but in this case her identity as a murder victim was established almost immediately and remained stable throughout the investigation.

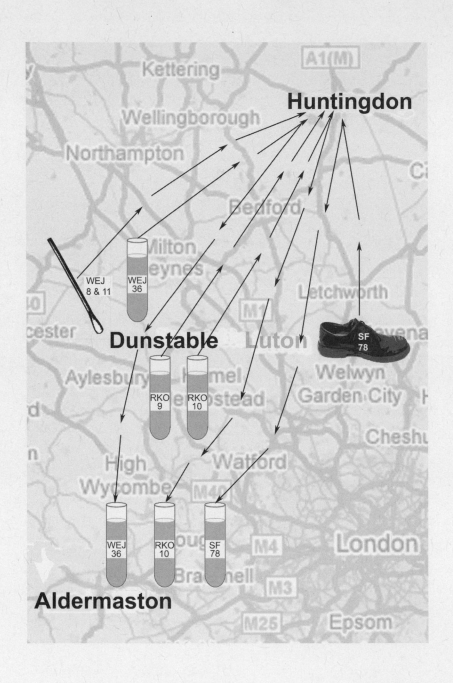

a translucent stain (presumed to be semen) found near the victim's genital area. These items of evidence were labeled (WEJ1 to WEJ31—the letters denote the initials of a forensic scientist who reported on the samples, and the numbers identify the series of separate items collected).[13] The victim's body was then moved to a police mortuary, where further items of evidence (exhibits WEJ32–39) were collected, recorded, and packaged. All of these exhibits were sent to the Forensic Science Service (FSS) Laboratory in Huntingdon (fig. 4.2). The important samples for our story are WEJ8 and WEJ11 (swabs), and WEJ36 (the victim's blood).

The police and pathologist recorded the extent of the victim's injuries, determined the possible cause of death, and began to reconstruct the events leading to her death. The summary of evidence states that she apparently had been struck with a fist, and possibly knocked unconscious, in a car park behind a bowling alley. A jacket and a bloodstain were found near the rear wall of the bowling alley, and the police found tracks indicating that her body had been dragged a distance of seventy-five yards to the base of a tree near the edge of the football pitch. There, her assailant apparently tore off some of her clothes and attempted a rape, but ejaculated on the surface of her body. Either before or after this act the assailant apparently tried to strangle the victim, and he forcefully kicked or stomped on her head, creating severe injuries to the face

13. The initials and corresponding names of the forensic scientists have been changed in this chapter, in order to maintain confidentiality.

FIGURE 4.2. Movements of Samples. This schematic map roughly traces the movements of crime samples to various laboratories in the south of England during the investigation. As the arrows indicate, the swab evidence (WEJ 8&11) was moved from Dunstable (crime scene) to Huntingdon for extraction and SLP analysis. SF 78 (the shoe) was moved from Dunstable Police Station to Huntingdon, where a blood sample was extracted. The sample was then sent to Aldermaston FSS lab for STR analysis. The suspect's blood (RKO 9&10) was moved from Dunstable police station to Huntingdon for SLP analysis, and a portion of RKO 10 was sent to Aldermaston for STR analysis. The victim's blood (WEJ 36) was extracted by a pathologist and moved to Huntingdon for SLP analysis, and then to Aldermaston for STR analysis. Note that some of the samples that originated in Dunstable were sent from Huntingdon to Forensic Science Services labs in London and Birmingham. This map does not trace pathways from Huntingdon, except in the cases of three samples (SF 78, the extract from the shoe); RKO 10, one of the blood samples from the suspect; and WEJ 36, the victim's blood sample). These samples followed a more complicated overland route than the arrows indicate, as they were moved by a series of vehicles operated by a courier service from Huntingdon to Aldermaston on 7 February 1995. Image by Brian Lynch.

and internal bleeding which the pathologist's report said led to the victim's death by asphyxiation.

While the body was being examined in the evidence tent, and later at the mortuary, the police investigation turned up further evidence. The mother of the family that had employed the victim told the police that the night before the murder she had driven the young woman to a nightclub near the football pitch. The police obtained videotapes taken that night by a security camera mounted above the nightclub entrance, and the victim's employer viewed the videos and identified her entering the nightclub and later leaving with a man described as a white male with short dark hair, dressed in a light-colored shirt and trousers. This man—later identified as Smith—became the prime suspect. As a result of media coverage—the video segment was aired on local television for the week after the murder—a man who identified himself as a long-time friend of Smith's contacted the police. According to this witness, on the weekend of the murder, Smith was on home leave from prison, where he was serving a sentence for a sexual assault conviction. The witness told police that he had accompanied Smith to the nightclub on the fateful Saturday evening, and had seen him with a woman fitting the victim's description. The witness also said that he lent Smith the shirt, trousers, and shoes he wore that evening, and that Smith returned them the next day. The police collected these items from the friend's residence, and noted that the shoes remained soiled and bloodstained. These items were sent to the FSS in Huntingdon.

On Wednesday, 11 January, the police arrested Smith at his brother's house in Luton. He attempted to flee through a window but was apprehended. At Dunstable Police Station a police surgeon examined Smith and drew two blood samples from him. The blood samples were labeled (RKO9 and RKO10) and forwarded to the FSS lab at Huntingdon (see fig. 4.2). On Friday, 13 January 1995, Smith was charged with murder, and after being cautioned he replied, "I'm innocent and I will prove it without a shadow of a doubt."

Preston, a scientist at the FFS lab in Huntingdon, examined various items: the victim's blood sample (WEJ36), various items of her clothing and swabs taken from the surface of her body; Smith's blood sample (RKO10), his borrowed shoes, shirt, and other items of clothing. Preston found bloodstains on the shoes, blood and semen stains on the victim's clothing, and blood and semen on the swabs. He performed routine blood tests on the shoes, and said that the blood could have come from approximately one out of every sixteen persons in the population, including the victim. He stated in his report that he transferred the blood samples from Smith (RKO10) and the victim (WEJ36) and a sample of blood from the right shoe (SF78) to Drake of the Aldermaston Forensic Science Service Laboratory.

Samples from two swabs (WEJ8 and WEJ11) recovered from the victim's body, together with her blood sample (WEJ36) and Smith's blood sample (RKO9), were analyzed by another forensic scientist at the Huntingdon laboratory, Dr. Jones, for the purpose of developing DNA profiles with the SLP (single-locus probe) technique. Dr. Jones stated that he had obtained DNA profiles from the swabs and blood samples, and that the profiles from the swabs matched those developed from Smith's blood, but not those from the victim's blood. He stated that "the chances of a person taken at random who is not related to [Smith] having a DNA profile matching the DNA profiles taken from the swabs is about 1 in 70 million." Dr. Jones concluded that these findings showed that the stain on the swabs could be wholly attributed to the presence of semen from Smith.

At Aldermaston, Drake performed tests "to determine with greater certainty the origin of the blood staining on exhibit SF78 [the right shoe]." She used the PCR (polymerase chain reaction) to amplify the DNA in the sample, and an STR (short tandem repeat) DNA typing system to compare DNA profiles from SF78 with the victim's (WEJ36) and suspect's (RKO10) blood samples. The STR system she used at the time (the "Quad") did not have quite as much discriminating power as the older SLP technique, but it could be used to analyze much smaller samples.[14] However, PCR also is known to be highly susceptible to contamination from "foreign" DNA molecules, which can lead to confusing or misleading results. After performing the analysis and determining a match between the victim's blood sample and the blood on the shoe, Drake stated that in her opinion, "the STR typing evidence is 7,000 times more likely" under the assumption that the bloodstain on the shoe (SF78) came from the victim rather than from an unknown person who was not related to her.

The defendant contested this evidence when his case came to trial at London Central Criminal Court (the Old Bailey) in December 1995.

14. The British Forensic Science Service began developing the Quad system in 1990, as the basis for a national DNA database.

From interviews with the police we learned about a small, but trouble-some, complication that came up during the trial. This complication sheds light on an organizational production of "chains of custody." As noted earlier, a chain of custody extends well beyond the laboratory and includes personnel besides laboratory technicians and staff scientists.[15] Viewed retrospectively and inferentially, a chain of custody reaches back from the lab all the way to the crime scene. The chain also extends pro-spectively into the courtroom, and it encompasses an entire series of ma-terial, clerical, and other referential practices that make up the local his-tory of a sample.[16] Scenes-of-crime officers act wittingly and unwittingly as agents of science, and their actions can become accountable in court by reference to the identity and integrity of DNA profiles.

Exploding the Linkages: A Story of a Shoe, a Signature,
and Four Lorry Drivers

What is impressive about chains of custody is that they are held together by heterogeneous practices for certifying identity and establishing cred-ibility. The very elements that compose a particular chain can occasion-ally be shown to be sources of disruption and dissolution. To use a hu-man analogy, the chain can be likened to a relay race in which a baton is passed from one runner to the next in a series; only, in the case of a chain of custody, there are numerous overlapping races running on different tracks; each with its own characteristic baton and each with its own pos-sibilities for fumbling the handoff. A further complication, which we dis-

15. This point is made in Halfon (1998).

16. Alfred Schutz (1962) developed a sociological conception of phenomenological time-consciousness: the "retrospective-prospective sense of occurrence." Viewed as an es-sential property of time-consciousness, this concept refers to how a momentary percept (for example, the apprehension of a musical note) is seen or heard as a moment-in-a-series with both past and future horizons. This involves both a retrospective grasp of the series in which the moment occurs (even if it was not concretely seen or heard) and a prospective anticipation of what will happen next. Schutz (1964c) points out that this structure of con-sciousness is intersubjective, and has important coordination functions in social activities such as playing music together; indeed, recruitment into such coordinated activities can be said to have precedence for establishing the "structure of consciousness" that a social-ized member embodies. A further step beyond individuality, and even consciousness, is implied by ethnomethodological conceptions of local historicity (Garfinkel et al., 1981). In our case, the local historicity of an evidence sample is materially embodied in the form and packaging of a sample, and inscribed in, and contingently retrievable from, the paper trail that records and certifies its handling and movements.

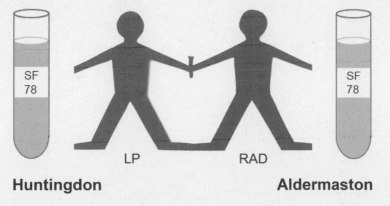

Huntingdon   Aldermaston

FIGURE 4.3. The blood extracted from the Shoe (SF 78) is transferred from Mr. Preston (LP, Huntingdon) to Ms. Drake (RAD, Aldermaston). Image by Brian Lynch.

cuss shortly, is that it is not always clear from the outset who is involved in the running of the race. Like a marathon in which runners sneak into the race along the way, ambiguities and retrospective inquiries can arise about whom the significant actors involved in a chain of custody really are, or were.

When attempting to reconstruct the careers of samples in the *Smith* case, we found that it was difficult to specify in advance which links, strands, relays, and sources of fragility would become significant. Moreover, our initial sense of the actions and agents that made up particular links in the chain was subject to doubt. The adversary situation in the courtroom trial disclosed previously hidden linkages that did not seem significant, or even visible, prior to the trial. For example, during the trial of Smith, a problem arose about the transfer from Huntingdon to Aldermaston of the blood sample taken from the shoe (SF78). The account of this transfer in the police summary of evidence can be schematized with a graphic device known to young children: "paper dolls." These are cut-out figures, joined together and folded accordion-style, so that a stack of figures can be folded out into a chain (see figs. 4.3–4.6).

During the trial, other agents and transfers became salient. An alert defense took note of the following set of contingencies:

1. Mr. Preston placed the evidence (SF78) in a tamper-evident bag (a sealed bag, designed to reveal any attempt to open it).
2. A clerk in Huntingdon was responsible for witnessing the transfer of the

FIGURE 4.4. Break in the chain as Smith's defense makes an issue of a photocopied signature, introducing a clerk into the evidence transfer. Images 4.4–4.6 by Brian Lynch.

FIGURE 4.5. "Paper doll" analogy of exploding number of agents in the transfer of the sample from Huntingdon to Aldermaston. Four lorry drivers who were, as though, hidden beneath the initial transfer from LP to RAD, mediated by the Clerk, now "fold out" into the space between them. The shadows (initially an artifact from photographing the three dimensional "paper dolls" for a poster exhibit), are significant for our purposes, as they indicate to the shadowy possibility of further agents beneath those that currently are exposed.

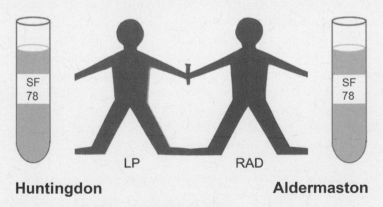

FIGURE 4.6. Restored Links in the transfer of evidence.

evidence bag to a courier service, and certifying the continuity of the samples transferred to Aldermaston.

3. The clerk (who shall remain anonymous) used a shortcut. He routinely completed "sample sheets" certifying evidence transfers. A sample sheet is a standard form used for purposes of recording and denoting the contents of evidence samples whenever they are transferred from one laboratory to another. In this case, several samples were moved in the same batch, and rather than sign a fresh sample sheet for each item, the clerk photocopied a form with his signature on it and then filled in the identifying details for the particular items.

When the defense attorney discovered the photocopied signature, he demanded the original. None could be produced. The trial was delayed while the prosecution attempted to recertify the transfer in a more laborious fashion. The prosecution contacted the clerk and the courier service. According to the courier service's records, the sample in question was moved by four different lorry drivers on different legs of its journey from Huntingdon to Aldermaston (fig. 4.5). The Crown attorney's office located the four lorry drivers and two controllers working with the courier firm and arranged to have all of them come to the Old Bailey in London, along with the clerk who authored the photocopied signature. This now-expanded group of agents was prepared to testify that the evidence sample moved through an unbroken chain of transfers on its way to Aldermaston. As it turned out, Preston's and Drake's testimony, along with the silent testimony provided by the intact tamper-evident bag, persuaded the defense to accept that the sample that left Huntingdon was identical with the one that arrived at Aldermaston, and the clerk and lorry drivers no longer were required to play their cameo roles at the Old Bailey. The break in the chain was repaired (fig. 4.6), the lorry drivers were folded back into an unbroken link between Preston and Drake, and at the end of the trial Smith was convicted. The "monster," as the tabloid press dubbed Smith, is now enchained in jail without any prospect of weekend leave.

The Contingency of Referential Matches

When DNA profiling is used successfully to identify and convict a perpetrator, it is customary to speak of a referential correspondence—a match between two samples—as the key item of evidence. In the *Smith* case, as we noted, a match was declared between the DNA profile developed from bloodstains on a shoe allegedly worn by the suspect (SF78) and the profile developed from a blood sample taken from the victim's body (WEJ36). However, when we retraced the careers of the particular samples, we were able to notice that the match in question was produced and made accountable by means of an extended chain of custody. The integrity of the samples, and the referential match between them, held only as long as the chain remained intact. The defense attorney's skeptical examination of a mundane clerical form momentarily expanded the chain to reveal a subchain of previously hidden agents and actions, whose role in the story suddenly became relevant and problematic. It is imaginable that a persistent defense would seek further expansion of this seemingly trivial link by investigating the actions and agents involved in loading and unloading each vehicle throughout the journey, or seeking assurance that the vehicles and their cargo of precious bodily fluids were secure and under surveillance at all times.

In our "paper doll" reconstruction (figs. 4.3–4.6), the figures of the clerk and lorry drivers are initially stacked underneath the transfer from Preston to Drake, but after the defense counsel's discovery of the photocopied signature, the hidden agents are folded out and the chain is evidently (and problematically) extended through their agency. Then, after the prosecution undertakes a laborious effort to certify the transfers through direct testimony, the lorry drivers fold back in, and out of sight, so that the transfer between Preston and Drake is once again direct and intact. This banal link was but one of many transfers that took place over the careers of the samples. A crucial aspect of these transfers is that numerous bureaucratic devices were used to attest to the continuity of the evidence. The credibility of the DNA match presented at the trial rested on an acceptance of the various statements, attestations, and official certifications, embedded in, and embodied by, the paper trail. And so, while it might seem as though *science,* after being subjected to challenge, triumphed at the end of the day, its victory was sustained and supported at many points by a small, mostly hidden, labor force of bureaucratic officials, clerks, and lorry drivers.

Consider how the evidentiary link between Smith and the crime scene was accomplished. Even a restricted, schematic account of DNA profiling gives us a picture of a multifaceted chain (fig. 4.7). The chain is made up not of tandem repeats of identical elements, but of representations of many different kinds: denatured material extracts, designer frag-

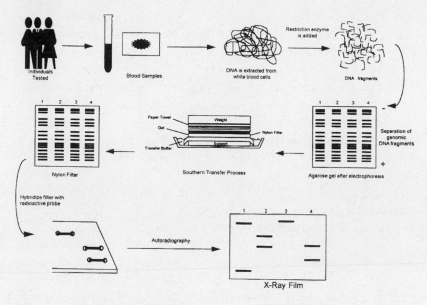

FIGURE 4.7. Schematic representation of restriction fragment length polymorphism analysis using a single locus probe. Source: Lee et al. (2004, fig. 5): 273. Reprinted with permission of WoltersKluwer Health.

ments, gel images, paper transfers, radioactive markers, x-ray exposures, visual matches, numerical codes, and statistical estimates. Expanding the chain so that it extends beyond the lab, we find such items as tamper-evident bags, courier services, forms, signatures, and oaths.[17] And in addition to metaphoric bar codes inscribed on autoradiograms, we find real bar-code stickers fastened to test tubes, records, and other items. These bar codes are organizational artifacts used for tracking the identity and assuring the integrity of samples. They are rails of justice: materials that facilitate the legally accountable transport and transformation of evidentiary cargo from one designated staging point to another.[18] Consider the highly detailed nature of the practices and inquiries that bolster or undermine the credibility of these representational devices. In forensic laboratories, protocols are set up so that one laboratory worker officially "witnesses" the work of another. Clerks also "witness" trans-

17. See Halfon (1998) for an analytic expansion of such a schematic account.

18. See Riles (2006) for studies of the role of documents for constructing accountable knowledge in law and bureaucratic organizations. For an illuminating account of such rail-building work in another organizational context, see Bowker (1994).

fers of evidence from one laboratory to another. Throughout the process, bureaucratic records stand as documents of identification, certification, and organizational memory: a signature, appropriately inscribed on a form by the relevant official, performs witnessing, independently of what the official may actually have seen or heard. The momentary glitch in the chain of custody we described earlier called into question an act of clerical witnessing because no "original" signature could be produced.[19] Representation unquestionably takes place, but not in accordance with a single rule or set of criteria, and never with the certainty demanded by the metaphysician.

In this case we see that the integrity of the chain is underwritten not by a natural identity that runs from one end to the other, because such an identity is a contingent product of such integrity rather than a cause. The integrity of the chain is forged and secured through a complex arrangement of technical devices and organizational practices that serve to record, certify, and practically ensure that samples remain identical with themselves, that signs signify their objects, and that a naturalistic reading of matching evidence coherently emerges from the assemblage. Reference, signification, and identity become accountable as relations between elements, but the elements and the referential functions are inscribed through the chain. As Goffman's (1961) metaphor of moral career suggests in the case of a mental patient, the career of a sample tracked through a chain of custody is likewise a moral career. In Goffman's antirealist conception of madness, the institutionalized mental patient suffers not from the revelation of a biological disease but from the biographical and social contingencies that make up the pathway to the asylum rather than to some other destination.[20] According to the same counterfactual logic, one could say that Smith's institutionalization was determined not by biological properties of the evidence but by the contingencies that forged the pathway the evidence traveled from crime scene

19. Readers who are mindful of literary theory may recall Derrida's (1977) closing line to "Signature, event, context" in which he professes to "forge" his signature, thereby confusing the conventions through which authorship is secured. When transposed to the imagery of a chain, the "forgery" of a clerical signature takes on a double (potentially duplicitous) sense: a link in the chain is manufactured (forged in a concrete sense), but the manner of manufacture is questioned for being unauthorized (a potential illegality, a forgery).

20. In a famous line, Goffman (1961: 135) states, "In the degree that the mentally ill outside hospitals numerically approach or surpass those inside hospitals, one could say that mental patients distinctively suffer, not from mental illness, but from contingencies."

to court. The either-or logic of such statements is highly dubious, but the point remains that the *accountable* integrity of samples, and of matches between DNA profiles developed from them, depends upon the regularity, propriety, and absence of prejudice ascribed to bureaucratic protocols, tracking mechanisms, systems of surveillance, and cross checks between multiple forms of indexing. Automatic recording devices such as bar codes and scanners and tamper-evident bags minimize the degree of trust invested in persons who staff the chain, but when challenged, these operations are underwritten (or, perhaps, underspoken) by personal testimony and credibility (we return to this point in chapter 7).

The minor crisis occasioned by the photocopied signature illustrates how previously visible elements of a chain of agents and actions can expand, and even explode, as hidden agents and actions are made to emerge from obscurity. Consequently, it becomes impossible to specify in advance just which "elements" make up the chain of custody. As long as it stood unquestioned, the signature certified the integrity of a transfer of the sample from one place (and agent) to another, but when its authenticity was challenged it became clear that it "stood for" (that is, it stood proxy for) a more complicated set of organizational agents, operations, and records. As noted early in this chapter, savvy lawyers know about, and exploit, such properties of chains when interrogating forensic witnesses and demanding detailed accounts of every possible step and substep through which samples pass on the way to becoming evidence.

A DNA profile match presently is regarded as a particularly strong form of physical evidence. Those who promote the use of DNA evidence to identify suspects, convict perpetrators, or prove the innocence of wrongly convicted suspects often draw invidious contrasts between the strong objectivity of such "scientific" evidence and subjective, highly fallible, forms of "common sense" evidence such as eyewitness testimony. However, when we take into consideration how DNA profile matches are embedded in chains of custody, we can appreciate that the objectivity of DNA evidence is subject to lines of attack similar to those that dismantle eyewitness accounts. In the words of philosopher Karl Popper describing a potential explosion of empirical (eyewitness) knowledge when subjected to relentless skeptical questioning, "you would in fact never arrive at all those observations by eyewitnesses in the existence of which the empiricist believes. You would find, rather, that with every single step you take, the need for further steps increases in snowball-like fashion."[21] As we have seen, when subjected to skeptical scrutiny, the chain of custody also ramified into further steps involving potentially relevant actions and agents. These actions and agents had been hidden— glossed over—in previous accounts of the chain of custody, which took for granted or trivialized their significance for the evidence they helped to deliver. A mere bureaucratic detail—a difference in the accountability of a photocopied versus handwritten signature—was used as leverage to bring these hidden agents back into the picture. However, we also should recall that the expansion was reversed, the hidden agents folded back into obscurity, and the original link was restored. That also was a contingent outcome, no less so than was the original folding out.

When we take the folding-out and folding-in of agents and agencies into account, the evidential weight of the DNA match—as a correspondence between two physical traces—becomes importantly connected to the administrative forms of accountability that testify to the bureaucratic regularity, routine competence, impersonality, and disinterestedness in the way the materials were handled. Such forms of accountability can be said to be objective matters, but this is the objectivity of an ideal-typical bureaucracy, specified and repaired in this instance by a complicated paper trail that certified that standard protocols were followed and proper procedures were accomplished, witnessed, and supervised. For the most part, the courts place a burden on defense attorneys to demonstrate prejudice, incompetence, or other sources of rupture in the chain of custody, and until such demonstration is made the paper trails produced by the administrative processes that make up the chain have presumptive credibility.[22]

The "scientific" objectivity of DNA profile results is thus inscribed on a platform of bureaucratic objectivity—an administrative accountability that imposes demands of purity, integrity, and continuity upon police agents and agencies as much as upon the work of forensic laboratories.[23]

21. Popper (1963: 22–23); quoted in Shapin (1994: 24).

22. See Redmayne (1995) on burden of proof in cases involving DNA evidence.

23. Historians of science have pluralized and historicized the concept of objectivity, pointing to the particular salience in nineteenth- and twentieth-century sciences of the category of "mechanical objectivity"—objectivity associated with the repetitive, reliable operations of machinery in contrast to the fallible "subjectivity" of human actions and judgments (Daston & Galison, 1992; Megill, 1994; Porter, 1995). Administrative objectivity overlaps with this category, but is founded in ideals of bureaucracy rather than of science and technology. It is no accident that the norms of science, which Robert K. Merton (1973) first formulated in the 1940s, overlap with the properties of Weber's ideal-type of bu-

Claims to administrative objectivity can be attacked, sometimes successfully, by a forceful skeptic, but administrative objectivity has presumptive standing in a system charged with the administration of justice. In an adversary system of justice, the role of "forceful skeptic" is not played by a philosopher, but by a cross-examining attorney. To a large extent, how well a chain of custody holds together in the face of efforts to "deconstruct" it depends upon the questions the adversary raises, the rebuttals the proponent provides, and the judge's and/or jury's reaction to the interplay of testimony and argument.

Just What Kind of "Thing" Is a Chain of Custody?

Although a chain of custody has little in common with the classic exemplars of physical objects favored by philosophy of science, we believe that it is appropriate to speak of it as an organizational object with interesting referential properties. Contrary to the classical philosophical concept of reference, which depicts a field of (subjective, discursive) references laid up against a field of (objective, material) referents, we can think of an interwoven, continuous, and temporally performed coherence of things and signs: "We never detect the rupture between things and signs, and we never find ourselves faced with the imposition of arbitrary and discrete signs upon shapeless and continuous matter. We only see an unbroken series of well-nested elements, each of which plays the role of sign for the previous and of thing for the succeeding" (Latour, 1995: 169).[24] The vernacular analogy with a "chain" can be misleading, however, if it leads us to suppose that a chain of custody is composed of a linear arrangement of repeating links composed of identical elements.

---

reaucracy. Whether or not they describe actual conduct, variants of Weber's and Merton's formulations recurrently appear in normative talk and writing about science and modern bureaucratic life. When Merton first formulated the norms of science, he (overoptimistically, perhaps) argued that they could fully exist only in a democratic society that allowed an autonomous scientific institution to remain free of political and religious ideology. He did not make much of the affinity between the norms of science and the idealized properties of bureaucratic decision-making: that it is disinterested, based on rules and criteria (i.e., rational), and universalistic (in the sense of treating persons equivalently, by merit). What interests us is the interplay between administrative or bureaucratic operations and the accountability of scientific results. The emphasis given by overseeing bodies on quality assurance/quality control audits provides some insight into the way administrative operations stand proxy for the reliability and validity ascribed to "science" and "scientific results."

24. For a detailed study of the sequential composition of activities and materials that eventuate in laboratory images see Lynch (1985a, b).

It is less like a chain composed of solid and repeated links, and more like a rope of many strands, no single one of which runs the entire length.[25] Moreover, the strands are not all made of the same stuff, and the rope may harbor bugs, patches of rot, and other potentially destructive elements. It also would be misleading to liken it to a continuous wire that acts as a conduit for passing information, knowledge, or truth back and forth.[26] What counts as a significant element of the chain is tied to assessments of the identity and validity of what passes through it. As we have seen, a given link can ramify to expose constituent (or even contradictory) linkages hidden by a record or signature. Moreover, a chain, with its associated paper trails, is not a single, linear series of stages, as it is likely to be supplemented by parallel, intersecting, and cross-cutting chains and trails.[27] Such complexity can be a source of confusion or contamination, but it can also provide a source of corroboration, as we saw when courier company records were used to repair a minor rupture in a forensic science paper trail. A selective attention to instances of adversary courtroom "deconstruction" may lead one to think that chains of custody are essentially fragile—constructed from rather flimsy stuff—

25. The analogy of a rope with many strands is famously used by Ludwig Wittgenstein (1958: sec. 67) to refute the idea that an identical meaning requires one or more essential features to be common to all referential instances. In the rope analogy, overlapping "family resemblances" are the source of identity.

26. This analogy is suggested by Latour (1995: 180): "Truth circulates here like electricity through a wire, so long as it is not interrupted." Our example of the clerk and lorry drivers points to a broader criticism of actor-network theory (Latour, 1987; 1990; 1992; 1999; 2004a; Callon, 1986). The theory's major contribution to the S&TS field has been through its fecund vocabulary and suggestive metaphors, but its proponents sometimes operationalize it technically to represent the networks of "actants" in a narrative field. In the case of a chain of custody, such an effort would be confounded by the complexity of the arrays of traces, inscriptions, and paper trails that the notion of chain of custody glosses. Worse, as we have seen, a given specification of a link, and of its relevant agents and agencies, may expand and contract (temporarily repopulating the link with a temporary team of agents) in the course of an adversary inquiry. Moreover, Latour's conception of reference as a constancy running through a series of transformations identifies a contingent *claim* for proponents of evidence. Importantly, however, the witnesses in court can deliver only fragmentary elements of the chain. Unlike Benjamin Franklin in his public demonstrations, they cannot deliver the jolt of electricity in a courtroom demonstration. Truth never appears as a unitary force; instead credibility is delivered through a crowd of proxies, in the face of an adversary effort to redistribute the proxies around a different story.

27. There is rough affinity here with Adrian Cussins's (1992) theme of "cognitive trails," though the chains and trails we discuss are far more "concrete" and tied to specific organizations and their records.

but links that are exploded also can be repaired. Fragility is no less occasional and contingent than robustness.

Breaks or interruptions in chains of custody provide perspicuous occasions for raising questions about the identities and origins of samples, but *just where* these interruptions occur—or may have occurred—is itself an open question. Did the clerk's photocopied signature interrupt the chain? Yes and no. As long as it remained unquestioned, it was unremarkable, and after it was questioned, it was repaired through a bypass operation using other records and testimonies.

Continuity and Object Constancy

Although we have tended to use the American expression "chains of custody" in preference to the British "continuity of evidence," the latter term can also be used to open up distinctive conceptual associations. Continuity of evidence can be construed as a variant of what is sometimes called "constancy" or "accordance" in gestalt and phenomenological theories of perception. The problem addressed by these theories is how separate moments of perception can be experienced as being related to an identical object or constant field, despite variations in perspective and appearance. Conversely, the theme describes how a preconception that the object is a self-identical thing structures the continuity and coherence of its perception.[28] Aron Gurwitsch (1964: 209) provides an apt summary: *"[I]f there is to be the consciousness of an identical thing . . . there must be accord and harmony between the multiple acts through which the thing in question is imagined as to the aspects under which it presents itself from various stand-points and sides, as to the qualities and attributes progressively displayed and as to the behavior which it exhibits under various circumstances, etc."* (italics in original). There is a key difference, however, between the continuity of evidence, and Gurwitsch's formulation of the gestalt/phenomenological continuity of "acts" of perception. In police investigations and courtroom testimony, constituent "acts" are not limited to perceptions; importantly, they include material practices and communicative actions performed by many different

28. See Garfinkel et al. (1981) and Lynch et al. (1983) for ethnomethodological analyses of how gestalt themes feature in the performance of experimental practices.

agents. These are public, as well as private, acts, and their coherence does not reside in the mentality of an individual "perceiver"; instead, the coherence is inscribed, literally, in organizational records that track the history of samples and in the form and packaging of the material evidence. The "accord and harmony" between acts is administered through the use of standard protocols, the performance of organizational routines, and the recording of paper trails. To borrow Latour's (1995: 180) language, the integrity of evidence moved from one staging point to another in a chain of custody depends upon its "traceability" (also see Lezaun, 2006). However, the paper trail through which it is possible to retrace the evidence chain does not simply provide access to the original movement, it lends its own qualities to the chain. The trail is not simply a trace on paper, because the *accountability* of the records resides in a local variant of what historian Adrian Johns describes as a "print culture." In a study of the largely unregulated book trade in seventeenth-century London, Johns shows how questions of literary identity turned on legal and practical protections against piracy.

> Could a printed book be trusted to be what it claimed? Perhaps a reader would be prudent to reserve judgment. On the most obvious level, whether a *Sidereus Nuncius* printed in Frankfurt was really Galileo's text, or an *Astronomiae Instauratae Mechanica* produced in Nuremburg was really Tycho's, could justifiably be doubted. More broadly, the very apprehension that printed books might not be self-evidently creditable was enough to rule out any possibility of their bearing the power attributed to them by most modern historians. . . . Unauthorized translations, epitomes, imitations and other varieties of "impropriety" were, they believed, routine hazards. Very few noteworthy publications seemed to escape altogether from such practices, and none at all could safely be regarded as immune a priori. (Johns, 1998: 30)

The printed object (the book) and the technology of printing did not guarantee that the same text would be disseminated under a title and author's name. Johns emphasizes the variability of reading and interpretation, but he also indicates less often discussed *sources* of variability. The apparent identity and identicality of a book was used (literally?) as a cover for subversive operations that went under the name of piracy.[29]

---

29. It is worth mentioning (or, perhaps, it is not) that one of the authors of the present book has for many years owned a pirated copy of a book (a copy of the English translation

So, what does this have to do with criminal evidence? Consider the following case: the prosecution in a criminal trial adduces evidence that a fingerprint found at a crime scene matches a suspect's fingerprints. In other words, two "texts" or "signatures" are shown to be identical. However, if the defense persuades the court that a policeman forged the evidence by, for example, picking up the suspect's fingerprints with cellophane tape and transferring them to objects collected from the crime scene, then the relationship between the two fingerprints becomes analogous to that of a book and an unauthorized copy. Even if they are identical, the identity does not have legal authority, and suspicions of authorship turn from the suspect to the police.

As noted early in this chapter, chains of custody are inviting targets for cross-examiners. Given the contingent linkages and the potential for expansion, one can wonder how they ever hold together under the onslaught of "deconstructive" interrogation. And yet, for the most part, they do hold together. We shall examine both sides of this issue in later chapters. In chapter 5 we go into further detail about interrogations of expert witnesses, focusing on techniques for building up and attacking credibility. In chapter 7, we discuss "administrative fixes": regimes designed to protect forensic evidence from common lines of courtroom attack, and we shall argue that the scientific credibility that DNA typing currently enjoys has much to do with the administration of objectivity.

---

of Heidegger's *Being and Time*). We assume that the pirated copy printed in Taiwan is essentially identical to the authorized translation. There are material differences: the paper is thin and cheap, the print and even the dust jacket is a faded photocopy of the original, and the binding is falling apart. Independent of these material discrepancies, for certain commercial and legal purposes the pirated copy is not identical with the authorized translation. Even if the text was "the same" and would make no difference for an individual reader, it would not be an authorized copy, and certain things could not be done with it. To relate this to the case at hand, without proper authorization certain serviceable bits of evidence are likely to be excluded from legal accountability.

The U.K. National DNA Database

Launched on 10 April 1995 at the Forensic Science Service (FSS) laboratory in Birmingham, the U.K. National DNA Database (NDNAD) of England and Wales is the oldest and largest national DNA intelligence database.[1] Although other nations have DNA databases, in this interlude we focus on the United Kingdom, which has the most enabling DNA databasing legislation in the world and is adopting the most privatized model for the provision of forensic sciences to the criminal justice system.[2] (Elsewhere in this book we discuss other systems, such as CODIS [Combined DNA Index System] recently developed in the United States by the FBI.)

The day-to-day running of NDNAD is performed by FSS Ltd. Formerly an executive agency of the Home Office, the FSS formally became FSS Ltd. in December 2005, although it is also referred to by its trading name "FSS." FSS Ltd. is a profit-seeking, government-owned company under contract to the Home Office. It was formed as a transitional stage toward becoming a private company in a fully commercialized market in forensic services to the police (House of Commons Science and Technology Committee, 2005).[3] NDNAD integrity, the

1. Scotland and Northern Ireland, which have their own DNA databases, submit profiles to the NDNAD.

2. For more detail about the NDNAD see the report by Williams et al. (2004); National DNA Database Annual Reports (2003–4), (2004–5), and (2005–6); House of Commons Science and Technology Committee (2005); Nuffield Council on Bioethics (2007); Parliamentary Office of Science and Technology (2006). For regular updates and critical analysis of the NDNAD, and for information on Scotland, where the law is significantly different, see the Web site of GeneWatch U.K.: http://www.genewatch.org/.

3. On the recommendation of the McFarland review of the FSS (Home Office, 2003).

setting of standards for procedures and profiling, and approving and monitoring laboratories that supply the profiles is the responsibility of the NDNAD custodian, formerly chief scientist at the FSS. With the partial commercialization of FSS Ltd. in 2005, the NDNAD Custodian Unit has been transferred to the Home Office in order to avoid conflicts of interest with FSS being both database supplier and custodian. In 2007 the Custodian Unit was moved from the Home Office to the National Policing Improvement Agency. The NDNAD is governed by the National DNA Database Strategy Board, comprising representatives of the Home Office, the Association of Chief Police Officers and the Association of Police Authorities, with ethical oversight and a lay view of the board's decisions being provided by two members of the Human Genetics Commission. In 2007, members were appointed to the independent NDNAD Ethics Group to advise the Database Strategy Board.[4]

The profiles on the NDNAD (and the samples they are derived from) belong to the individual police forces from where they originated. The majority (80 percent) of samples represented on the NDNAD have been profiled by FSS, with the remainder being profiled in the accredited laboratories of five private organizations (National DNA Database Annual Report, 2005–6).

The NDNAD contains DNA profiles from three different sources:

1. Casework profiles from unknown persons, known as scenes-of-crime (SOC) profiles, derived from bodily samples, such as blood, semen, sperm, and saliva, collected from crime scenes.
2. Criminal justice (CJ) profiles, which are reference profiles from known people arrested on suspicion of involvement in a criminal offence.
3. Elimination, or volunteer, profiles, which are another source of reference profiles, from known volunteers who provide samples to eliminate them from a specific criminal investigation, for example as part of an intelligence-led "mass screen" in a particular area or workplace.

---

4. The Human Genetics Commission (2002) and the Select Committees on Science and Technology in both Houses of Parliament recommended that there should be an independent oversight body for the NDNAD with lay membership (see Parliamentary Office of Science and Technology, 2006). The formation of the Ethics Group was the response.

Each CJ record on the NDNAD includes the following information (GeneWatch U.K., 2005: box 1):[5]

- unique barcode reference number linking it to the stored DNA sample;
- Arrest Summons Number, which links it to the record on the Police National Computer (PNC) containing criminal records and police intelligence information;
- the person's name, date of birth, gender and "ethnic appearance" (as assigned by a police officer);
- information about the police force that collected the sample;
- information about the laboratory that analyzed the sample;
- sample type (blood, semen, saliva, etc.);
- test type;
- DNA profile as a digital code.[6]

Every day, all of the profiles in the NDNAD are searched against each other to look for matches. The objective of these daily "speculative" searches is to identify "hits"—a match or partial match between a newly entered profile and an already databased one. Such matches are reported back to the relevant police forces as intelligence matches. Matches between profiles from two different crime scenes could mean that the same person was present at both crime scenes. Matches between two CJ profile entries could mean that the same person has been profiled twice.[7] A match between a CJ profile and a SOC profile may identify a suspect for the crime.[8] A suspect may be charged on the basis of an NDNAD intelligence match only when there is further supporting evidence.

Technically, there are two types of profiles on the NDNAD. From 1995 to 1999, the standard profiling system was the Second Generation

5. Records for SOC profiles contain information about the crime rather than the (unknown) individual.

6. The code comprises twelve numbers for SGM profiles and twenty for SGM Plus profiles, plus a gender indicator (see fig. A.6).

7. Based on an estimated replicate sampling rate of 11 percent, the 4.251 million CJ profiles on the NDNAD on 20 November 2006 belong to just 3.783 million individuals (Parliamentary Question, 2006b). According to Cole (2001a: chap. 1), one of the original incentives in the late nineteenth century for collecting systematic information about criminals and suspects was to track recidivists, who often disguised their identities with aliases. More than one hundred years later, false names and aliases continue to be a problem, although it is unclear how duplicate samples are distinguished from adventitious matches (Williams et al., 2004: 125).

8. Such matches can open up reinvestigation of an old, unsolved ("cold") crime.

Multiplex (SGM), with an average discrimination power of 1 in 50 million.[9] SGM was replaced in 1999 by SGM Plus, the current database standard, to reduce the chance of an "adventitious match" as the size of the database increased. An adventitious match is a chance match between the profiles of two different people and it gives a false positive result. The risk of an adventitious match between two SGM profiles has been calculated to be 26 percent.[10] It is claimed that no adventitious matches have ever occurred between full SGM Plus profiles (Nuffield Council on Bioethics, 2007). However, Sir Alec Jeffreys, the original inventor of DNA profiling, warns that when a database the size of the NDNAD undergoes large numbers of speculative searches, even extremely rare matches will occur. Furthermore, the probability of an adventitious match increases when partial profiles or the profiles of related individuals are compared.[11] Dissatisfied with the discriminatory power of SGM Plus, Jeffreys recommends that following the identification of a suspect, the authority of the match should be tested by reanalyzing the sample at six additional loci (House of Commons Science and Technology Committee, 2005: 40–41). However, while reanalysis with more rigorous systems has the potential to exonerate the innocent, it requires the retention of the biological samples after they have been profiled for the database, a practice that raises its own concerns, given that profiles and the associated samples of NDNAD inhabitants, including those of volunteers and "postsuspects" (those arrested but not subsequently charged

9. This analyzed six short tandem repeat (STR) loci plus a locus for sex designation.

10. Twenty-two thousand CJ profiles were upgraded from SGM to SGM Plus because they were reported as matching crime scene profiles. Six thousand of these were recompared to the crime scene samples they originally matched. Of these, 52 percent confirmed the original match; 19 percent showed that the previously reported SGM match was adventitious and eliminated the suspect; and no comparison was possible for 29 percent because the crime sample profiles were no longer on the database. On the assumption that 19 percent of the latter would also have been false positives, it was concluded that 26 percent of the matches reported using the SGM would have been adventitious (DNA National Database Annual Report, 2003–4: 16).

11. Fifty percent of SOC profiles are only partial because the source samples are degraded or involve low quantities of DNA (presentation by Bob Bramley, then custodian of the NDNAD, Techniquest, Cardiff, 19 July 2005). The standard for including partial profiles in routine speculative database searches is that they must comprise genotypes from at least four loci. At the end of March 2003, of 168,308 scenes-of-crime profiles on the database, 22,849 were partial (Williams et al., 2004). The probability of a chance match between full SGM Plus profiles of full siblings is 1 in 10,000 compared to the conservative estimate of 1 in 1,000,000,000 for unrelated individuals (National DNA Database Annual Report, 2004–5).

with an offence; see below), can be used in forensic research without the explicit consent of the donors.

Toward a National DNA Database

The realization of the NDNAD in its current form has been both the driver for, and the consequence of, a series of interdependent developments in forensic science, government policy, policing practices, and the law, coproduced with a model of the typical criminal career. The FSS first explored the potential of forensic DNA database searches through trials of a small database comprising 3,000–4,000 "single-locus probe" (SLP) profiles (Allard, 1992). This pilot study highlighted technical and legislative barriers to the efficient operation of a DNA database in criminal detection and conviction. On the technical side, the SLP system was labor-intensive, slow (days or weeks per profile), and required larger amounts of biological sample than were obtainable from many crime scenes and some suspects. On the legislative side, the law governing the taking of bodily samples from suspects was sections 61–65 of the Police and Criminal Evidence Act of 1984 (PACE). PACE classified blood or scrapings of cells from inside the mouth as "intimate" body samples, which therefore could not be taken from a suspect without his or her consent. Hair samples (apart from pubic hair) were classified as "nonintimate," and therefore did not require consent, but hair cut by scissors rarely contained sufficient DNA for SLP profiling, which generally required a clump of hair pulled out by the roots. Furthermore, suspect samples could be taken only when it was believed that the sample would be relevant to the investigation of a serious arrestable offence, such as treason, murder, rape, or kidnaping.

By the late 1980s, it had been demonstrated that using the polymerase chain reaction (PCR) it was possible to amplify and type DNA from a single hair. Forensic organizations throughout the world tailored the PCR approach to their needs, including the FSS, which in 1990 began developing the Quad PCR DNA profiling system (Werrett, 1995).[12] Designed to develop profiles from "easily available substances of the body" which were "capable of being reduced to a numeric value and easily cap-

12. The Quad was a multiplex system that amplified four STR loci and had an average discriminating power of 1 in 10,000. It was succeeded by the Second Generation Multiplex (SGM) system.

tured on a computerized database," PCR profiling featured in the Association of Chief Police Officers' (ACPO's) Police Service Business Case for a national DNA database (Dovaston, 1994a, 1994b: 1). In 1993, having received evidence from across the criminal justice system, the Royal Commission on Criminal Justice (1993) recommended that a national forensic DNA database be established, and the following year the Criminal Justice and Public Order Act 1994 (CJPOA) introduced the necessary amendments to PACE.

One of the key changes to PACE was the reclassification of mouth and hair samples (apart from pubic hair) as "nonintimate" samples which could be taken without consent. Another was lowering the threshold for sampling, making it lawful for the police to obtain a nonintimate sample from anyone in police detention or custody who was charged with, about to be reported for, or convicted of a "recordable" offence (rather than a "serious arrestable offence") regardless of the relevance of DNA evidence to the particular offence concerned.[13] One rationale for lowering the severity threshold was research findings from the Derbyshire Constabulary CATCHEM (Centralised Analytical Team Collating Homicide Expertise and Management) Project, which "demonstrated that in more than half the cases of murder where the victim is a child or young woman, the offender had a previous conviction for assault or a sexually oriented minor crime before going on to commit the offence of murder" (Dovaston, 1994b: 2). ACPO reported these and other studies to the Royal Commission, citing them as evidence of the typical trajectory of a criminal career characterized by recidivism and crime "progression," for example from vandalism, auto theft, and burglary, to armed robbery, rape, and murder (Gaughan, 1996: 12–13). Such studies formed the basis for ACPO's recommendation that the threshold for taking samples should be lowered in order to get as many offenders as possible onto the database early in their criminal careers.

A third key change to PACE allowed samples, or the information derived from them, to be checked against others. In support of this ACPO cited well-known criminal cases where "speculative searches" would have made a difference. One was the "Black Pad murderer," Colin Pitchfork, convicted of murdering two young women in 1983 and 1986. ACPO

---

13. Recordable offences are those that have to be included on the Police National Computer. They include begging, being drunk and disorderly, and participating in illegal demonstrations.

argued that Pitchfork's history of prior convictions for minor offences meant that the combination of compulsory profiling and a searchable, national DNA database would have identified him during the investigation of the first murder and thus saved the life of his second victim (Dovaston, 1994b: 6).

Populating the NDNAD with the "Active Criminal Population"

The CJPOA came into force in April 1994, and in 1995 the NDNAD was launched. ACPO's requirement was that the database should process around 135,000 samples in the first year (Dovaston, 1995: 5–6, 15). However, at the end of the first year only 36,000 CJ profiles had been loaded on the database, and an annual loading rate of 135,000 was not attained until 1997–98 (National DNA Database Annual Reports, 2003–4; 2004–5). Failure to meet targets was initially attributed to practical and technical "teething problems." Police officers failed to seal evidence bags, misused sample identity labels, neglected to refrigerate samples, delayed in sending them to the laboratory, and did not ship them appropriately. Furthermore, police forces did not fully exploit the potential of the new legislation due to ignorance and reluctance to spend the requisite time and money.[14] Shortfalls were also due to a backlog at the FSS, where samples sent in by police forces accumulated while awaiting analysis and database entry. To alleviate the backlog, a second database unit was opened in November 1996 at the London Forensic Science Laboratory in Lambeth, and some of the more labor-intensive tasks for DNA extraction, quantification, amplification, and gel setup were automated (Gaughan, 1996).

In spring 1997, the Labour Party came to power, and in autumn 1999 Prime Minister Tony Blair announced the government's intention to expand the NDNAD to 3 million profiles by April 2004 (Home Office, 2000). In April 2000, when the database population was 750,000, the Home Office DNA Expansion Programme was launched. The goal was to capture the DNA profiles of "all active offenders" by March 2004. Between the years 2000 and 2005, the government invested £240 million in the DNA Expansion Programme to increase the rate of sampling

14. According to interviews in 1995 with a London police inspector and scenes-of-crime officers, when the system of charging police forces for FSS services was first instituted, police were sometimes reluctant to submit DNA samples unless the crime seemed significant and they had a very clear expectation that the FSS analysis would lead to a conviction.

from individuals and from the scenes of high-volume crimes (e.g., burglary and vehicle crime), and to fund DNA awareness training for police officers and provide scientific support personnel.

At the end of March 2005, only a year behind schedule, the database had reached the Expansion Programme 3 million target, with the retention of 3,072,041 CJ profiles, 12,095 volunteer profiles, and 230,538 SOC profiles (National DNA Database Annual Report, 2004–5). A small, but significant contribution to this growth was an annual doubling of the number of crimes from which profiles were loaded, from around 25,000 in 1999–2000 to 49,000 in 2004–5 (Home Office, 2006: 4). However, SOC profiles constitute only a small proportion of the profiles on the database. The vast majority are CJ profiles, whose numbers on the NDNAD increased by 0.4–0.5 million annually from 2000 to 2005 (NDNAD Annual Report, 2004–5: 6). This increase was due not only to investment under the DNA Expansion Programme, but also to legislative amendments which substantially broadened the net for entering and permanently retaining CJ and volunteer profiles on the NDNAD.

Initially, only the profiles of convicted offenders and detainees under the Prevention of Terrorism Act became permanent residents of the database. However, a 2001 amendment to PACE meant that once profiles were entered on the database they stayed there by permitting the indefinite retention of samples and profiles even in the absence of prosecution or acquittal.[15] By the end of March 2005, this amendment is estimated to have accounted for the retention of the profiles of 186,900 individuals, rising to 200,300 a year later (National DNA Database Annual Report, 2004–5: 9). In 2004, the threshold for entry onto the database in the first place was dramatically lowered by the extension of sampling powers to include "arrestee sampling" under the Criminal Justice Act 2003. This act empowers the police to take nonintimate database samples, without consent, from anyone arrested and detained in a police station on suspicion for a recordable offence, even if not charged. Furthermore, once on the NDNAD, their profiles can be retained indefinitely. During 2004–5, arrestees accounted for the taking of an additional 71,600 CJ samples for database profiling, with an estimated 125,000 arrestees retained on the database without charge by December 2005 (National DNA Database Annual Report, 2004–5: 10; 2005–6: 32). Other new permanent residents

15. Under section 82 of the Criminal Justice and Police Act 2001. Even after death the sample and profile may be retained for one hundred years.

of the database are volunteers who provide samples in order to eliminate themselves from the investigation of a specific crime. Since 2001, if
volunteers give their written consent, their profiles become permanent
members of the database population and participate in the daily speculative searches which may match them to new SOC profiles.[16] At the end
of October 2007, the NDNAD contained over 26,000 volunteer profiles
(Parliamentary Question, 2007).

Parliamentarians, academics, and civil liberties groups have expressed concern that the progressive widening of the net for inclusion
and retention of profiles in the NDNAD transcends its original objective of crime prevention and detection. The matter has even been the
subject of judicial review of two cases in which the claimants argued that
to retain their fingerprints and DNA samples after they were cleared of
criminal charges was a breach of the European Convention on Human
Rights. The claimants' appeals were dismissed by the House of Lords.[17]
The case subsequently went to the European Court of Human Rights
(Nuffield Council on Bioethics, 2007: 6).

Profile of the NDNAD

The DNA Expansion Programme was based upon a claim that once
the population of the NDNAD approached 3 million, every "active offender" would have been profiled. Having reached this target, the goal
is to keep the database up-to-date by adding the profiles of "newcomers
to crime" as soon as possible (Home Office, 2006: 3). Aided by legislative changes, by the end of 2005, the profiles of 40,000 individuals were
being added to the NDNAD per month. With 3 million profiles at the
end of March 2005, an estimated 5.2 percent of the U.K. population was
on the database (Nuffield, 2006: 11). This compares with 1.0 percent of
the population in Austria (the next highest and next most "liberal" regime) and 0.5 percent in the United States (Home Office, 2006: 4–5). Although the rate of increase reduced to 30,000 per month, by the end of
2007 the database contained records of approximately 4.5 million people
(Cockcroft, 2007). If current legislation remains unchanged, it has been

16. Under Criminal Justice and Police Act Order 2001.
17. *Regina v. Chief Constable of South Yorkshire Police* (2004). See Williams et al.
(2004).

estimated that eventually the NDNAD will expand to 25 percent of the adult male and 7 percent of the adult female population (Williams and Johnson, 2005).

While the database has always contained the profiles of individuals who have not yet been convicted of a criminal offence, recent legislative amendments are increasing the proportion of NDNAD profiles belonging to such persons. Their numbers have to be deduced, because the database record does not include information regarding criminal proceedings. It is estimated that the database contains the profiles of at least 1 million people who have not yet been convicted or cautioned, and who may never be, at least not for the offence for which they were originally arrested (Parliamentary Question, 2006c; GeneWatch and Action on Rights for Children, 2007). In addition to a criminal database, the NDNAD is a police register of people who, by virtue of having come under police suspicion, are deemed more likely to offend in the future than the population at large.

Analysis of the profile of the NDNAD population itself has raised concern that many entries are the "usual suspects" who disproportionately attract police attention. According to the NDNAD Annual Report 2005–6, on 31 March 2006, the database contained 3.7 million CJ profiles, which, with an estimated 12 percent replication rate, represent 3.3 million individuals. Almost half of those loaded during 2005–6 were from people under twenty-five. Eight percent were below the age of fourteen. Eighty percent of the profiles on the database are from men, of which 76 percent are "white skinned European," and 7 percent "Afro-Caribbean." Using Home Office statistics (racial category selected by arresting officer) and census data (racial category self-selected) the *Guardian* newspaper has calculated that 37 percent of black men and 13 percent of Asian men in the nation are contained in the NDNAD, as compared with 9 percent of white men (Randerson, 2006; see also Leapman, 2006).

Although the NDNAD was originally championed and introduced by the Conservative Party, in January 2006 it became an object of political debate with Conservative MPs running a campaign on the pages of the *Daily Telegraph* accusing the Labour government of "Backdoor DNA" and compiling a database "by stealth." Of particular concern were the profiles of "juveniles" between the ages of ten and eighteen, 230,000 of whom were alleged to have been added following legislative changes in

2004, and of whom 24,000 were taken from "innocent children" against whom no charges had been brought (Johnston, 2006).

The justification for lowering the threshold for entry onto the NDNAD is that it "will allow offenders to be detected at an earlier stage than would previously have been possible, prior to any charges being brought, with corresponding savings in police time and cost" (NDNAD Annual Report, 2003–4: 6). Yet, although the Home Office (2006) has published an evaluation of the DNA Expansion Programme, an independent assessment of the cost-effectiveness of DNA collection and analysis has not been published.[18] Interestingly, official figures suggest that it is the number of crime scene profiles rather than CJ profiles which makes a significant difference to the number of crimes detected (Home Office, 2006). Although 1.5 million people were added to the database between April 2003 and April 2006, the DNA detection rate has remained constant at about 0.36 percent (GeneWatch, 2007). Moreover, being on the database is not without consequence. Unlike other citizens, database inhabitants are permanent "statistical suspects" or "pre-suspects" for all subsequent crimes that are searched on the NDNAD (Cole & Lynch, 2006). Furthermore, it is not just legislative amendments that are widening the net of suspicion, but technical ones too. A new FSS database application called "familial searching" is able to identify suspects for serious crimes from among the *relatives* of people on the database with profiles which are near matches to SOC profiles (House of Commons Science and Technology Committee, 2005: 38–39). Another cause of concern is that the profiles and associated samples of NDNAD inhabitants, including those of volunteers and individuals who have never been charged or convicted, can be used in forensic research without their explicit consent.

Looking toward the future, innovations in sampling and analysis pose new challenges to the quality and integrity of the database and to civil liberties. With regard to sample collection, some employees in the public transport sector who are at risk of public harassment involving spitting have been issued with "spit kits" to swab for samples (Williams et al., 2004). At present, no consent is required from the donors of these "crime" scene samples, and profiles derived from this "lay" sampling method can be added to the database and searched for suspects to link

18. For some recent figures on the costs of the NDNAD, see Parliamentary Question (2006a).

to the "crime," and, if no match is found, remain for future speculative searches.[19] While the quality of such profiles as evidence may fall short of formal standards of validity and admissibility, for example with respect to contamination or chain of custody, their entry onto the system is not without significance given that it is intended to direct police attention toward, and even intervention in, the lives of database inhabitants with matching profiles.

With regard to suspect profiling, the prototype "lab in a van" (or "forensic response vehicle"), which is a mobile laboratory, and the proposed "lab on a chip" (a miniature, automated laboratory on a plastic chip), are each designed to bring DNA analysis closer to policing practices in time and space, and in the case of chip technology, to place analysis and interpretation in the hands of nonexpert police officers rather than forensic scientists. The vision is toward the comparison of an arrestee sample against the NDNAD in a matter of minutes (as is the case for fingerprints). However, this poses another type of challenge to the integrity of the database. At present, access to the NDNAD is restricted to a small number of authorized people. Widening remote access to police officers raises concerns regarding the confidentiality and security of personal information stored on the database.

Aside from these future developments, such is the potential of current practices to infringe the civil liberties of those on the database that, in the opinion of Sir Alec Jeffreys, databasing the profiles of the entire population would be preferable in that it would at least be equitable (House of Commons Science and Technology Committee, 2005: 32). Controversial support for extending the database also came from a judge, Lord Justice Stephen Sedley, who advocated the inclusion of all residents and visitors to the United Kingdom as a measure to redress the bias in current profiling practices under which ethnic minorities are overrepresented (BBC News, 2007). Universal inclusion in the NDNAD is also supported by the Police Superintendents' Association, although for different reasons. Their argument is that early identification of suspects could prevent subsequent serious crimes, and even save lives (Williams et al., 2004: 121). In what some regard as a long overdue public consultation on the scope of the NDNAD and its uses, the Nuffield Council on Bioethics appears supportive of the benefits of a universal database. It finds that while the

19. Spitting directly at someone can give grounds for arrest under public order legislation (Williams et al., 2004: 105).

balance of argument and evidence presented in the consultation was against the establishment of a population-wide database, it recommends that the possibility should be subject to review, given its potential contribution to public safety and the detection of crime, and its potential for reducing discriminatory practices.[20]

20. See Nuffield Council on Bioethics (2007). Interestingly, the report links controversies about digitizing, storing, and searching for matches of fingerprints on databases with debates on the collection and retention of other bioinformation, such as DNA.

CHAPTER FIVE

Deconstructing Probability in the case *R. v. Deen*

THE LORD CHIEF JUSTICE: It makes it very difficult, even if the scientist gets it right, and the judge gets it right—if this is what is right—what on earth does an ordinary jury make of it.— *R. v. Deen*

Starting in the late 1980s, and proceeding through the early '90s, a series of appeal cases in English courts raised challenges to DNA evidence. DNA fingerprinting was the main, and almost the sole, form of evidence used to convict the defendants in these cases, and in each case the appeal involved problems with the interpretation of such evidence. The best known of them was *Regina v. Deen*,[1] an appeal of a conviction for three counts of rape in which a DNA profile match was the most important item of prosecution evidence. This was the first case in which DNA evidence was successfully appealed in the United Kingdom, and it raised numerous issues of general interest. *Deen* and

---

1. The oral arguments from *Regina v. Andrew Philip Deen* (1994) discussed in this chapter were drawn from a transcript of the testimony of three expert witnesses for the defense at the hearing before the Court of Appeal, Criminal Division, Royal Courts of Justice, The Strand, London (7 December 1993). The transcript was a personal copy that was made for us by a participant we interviewed. Other key appeal cases were *Regina v. Gordon* (1995), *Regina v. Doheny and Adams* (Gary Andrew) (1997); and *Regina v. Adams* (Denis John) (1996; 1997). The last of these cases is the focus of chapter 6. We are grateful to Peter Donnelly for meeting with Michael Lynch and Ruth McNally for an extended interview, and for supplying a copy of the transcript of his and other expert witness's testimony in the *Deen* case. References to the transcript in this chapter are to sections, identified by the witness's name, and page numbers for the respective section.

the other cases highlighted problems that also were debated in court-room hearings in the United States, and were aired in the pages of the international science press in the early 1990s. In this chapter, we focus on *Deen*, and in chapter 6 we focus on another of the British cases, *Regina v. Adams*. We chose to examine these particular cases because they bring into relief key problems with the quantification of DNA evidence that were featured in controversy in the early 1990s. Before the end of the decade, it was widely presumed that these problems were resolved through a series of technical and administrative changes. That presumption helped establish the extraordinary credibility now assigned to DNA fingerprinting. We discuss the British cases because we had relatively good access to documentary materials and some of the key participants.[2]

The *Deen* case became notable for the "prosecutor's fallacy," which applies to the way probability estimates are presented in testimony (Thompson & Schumann, 1987; Balding & Donnelly, 1994; 1995; Koehler, 1996), but it also exhibited a broader set of problems with interpreting DNA evidence and developing probability estimates. Many critical discussions of forensic DNA evidence treat technical problems with the determination of DNA matches separately from statistical estimates of how often such matches should be expected to occur in a relevant population. Proponents of DNA profiling have persuaded courts in the United Kingdom and United States to separate calculations of practical error from calculations of random match probabilities. An examination of the transcript of the *Deen* appeal indicates how deeply questions about just how the material evidence is interpreted bleed into calculations of match statistics. It also shows that such calculations depend upon the resolution of ambiguities in witness stories at different levels of detail: ambiguities in the victim's eyewitness account about the appearance (specifically, the "mixed race") of the perpetrator, and in the forensic scientists' "eyewitness" accounts of faint and blurry bands (specifically, "gray" areas) in the lanes of an autoradiograph.

2. Much of the research for this and the next chapter was performed by Lynch and Mc-Nally at Brunel University in West London, with funding from the ESRC, entitled "Science in a legal context: DNA profiling, forensic practice and the courts" (ESRC R000235853), 1995–98.

Practical and Statistical Errors

The *Deen* case elucidates some of the complications that arise with uses of *statistics* in the original, etymological sense of the word as a *science of the state* or a technology of governance. Numbers have become ubiquitous in modern institutions, and in many ways they lend authority to administrative decisions. Starting in the late eighteenth century, and accelerating through the nineteenth century, vigorous attempts were made to establish methods and develop the political means to quantify social and economic activities (Hacking, 1975; MacKenzie, 1981; Turner, 1986; Gigerenzer et al., 1989; Porter, 1995). Quetelet, Galton, Pearson, and two generations of Bertillons were some of the notable figures associated with this movement to press measurement, standardization, and quantification into ever more detailed domains of public and private life.[3] It is important to understand that this movement is not only a matter of *expressing* pre-existent social and natural properties in quantified form. Often, and to some extent always, measurements and measurement regimes are key elements of social reform movements and government schemes. As faculty members of British universities are well aware, after being subjected to repeated waves of Research Assessment Exercises (RAE), measures and standards become criteria and organizational goals as well as performance indicators. The numbers become reified, and are used as proxies for what they purport to measure. Economies form around them, and political reforms are leveraged by means of them; some analysts go so far as to characterize such regimes as audit societies (Power, 1999) or audit cultures (Strathern, 2000). One aspect of the Foucauldian theme of "discipline" is the way numbers become bound up with systems of surveillance and control, in the modern reorganization of life, labor, and love. Numbers become key elements of a discursive and material integration of the "capillaries" of civil society within a massive regime of power/knowledge. Although often associated with the sciences and the inexorable march of technique, such discipline—and the specialized disciplines that support and surround it—is spurred by numerous movements within a broader political economy

3. "Bayesianism" is currently a focus of much discussion and debate. There also is an important tradition of legal calculus, from Leibniz to Bentham and Leibniz, tracing through Wigmore to the present (see Golan, 2004a).

inspired by Enlightenment ideals and integrated with austere goals of efficiency, control, and conformity (Foucault, 1979).

The history of forensics is deeply intertwined with this compulsive movement to measure and standardize, but the general drive to objectify and quantify is not an unbroken progression. The Bertillon signaletic system and the Galton-Henry fingerprint system represent distinct "paradigms" in criminal forensics (Ginzburg, 1989). The former provided a multifactorial record correlated with identity, whereas the latter became an essential *sign* of identity. And, as we shall see in chapter 9, although DNA typing was initially linked to fingerprinting, the two methods of identification became differentiated, associated with very different organizational and analytical procedures. A key difference between the two is the fact that DNA evidence is expressed probabilistically and fingerprinting is declared as an experiential judgment.

Even within the domain of forensic DNA analysis, there are distinct techniques, each requiring different instruments, protocols, forms of quantitative analysis, and modes of interpretation. Just how to express DNA evidence probabilistically had been a source of extended debate, and the debate involves bio-socio-legal as well as mathematical questions.[4] The questions concern the distribution of genetic patterns in human populations, the linkage between separate alleles in the human genome, the composition of reference groups used for estimating the distribution of specific alleles, and the distribution of possible suspects in the sociogeographical environment of a specific crime.

Criticisms of forensic DNA typing focus on two domains of possible error. One domain is practical, involving mistakes or omissions occurring at (or impinging upon) any point in a chain of custody. Such errors can result in misleading or uncertain analytical results. Deliberate efforts to plant or manufacture evidence also can be included in this "error" category, although systematic efforts to conceal such fraudulent practices

4. The unwieldy hyphenated term extends the notion of "biosociality" popularized by Rabinow (1992) to point specifically to legal implications. Biosociality signals the way genetic information becomes wedded to personal identity and catalyzes the formation of collective interests, lobbies, and movements. In the specific context in which we discuss it in this chapter, the term *bio-socio-legal* signals an institutionalized biological method, which interacts with (and to an extent disrupts) legal definitions and procedures having to do with suspect identity, the presumption of innocence, and the weight of evidence. In addition, the term draws attention to presumptions about ordinary society (how crimes normally occur, who is a possible suspect, how bodily evidence may have been left at the scene of a crime, and so forth). Also see Rose (2007) on "biopolitics."

make it even more difficult to detect their occurrence and assess their frequency. The second type of error is statistical, and is related to a technical use of the term "error" to describe random variation from an expected or average value.[5] In forensic science, the term "error" is a quantitative measure of the discriminating power of an identification technique. When analyzing DNA matches, forensic scientists calculate random match probabilities (RMPs): odds, frequencies, or likelihood ratios that estimate the probability that a DNA profile taken from a randomly chosen person from the relevant population would match the criminal evidence. An RMP figure is an estimate of the probability of "coincidence error" or "adventitious match" between an innocent person's DNA and the DNA recovered from samples collected at a crime scene.[6]

For many kinds of forensic comparison evidence—fingerprints, human hairs, threads, handwriting samples—matches are declared without giving an RMP. As we shall see in chapter 9, proponents of fingerprinting—the traditional gold standard of forensic science—sometimes claim that there is a "zero error rate" for that technique. We understand this claim to mean that *if* one assumes that two individuals never have exactly the same fingerprint patterns, then "zero error rate" is simply a way to express the assumption that there is no chance that one person's fingerprints should be mistaken for those of another. This is separated from "practitioner error rate": the (often unmeasured) rate at which fingerprint examiners do in fact commit mistaken identifications. The problem is that the court needs to know *that* error rate, even if it goes along with the dogma that no two individuals should ever have exactly the same fingerprints. Although some proponents of early DNA fingerprinting techniques made similar claims about the potential for exact identification, courts thus far have required proponents of such evidence to present quantitative RMP estimates. Some proponents of the latest versions of STR (the short tandem repeat sequence method) are again making claims of unique individual identification (making exceptions for very close relatives), but the courts continue to insist on RMP estimates.[7]

5. See Simon Schaffer's marvelous essay (1988) on the disciplinary organization of the systematic marking and measurement of observational errors in astronomy.

6. It is necessary to keep in mind that talk of suspect and crime scene "DNA" is a shorthand reference to particular DNA *profiles* (or other graphic "tests" or visual displays of DNA "type"), which provide an approximate and indirect measure of small portions of the DNA in a given sample.

7. Budowle et al. (2000) claim that currently used STR profiles effectively reduce the RMP to zero. To simplify the presentation of evidence, the British Forensic Science

The practice of expressing forensic evidence probabilistically did not begin with DNA typing. It was already established with older serological techniques such as ABO blood group typing and serum protein analysis (see Aronson, 2006). When, for example, the suspect's blood type matches that of the blood or semen recovered from a crime scene, an estimate is given of the probability of finding the same serological type when examining a randomly chosen individual from a relevant population.

DNA profiles have much greater complexity, and correspondingly greater discriminating power, than older forms of serological evidence, but the basic procedure for quantifying comparisons is similar: for each matching band in a comparison between crime scene and suspect samples, an estimate is given of how frequently the specific allele occurs in the relevant population.[8] The reference figures are compiled by forensic organizations, and are based on estimates made from samples taken from blood banks, volunteers from police forces and forensic labs, and other sources. Just as with blood serum analysis, when DNA evidence is presented in court the proponent is asked to estimate how frequently a given profile occurs at random in a relevant population. When the profile in question combines multiple probes, the resulting RMP estimate can become vanishingly small; leaving aside possibilities of false-positive error, the chance of getting the same profile from another individual (excluding closely related persons) reduces to one chance in billions or more.

The discriminatory power of DNA typing depends upon a multiplication procedure that presumes the independence of the various alleles that compose a specific profile. Challenges to this assumption (known as the Hardy-Weinberg equilibrium assumption) led advisory panels to develop "conservative" estimates, in accordance with the "ceiling principle" (NRC, 1992: 80–85). The recommended procedure involved calculating allele frequencies for different "ethnic" populations and using the highest probability for the surveyed populations when estimating the RMP for the combination of alleles in a given profile.

It was difficult to get agreement about specific probability estimates

---

Service sets a lower limit for RMP estimates, at one chance in a thousand million (one in a billion) (Ian Evett, Forensic Science Service, interview with M. Lynch and R. McNally, 19 March 2002).

8. There are numerous complicating factors that we are not mentioning for the moment: for example, an SLP or MLP "band" does not necessarily correspond to a single genetic "allele," though the terms were used interchangeably in forensic testimony, and it is necessary to include consideration of homozygous and heterozygous alleles when calculating probability.

(whether "conservative" or not), because a whole series of judgments come into play for calculating such estimates: taking into account variations in the frequency of specific alleles in different population subgroups, establishing the relevance of one or another population subgroup for the case at hand, and articulating highly specific assumptions about possible suspects and their genetic relationships to the unknown person who was the source of the criminal evidence. RMP estimates expressed in terms of odds, frequencies, or likelihood ratios were presented separately from results of proficiency tests or other estimates of practical error, even though both were intertwined with the analytical results and their significance for the case at hand. As we shall see, however, when examining the evidence in the *Deen* case, contested procedures of analysis and interpretation fed directly into estimates of the likelihood that someone besides Andrew Deen could have "donated" the semen sample recovered from the victim. Although debates about these matters were highly technical, some of the most intractable issues did not fall strictly within the province of population genetics or mathematical statistics. As Richard Lewontin (1992: 78) has argued, "intuitive arguments about the patterns of people's everyday lives" should be crucial for determining random match probabilities in criminal cases. Urban neighborhoods and isolated villages often include concentrations of persons of similar ethnic origin with extended family ties, persons who are more likely to share genetic alleles with one another than with randomly chosen individuals from a more broadly defined population group or "racial" subgroup. When a crime occurs in such a locality, the most likely suspects tend to be closer—geographically and genetically—to one another than they are to randomly chosen members of a larger population. Consequently, when the police search for suspects, they search within a narrow band of the general population for a "pool" of possible suspects.[9]

9. Lewontin (1993: 78; 1994a: 261) cites a case that was tried in Franklin County, Vermont (*Vermont v. Passino,* 1991). This was a murder trial in which the accused was a half Abenaki Indian, whose DNA profile matched blood from the crime scene. The immediate area had a high concentration of Abenaki Indians, many of whom lived in trailer parks. The victim, an Abenaki, was assaulted in the trailer park in which she lived, and many of her acquaintances also were Abenakis living in the area. The defense persuaded the judge to exclude the DNA evidence, because there was no database on the Abenakis, and it seemed reasonable to assume that the most likely alternative suspects were Abenakis from the local area who, because of relative genetic similarity with the defendant, would have been more likely to match the criminal evidence than randomly chosen individuals from the population at large. Lewontin then observes (1994a: 261) that the prosecution could have argued that a larger population was the appropriate reference group for as-

When they apprehend a suspect, and that person's DNA pattern matches the profile developed from the crime sample, an estimate of how likely it would be for another possible suspect's DNA to match the crime sample depends upon who else might count as a suspect. The pool of most likely suspects depends upon the story of the crime and how it jibes with presumptions about what Lewontin calls "patterns of people's everyday lives." Like actors auditioning for a role in a play, potential suspects in the pool may be highly similar to one another in terms of age, sex, ethnicity, regional accent, consanguinity, and overall appearance.[10]

Consequently, the specific comparison groups used for generating the probability of a false-positive match between a suspect and crime scene DNA profile should depend upon *who counts as a possible suspect* for the specific crime in question. Could a close family member (for example a brother) have been in a position to commit the crime? Is it possible that an unrecorded but closely related person also lives in the neighborhood— for example, as a "hidden half-brother" (an unidentified individual who shares the same genetic father as the suspect)? Such closely related persons may also appear similar to the perpetrator described in an eyewitness account. Although DNA evidence is sometimes presented as a means of identification that is not subject to the vagaries of eyewitness evidence and local communal knowledge, the *probabilities* assigned to such evidence depend upon an array of *possibilities* inferred from eyewitness descriptions, neighborhood makeup, access to the crime scene, and other singular features of the case.

---

sessing the RMP, because the trailer park was near a major road, and so a large number of people traveling that road could have had ready access to the victim. His point was that the RMP calculation depended on a story that constitutes a pool of possible suspects. See Roeder (1994: 273) for a rebuttal.

10. The interdependence of DNA profile analysis and the makeup of suspect pools points to a connection between ethnic profiling and DNA profiling (Rabinow, 1996b). Ethnic or racial profiling is widely condemned for being prejudicial, but at a more subtle level, police investigation almost inevitably involves a pursuit of individuals within a combination of age, sex, ethnic, regional, and other categories (Meehan & Ponder, 2002; Rubinstein, 1973). A key issue is whether one or another racial or ethnic stereotype becomes an omnirelevant criterion for virtually any search for possible suspects in a particular criminal category. At a more profound level, Lewontin's conception of suspect pools is congruent with a distinction that Harold Garfinkel (2002: 184) draws between demographic populations and "endogenously exhibited populations." A suspect pool is not a stable configuration of individuals classified into age, sex, race, and other conventional categories; instead, its composition is case-specific, and contingent upon the elements of a story that develops as the case unfolds.

Further disputes have to do with the question of whether the probability of practical errors at any point in the chain of custody should be combined with estimates of the probability of coincidence matches in the relevant population. When the latter probability is extremely small, it seems reasonable to assume that practical errors are more likely to occur than coincidence matches (Koehler, 1996; 1997).[11] As we have seen in chapter 4, however, the indefinite and expandable arrays of agents and practices that compose a chain of custody make it impossible to contain the *possibility* of error or fraud within stable boundaries. Proficiency tests become a proxy measure of practical error rates, but rarely are they conducted in the recommended blind manner with unannounced visits (Risinger et al., 2002), and they invariably assess limited phases of the entire chain, from evidence collection through presentation in court. Consequently, if the courts and advisory panels were to insist that practical error rates should be used as the lower limit of a given probability estimate, then such estimates would become highly unstable and subject to potentially endless disputes. To make matters worse, in jury trials problems with expert calculation are compounded by real or alleged problems with juror understanding. Moreover, jury understanding is not a fixed reference point for presenting evidence, since it can depend upon how the evidence is presented. "Frequentists" argue that jurors can be led more easily to understand data expressed as frequencies or odds (Gigerenzer & Hoffrage, 1995), whereas Bayesians insist that likelihood ratios provide more comprehensive and accurate accounts of probabilities.

In our view, the various problems with quantifying DNA evidence interact with one another—they are not a simple list of discrete problems to be solved one by one. Moreover, the technical practices of analyzing DNA evidence themselves include intuitive arguments and eyewitness

11. Jonathan Koehler was reported to have conducted a study of DNA testing in which forensic laboratories were asked to examine a series of samples to see if they matched. According to the study, the labs falsely matched samples, or failed to notice a match, in 12 of every 1,000 cases. The report of this finding raised alarm among forensic scientists, who had assumed (and claimed) that error rates were much lower, to the point of being completely controlled. Koehler was quoted in an article in the *London Observer,* as saying in reaction to criticisms of the study, "The controversy over error rates is not over my calculations, but over the concept of an error rate" (Walsh, 2002). Note that this particular measure of error pertains only to a particular judgment at one phase of the collection, handling, and analysis of the evidence.

accounts of visual patterns. To demonstrate this point, we turn to the record of the 1993 U.K. appeal case *R. v. Deen*. We shall describe how analytic procedures, particularly involving visual inspection of autoradiographs, interacted with statistical assumptions, forensic practices, and tacit understandings of the events and circumstances of the case.

The Case *R. v. Deen*

The *Deen* case involved a series of rapes of young women that occurred in ground-floor apartments in the South Manchester area between May 1987 and October 1988. Andrew Deen was arrested in connection with these crimes, and the main evidence linking him to the crimes was a DNA match (analyzed with the multilocus probe method) and blood group match between Deen's blood and the semen extracted from a high-vaginal swab of the victim (in the trial and appeal transcripts, the semen evidence was referred to as the "crime stain" or "swab" evidence). The police report of the crime scene investigation also mentioned that a "Negroid hair" was found on the victim's bed, but the victim gave uncertain testimony about the race of the man who assaulted her. As we shall see, this was not an isolated instance of synecdoche: vernacular accounts of other evidentiary fragments and traces also inscribed them with personal and racial identities.

Expert witnesses for the prosecution stated that the combined occurrence rates of the DNA profile and blood group evidence was 1 in 3 million. This figure was an estimate of the odds that the specific combination of DNA and blood group evidence would show up in an analysis of a blood sample taken from a randomly selected, unrelated individual in the relevant population. The prosecution experts derived this figure by multiplying the frequency assigned to the blood group evidence with the RMP assigned to the specific DNA profile.

Deen was convicted of three counts of rape on 23 February 1990 in Manchester Crown Court, and sentenced to sixteen years of imprisonment on each count, to be served concurrently.[12] Following the conviction, Deen's lawyers filed an appeal, claiming that the prosecution witnesses and the trial judge had misled the jury in the way they presented

12. Although DNA evidence linked him to only one of the victims in a series of rapes involving at least two perpetrators, other evidence linked him to at least two others.

the DNA and blood group evidence. The court of appeal agreed to review the case, and the appeal was heard in 1993. During the appeal
hearing, Deen's defense (his solicitor was Robert Platts, and his barrister was Michael Mansfield) attacked the prosecution's evidence on
several fronts.[13] First, the defense challenged the statement that there
was a DNA match between Deen's blood sample and the semen sample recovered from the victim. Second, the defense disputed the procedure through which probability figures were derived from the analysis of
the DNA profile and blood group evidence; and, third, the defense contested the way the prosecution's experts presented the evidence to the
trial court. It is worth going through each of these phases of argument,
because all three had specific implications for the numbers assigned to
the match.

(1) Was There a Match?

In Deen's trial, the prosecution's experts used the multilocus probe
(MLP) method for analyzing the swab (semen) and suspect (blood) evidence. MLP autoradiographs are often likened to bar codes, as they display many bands of variable width arrayed in vertical lanes. One of the
prosecution's expert witnesses, Mr. Davie, a forensic scientist working
for the Home Office Forensic Science Laboratory in Chorley, declared
ten matching bands on the autoradiograph.

During the 1990 trial, a defense witness (Professor Roberts) observed
that there were two nonmatching bands that Davie did not count among
the ten matching bands in the region of the autoradiograph inspected.
According to standard recommendations, nonmatching bands should result in an exclusion; a declaration that Deen's blood did not match the
crime stain. In this case, however, the prosecution experts declared a
match, and explained away the apparent anomalies. After inspecting
the evidence during an adjournment, Professor Roberts testified that
one of the discrepancies could be explained as a result of an artifact of
the preparation known as "excessive stringency."[14] The other anomalous

13. In the U.K. system there is a division of labor between solicitors, who make contact
with the defendant and prepare the case, and barristers, who present the case in the courtroom and conduct the examination of witnesses.

14. This has to do with a phase of the Southern blotting procedure, in which a membrane
is laid on the gel apparatus in order to pick up the radioactive probe pattern in preparation

band, according to Roberts, could not be so easily explained away, and he testified that an exclusion should have been declared.

In an institutional setting in which variations on the theme of Popperian falsification provide a common argumentative resource, the defense and its expert witnesses had a field day with the prosecution's interpretive procedures.[15] The defense suggested that the prosecution experts gave ad hoc interpretations of the anomalous bands in order to preserve the evident visibility of a match. Although Imre Lakatos and many historians and philosophers of science have pointed out that it is commonplace (indeed, necessary) for experimenters to discount apparently falsifying results by citing (sometimes unspecified) equipment malfunctions and laboratory errors,[16] strict rules for declaring forensic matches only when there are no mismatching details have a legal as well as an epistemological rationale. Under the presumption of innocence, it can be held prejudicial to the defendant to presume that a potentially incriminating match exists, despite ambiguous and anomalous details. This was a crucial point for the appeal. During the original trial, the judge accepted Davie's evidence (which was supported by the testimony of a second prosecution expert), and focused on the fact that ten bands matched even though two bands did not seem to have matching counterparts. The judge stated to the jury that the probability that Deen was the man who left the crime stain was "pretty well to certainty." This confirmation bias was mentioned in the appeal court's summary: "Nowhere did the learned [trial] judge specifically state [to the jury] that if there was even one discrepant band which was not satisfactorily explained, a match would not be established."[17] The defense also suggested that, even if the court were to accept that there was a match, the anomalous bands should have been given some quantitative weight in the estimation procedure. Instead, they were simply noted by the forensic analysts, but not included in the calculation of the RMP.

During the appeal hearing, the DNA profile evidence was described in great detail, and even the justices got involved in the inspection of sen-

---

for autoradiography. The membrane is rinsed to wash off excess probe material, but if the washing is excessively stringent, the trace left on the probe will be faint.

15. See Edmond & Mercer (2002) and Caudill & Redding (2000).

16. The best known study along these lines is Holton's (1978) account of Millikan's oil drop experiment. See Lakatos (1970) for the philosophical argument about auxiliary hypotheses (often involving technical contingencies) that complicate judgments about whether or not a negative experimental result falsifies a hypothesis.

17. *Regina v. Deen,* Judgment, *Times* (10 January 1994).

sual qualities of bands. At one point, during the testimony of a defense witness, Thomas Fedor (an American forensic scientist), the lord chief justice queried the witness about a particular autoradiograph in which a "greyish shadow" in the lane ("track") for the swab (semen sample) appeared to be aligned with a band in the suspect's blood sample ("the 6" is a reference point along the margin of the autoradiograph):

> THE LORD CHIEF JUSTICE. Opposite the 6 in each of these photographs that you get on the blood sample, you find on the swab not a sort of white/black but a sort of greyish shadow?
>
> FEDOR. Yes.
>
> THE LORD CHIEF JUSTICE. In each of them, if you compare it with other parts of the track where there is nothing, it is just white. What causes the greyness if it is not a band that just has not gone up the autoradiograph?
>
> FEDOR. One of the things that may have caused the greyness in the swab area is some degradation perhaps of the specimen.
>
> (*R. v. Deen*, Appeal, Fedor, 7 December 1993, 7)

Two things are notable here: first, bands are not black-and-white data points; the one in question is gray, and most likely smeared, in appearance; second, the problematic quality of the evidence evokes a practical account of how that artifact might have been produced. Even if one were to use Fedor's account to save the data (to decide that the bands do, after all, match), that account refers the visible appearance of the band to the possible circumstances of its production. In other words, just as the criminal meaning of forensic evidence turns on stories of possible events that such evidence is used to investigate, so does the technical meaning of evidentiary details turn on stories of their production.

(2) What was the Probability of Matching Evidence?

After deciding that there was a match between the DNA profiles developed from Deen's blood sample and the crime stain, the prosecution experts used a procedure for giving a "conservative" estimate of the likelihood that the matching evidence could have occurred with a randomly chosen individual who was unrelated to the source of the crime stain. As noted above, Davie declared ten matching bands. According to Davie, with this technique, one should expect about 25 percent of bands to match for unrelated individuals in the general population. For purposes of calculation, this figure was adjusted to 0.26. The 0.26 figure was multiplied for each matching band, resulting in a figure of 0.26 to a power of ten, rounded off and converted to odds of 1 in 700,000.[18]

A further procedure was used for generating the 1 in 3 million probability figure. Davie claimed that Deen's blood group (PGM 2+1+) occurs in one out of every four individuals, and so he once again used the "conservative" 0.26 figure to multiply the DNA profile estimate of 1 in 700,000, resulting in a round figure of 1 in 3 million. The defense witnesses objected that there is unknown independence between the two forms of evidence, and so multiplying the figures together had even less biological support than multiplying the probabilities assigned to each band in a DNA profile under the Hardy-Weinberg equilibrium assumption.

Consequently, if one were to reject the procedure of combining estimates from the blood group and DNA evidence, this would change the resultant figure from 1 in 3 million to 1 in 700,000 (the estimate given for the MLP evidence). Further, working with the prosecution's figure of 0.26 for each matching band, the size of the resulting estimate will change correspondingly with the number of bands counted as matching. The weight of the prosecution evidence would thus be reduced by a factor of 0.26 whenever an initially counted band is discounted for being mismatched or dubiously counted as matching.

Two experts called by the defense (Fedor and Bernd Brinkmann) testified that the two nonmatching bands and four of the ten bands that had been declared matching showed "illogical" characteristics—relative

18. Compared with the procedure used for quantifying single-locus probe (SLP) evidence, this method of giving the same average figure for each matching band is very crude, and this was one reason that the SLP method replaced the multilocus probe (MLP) method. In fact, the MLP method was no longer widely used in 1993, when Deen's appeal was heard. One of the defense witnesses in the *Deen* appeal (Prof. Peter Donnelly) used the following analogy for contrasting the estimates of band frequency in the MLP and SLP approaches:

> There is analogy . . . to try to estimate how common shirts are in a given situation, and the natural question, if you want to know how common a white shirt with a blue tie is, the natural thing to do is to go and look at people and see how many of them have white shirts and blue ties. That would be the single-locus approach. The multi-locus approach effectively: I will ignore the white shirt and blue tie, I will just ask how common it is to see two men wearing the same shirt and tie. In some situations some combinations of shirt and tie will be more common than average and some will be less common than average, so the average figure is only indirectly relevant, not directly relevant. (*Regina v. Deen,* 1993; Donnelly, 11)

patterns of darkness and faintness that were inconsistent with surrounding bands.[19] Nevertheless, the prosecution counted the four "illogical" bands among the ten matching bands between the suspect and crime samples. The defense experts argued that they should have been treated as "neutral"—neither matching nor nonmatching—for the purpose of calculating the RMP. Not counting the four bands would have greatly increased the estimated chance that a profile from a person other than Deen would match the crime stain. Instead of being 1 in 700,000 (0.26 to the power of ten, rounded off) it would become approximately 3 in 10,000 (0.26 to the power of six). Backed by a third defense witness, Peter Donnelly, a professor of mathematical statistics, Brinkmann and Fedor argued that this figure should further be qualified by taking account of the fact that mismatched lines occurred, perhaps because of an artifact.[20]

> MANSFIELD (QUOTING FROM BRINKMANN'S WRITTEN REPORT OF THE RE-ANALYSIS). "The frequency of artefactual bands as observed here is unknown and has not been documented in the literature." Can you just elaborate that a little bit further by what you mean by "the frequency of artefactual bands"?
>
> BRINKMANN. Well, if one considers them artefactual bands, then of course different sorts of artefacts can exist and one can only calculate the frequency of them if one makes side-by-side comparisons between samples coming from the same source, from the same individual. This experiment to my knowledge has not yet been performed, so we don't know about the frequency. Artefactual bands can occur due to partial digest, to stringency differences. They can also be due to contamination, especially of course in vaginal swabs and mixed body fluids, and differences can be due to the samples coming from different persons. (Brinkmann transcript, 4)

Brinkmann then recalculates the RMP after factoring in the anomalous bands, and this further reduces the estimated weight of the prosecution evidence. The prosecution had declared a match, while admitting that there were some discrepancies, and then estimated the frequency with which the ten matching bands, together with the blood group evidence,

19. At the time, Brinkmann was chair of legal medicine, University of Munster.

20. Fedor and Brinkmann also noted that one band that had been declared matching fell just outside the 4–6 kb window, but because the margin was so small they offered to allow that band to count in the calculation.

would occur in the population, but the procedure Brinkmann used did not separate the declaration of the match from the calculation of its probability. Instead, he used a Bayesian procedure, stating different "hypotheses" about the source and reliability of the evidence. Donnelly's testimony in the appeal hearing provides a concise and lucid account of the different approaches to the evidence:

> DONNELLY. Now, I think it is common ground in this case that there are two forms of explanation. One form of explanation is that Mr. Deen [who is the source of the DNA in the crime stain] and we have seen experimental artefacts [accounting for the discrepancies]; the other form of explanation is that it is not Mr. Deen, and the statistical [Bayesian] approach is designed exactly to compare these two different sorts of discrepancies.
>
> THE LORD CHIEF JUSTICE. . . . Supposing there had been no discrepancies [between Deen's profile and that of the crime stain]. There would still have been statistical analysis, would there not, in front of the Jury, as to the likelihood, or rather the odds, of the DNA being someone other than the defendant's?
>
> DONNELLY. Yes, if there had been no observed differences, a very relevant question is: how likely would we be to see an identical profile by chance? But it is also relevant, even in a case in which there weren't observed differences, it is relevant to ask: how likely would we be to see identical profiles if the defendant is the source of the DNA? Perhaps I might give an analogy which might be helpful. There is an eye witness who reports on it—I want to compare two different scenarios. In each case we have a red-haired defendant. In one case the crime was committed in the middle of the day. An eye witness with good eyesight and a perfect view said, "The person who committed the crime had red hair." In another case, the same eye witness, with the same ability and so on, but it was night time, they did not have their glasses and so on, and they say, "I think the person had red hair, but I can't be sure." Now, in each case the defendant matches whatever this trait is that has been observed at the scene of the crime— red hair for example—but you still have to ask how likely the true culprit would be to match. (Donnelly testimony, 3)

This analogy with an eyewitness account has interesting implications. Donnelly likens the "genetic witness" to an ordinary eyewitness, and a DNA profile match to a match between an eyewitness description and a suspect's characteristic ("red hair"). In other words, Donnelly explicitly treats the DNA match as an "eyewitness account" produced through the use of restriction enzymes, gel electrophoresis, Southern blotting, and so forth, with technical contingencies such as partial digestion, stringency, and band-shift being analogous to contingencies of lighting, vantage point, and memory for eyewitness accounts.[21] Both kinds of account can be more or less probable depending upon the presence or absence of the contingencies.

The analogy between eyewitness testimony about "red hair" and the testimony of the "genetic witness" also relates to ambiguities about race in the *Deen* appeal. Like the hypothetical witness in Donnelly's account who did not have a good look at the perpetrator and gave an uncertain account of his red hair, the victim in the *Deen* case described a man whose skin color was difficult to determine. In his testimony, Donnelly noted that she gave inconsistent descriptions at different times and expressed uncertainty about her attacker's skin color: "I got the impression of a face. It appeared white. It had a darkness about it"; " . . . probably half-caste. My eyesight and light were poor"; "He was light brown." According to the case record, Deen was a man of mixed race (Caucasian/Afro-Caribbean), and thus he "matched" the description. But this raised the further issue of whether the victim described accurately the face of a man like Deen, of mixed race, or whether she simply gave a vague description of a face she did not see very clearly. Her characterization "probably half-caste" might occasion doubts about whether she was revising her description in light of the suspect's characteristics. The indeterminate/intermediate race of the suspect left further ambiguity about the comparison group against which to calculate the RMP associated with the DNA match.

As noted earlier, forensic organizations compile allele frequency tables for broad "racial" subcategories of the general population: African American, Caucasian, and Hispanic in the United States; and Afro-Caribbean, (South) Asian, and Caucasian (Northern European) in the United Kingdom. They recognize that these broad groupings lack genetic integrity, but they use them to refine their estimates of allele frequency. The resulting tables show that frequency estimates for differ-

21. Incidentally, the analogy with viewing red hair is particularly salient, given the development by the British Forensic Science Service of a DNA test that allegedly can determine with a high level of probability that an unknown person who "donated" a sample had red hair (cf. Duster, 2004). When used in crime scene investigations, the DNA test thus lends itself for use as an "eyewitness" of a crime that declares "the suspect had red hair."

ent alleles found in DNA profiles can vary considerably between such "racial" categories, even as crude as they are.

In Deen's case, the issue was not that a different RMP figure would result for the "Afro-Caribbean" and "Caucasian" populations. The 1 in 3 million figure given by the prosecution was calculated independently of any population subgroup. The issue had to do with the relationship between that figure and different suspect pools. The prosecution put forward the assumption that the perpetrator was a "black" male between the ages of sixteen and thirty-four, from the greater Manchester area. Evidence for this included the victim's testimony that assailant had a local accent, and the police statement about the "Negroid" hair found at the crime scene. The prosecution cited a figure of 5,450 such individuals, according to census data. With only 5,450 potential suspects in the pool, and assuming the RMP of 1 in 3 million, it would seem that the probability that one of these suspects other than Deen would match the crime sample would be very low (5,450 over 3 million = 0.0018). When the prosecutor Shorrock suggested this figure during cross-examination, Donnelly gave a very interesting response:

> SHORROCK. The total is 5,450. Do you agree that would be a much more reasonable figure to take as the figure of possible perpetrators?
> DONNELLY. I would have some concerns. First of all the witness—the victim in this case—says that the assailant was a certain colour. We would want to try and assess the chance that she might be wrong. Now, that is not my expertise; that is clearly the jury's expertise, or the Court's expertise.
> MR. SHORROCK. Can I just stop you there. This census deals with black Caribbean, black African, black other.
> MR. JUSTICE OWEN. I have no doubt you are coming to it, but why should one assume Afro-Caribbeans are light brown?
> MR. SHORROCK. I suppose the pubic hair in so far as that link in this case indicates that that came from an Afro-Caribbean.
> DONNELLY. Can I ask whether it is known whether pubic hairs can uniquely discriminate races?
> THE LORD CHIEF JUSTICE. It is described as "Negroid."

Shorrock's line of questioning is premised on the assumption that the perpetrator and Deen are both "black." If Deen were coded as Caucasian, there would be two problems—one being to account for the "Negroid hair" at the crime scene, and the other being that there would be a much larger pool of potential suspects from which to generate a coincidence match. In reply, Donnelly disclaims any expertise in evaluating the victim's description, and defers to "the jury's expertise, or the Court's expertise."[22] As we understand the disclaimer, Donnelly is saying that he has no special expertise on judging the veracity of the witness's eyewitness account. He is a statistician. But rather than defer to another type of expert (for example, a psychologist who studies eyewitness recall), he defers to the "expertise" of judges and jurors, parties who are distinguished by their absence of expertise (judges are experts in law, but not on the specific matters marked for expert testimony). The makeup of the pool of suspects, and thus the calculation of the probability that another suspect's DNA profile would match the criminal evidence, would depend upon the judgments of the "lay experts." As the colloquy unfolds, and two of the justices get involved, the conventional terms used to fix the evidence into racial categories begin to unravel: can Afro-Caribbean men be "light brown," and can a pubic hair be uniquely identified as "Negroid"?

When interrogating Donnelly, Shorrock pressed for a resolution in (literally) black-and-white terms, which set up the use of census data to define the relevant suspect pool. If Deen were coded as "mixed race," intuitively we could figure that the pool of possible suspects would be quite small—but then we run into the ambiguities in the victim's description: was the perpetrator a mixed-race individual (like Deen), or an individual whose race the victim was unable to identify? When cross-examining Fedor, Shorrock tried a line of questions premised on the idea that the perpetrator was "half-caste":

SHORROCK. The defendant Deen is a half-caste, is he not?

FEDOR. If you say so.

SHORROCK. Well, in the evidence before the jury it was established that Mr Deen's mother was white and his father was black.

FEDOR. As you say.

SHORROCK. If the defendant's mother is white, and his father is black, by definition he himself is a highly outbred member of the population, is he not?

FEDOR. Of which population, Sir?

22. This is an interesting use of the word "expertise," akin to the term "lay expertise" sometimes used in science and technology studies to describe experiential knowledge that is not officially granted expert status. Examples of studies that identify instances of "lay expertise" include Wynne (1989b) and Epstein (1995).

SHORROCK. Of the Caucasian population and the Afro-Caribbean population?

FEDOR. Of those two populations as pure ethnic groups, then yes of course he is. I would perhaps not accept that he is an outbred member of the mixed ethnic group.

SHORROCK. I am not quite sure what you are saying here. Are you saying that because he comes from a mixed ethnic group the chances of linkage is greater?

FEDOR. No, I am saying it is unknown.

SHORROCK. It is unknown?

FEDOR. Yes.

THE LORD CHIEF JUSTICE. Does the word "outbred" have any specific meaning?

SHORROCK. As opposed to inbred. It is not meant to be a precise word—it is my word, perhaps an inappropriate choice.

Shorrock here attempts to turn Deen's "half-caste" status to the prosecution's advantage. In their arguments against the procedure of multiplying frequencies assigned to different alleles, Lewontin & Hartl (1991) had argued that linkage between alleles is more likely with inbred populations. Shorrock suggests that, far from being a member of an inbred population, Deen is a "highly outbred member" of the Caucasian and/or Afro-Caribbean populations. Fedor does not go along, and instead raises the possibility that Deen may be part of an inbred, mixed-race population for which there is no database. The irresoluteness of the accused's racial profile thus plays into indeterminacies about his genetic profile.

(3) The Meaning of Numbers: The Prosecutor's Fallacy

Although, as we have seen, the *Deen* case involved a whole series of contested procedures for arriving at the 1 in 3 million figure, the most notable problem that came up during the appeal had to do with how that figure was presented and understood in the courtroom situation. That problem is often given the name "the prosecutor's fallacy." Another name for it is "the fallacy of the transposed conditional."[23] This fallacy

23. According to Peter Donnelly (interview, 28 November 1997), the term "prosecutor's fallacy" was first used by William Thompson (Thompson & Schumann, 1987), but the idea had been used earlier by forensic scientists such as Ian Evett of the British Forensic Science Service, sometimes under the name "the fallacy of the transposed conditional." In

was attributed to the following excerpt of testimony in the 1990 trial, when Davie was questioned under direct examination:

> Q. Are you sure Andrew Deen committed this particular rape? On the figures which you have established, according to your research . . . the possibilities of it being somebody else being 1 in 3 million, what is your conclusion?
> A. My conclusion is that the semen has originated from Deen.
> Q. Are you sure?
> A. Yes.

The Times Law Reports (10 January 1994, 12) gave the following summary of the problems raised by this testimony:

> There were two distinct questions:
> 1. What was the probability that an individual would match the DNA profile from the crime sample given that he was innocent?
> 2. What was the probability that an individual was innocent, given that he matched the DNA profile from the crime sample?
>
> The "prosecutor's fallacy" consisted of giving the answer to the first question as the answer to the second.

In his testimony, Donnelly provided the following example, which he attributed to John Maynard Keynes:

> Imagine that you are playing poker against the Archbishop of Canterbury. The first thing that happens is that the Archbishop of Canterbury deals himself a straight flush. I don't know whether the Archbishop of Canterbury would be aware of this, but a straight flush is a very good poker hand. There are two questions, one might ask there. The first question is: what is the chance, if the Archbishop is playing honestly, that he would deal himself a straight flush. The answer to that is quite small; it is about 1 in 100,000, I think. So the first question is: if we assume honesty, what is the chance of a straight flush? A straight flush is very unlikely. The second question is the one I think you would want to ask yourself if you were in this position—you are playing poker against this person, they have just dealt themselves a straight flush—is: do I think they are playing honestly? Those two questions are very

---

the interview, Donnelly noted that "[t]he concept of the prosecutor's fallacy captured the imagination of the public more than the fallacy of the transposed conditional."

different. I think many people—I think almost everyone—would come up
with the answer to the second question: what is the chance that he is honest
given that a straight flush is much more than 1 in 100,000? Many people, because they have a strong prior belief in the Archbishop of Canterbury's honesty, would still believe that he was being honest. Your Lordship does not
look entirely convinced. (17)

Indeed, as Donnelly recognized at the end of this passage, the lord
chief justice was troubled by the analogy, and he raised questions about
the fact that Donnelly brought in an "extraneous factor" by identifying
the dealer as the archbishop, rather than, say, a card sharp. Donnelly replied, "That is exactly my point, my Lord. To answer the second question, you need to know the answer to the first question [the probability
figure] and extraneous factors, and you get a different answer to the second question depending on those extraneous factors" (17). Later, after
further exchange with the lord chief justice, Donnelly emphasized, "The
answer to question 2 depends on the extraneous factors, and I would argue that it is not the forensic scientist's job, or any expert's job, to assess those in a case" (18). Donnelly also offered the term "background
knowledge" when elaborating upon "extraneous factors" about the card
dealer, or in a forensic investigation, knowledge (or, presumably, prejudice) about a particular suspect.

The justices had difficulty accepting Donnelly's argument that (1) the
answer to question 1 provides no basis for answering question 2 in the
absence of background information; and (2) in a specific case there is almost invariably other information about crime and the defendant, which
should bear on the answer given to question 2. Donnelly drew an interesting distinction between "information" and "evidence," with the former being conveyed through the case description, appearance of the
defendant, and so forth, whether or not it is formally presented as evidence. Another way to understand this distinction would be in terms of
the contrast between "tacit" and "explicit" information in the case: with
some information being marked as evidence (given as statements in testimony or exhibits), while other information is "given off" (to borrow
Goffman's [1959] term) in the scenic unfolding of the trial. The lord chief
justice then raised a question about fingerprints:

THE LORD CHIEF JUSTICE. If it were a fingerprint, you would be able to say,
    would you not: that is this man's fingerprint. You would still be left with

the possibility that someone lifted it and planted it on the car, or whatever it was, but subject to that.

DONNELLY. Yes, I think it is very important to distinguish between genetic profile and dermal fingerprints. I think it is well accepted in the case of dermal fingerprints that they are unique. So if you have a matching fingerprint it automatically follows it is the same person. That is not the case with DNA. I don't think anyone would suggest that DNA profile—it is unfortunate that they were once called DNA fingerprints—no-one would suggest that they are known to be unique, so one has to assess them in terms of probability. (19)

As we shall see in chapters 8 and 9, this account of fingerprinting also could be subjected to the kind of unpacking that Donnelly accomplishes for DNA evidence, whether or not one accepts the dogma that every individual's fingerprints are unique. After further discussion and explanation, Donnelly and Mansfield, the barrister for Deen, turned to the key lines of testimony by Davie in the transcript of the 1990 trial. First, they picked out Davie's statement about the DNA profile evidence: "The probability of the semen having originated from someone other than Andrew Deen [is] 1 in 700,000" (19; citing trial transcript, 25). The first problem with this statement that Donnelly addressed had to do with "the form of the words": did the words "someone other" mean *anybody* other than Deen in the relevant population, or were they a reference to a particular person picked at random from that population? Donnelly asserted that Davie's statement would be correct only under the second interpretation, but would be committing the "prosecutor's fallacy" under the first (which he marked as the more "natural" interpretation to make). He explained the difference as follows:

It is one thing to say: if I pick someone off The Strand [a famous London street], the chance of them having the profile is 1 in 700,000. That is one statement, which is true if you accept the 700,000. That is not the same as saying: the chance that there is no-one out there who would match is 1 in 700,000, because even though each individual might be unlikely to match, if there are lots of individuals we should be thinking [that] the chance of one of them matching might be nontrivial.

Donnelly expressed less qualification about the conclusion that Davie stated: "My conclusion is that the semen has originated from Deen." He raised two issues. The first is sometimes formulated under the rubric of "jury usurpation", which in this instance would mean that the expert witness is pronouncing an opinion about the "ultimate issue" of guilt or innocence.[24] As Donnelly noted, this issue is a legal issue for the court to address. The second objection had to do with the meaning and relevance of the 1 in 3 million figure.

> The figure of 1 in 700,000 or 1 in 3 million is a chance that if we picked a particular person and examined them they would match. So if that statement is saying the chance that a particular man would match, it would be correct. Another very relevant question is: what is the chance that there would be somebody who matches? One might also interpret the statement [as] the chance of a man other than the defendant matching. In that way, that would be incorrect.

If one were to assume nothing else about the evidence, and were to search the population of adult males in the United Kingdom (for the sake of this example, assume it to be 30 million), the 1 in 3 million figure would suggest that there would be approximately ten other men besides Deen in the population whose profiles would match the crime sample.

One reason it is difficult to understand the difference between the two questions, and to adhere to the "correct" way of formulating the evidence, has to do with the matter mentioned earlier of "extraneous" evidence or "background knowledge." The fact that Deen had been apprehended by the police, that he was from the local area, and that his characteristics (ambiguously) matched the victim's description(s) of the perpetrator, are likely to enter into our understanding of the probability figure.[25] Accordingly, Donnelly's argument might seem "academic."

24. Donnelly mentioned that Deen did not put forward a consent defense, and so certainty that the semen originated from him effectively meant that he was the perpetrator of the rape. In an interview (18 November 1987), Donnelly stated that "[t]he DNA evidence only bears on whether the suspect left the crime stain. I think in all the cases I've been involved in, that's been the key issue. I guess in events like rape, it's possible that the defense will be consent, and so the issue of whether or not it's his semen isn't relevant. But in all the cases I've been involved in that hasn't been the case. So there is a technical distinction between guilt and innocence and whether the defendant left the crime stain, but it's only what I would call an academic distinction in all the practical cases I've been involved in." Note, however, that the difference becomes "academic" precisely because of a presumed agreement among parties to the case about which features of the defense and prosecution stories are contestable or not.

25. The term "defense fallacy" was used in reference to a statement made by one of O. J. Simpson's lawyers, the late Johnnie Cochran, who argued that approximately six

However, there are a number of reasons to treat it seriously—and it was treated seriously by the court of appeal, which quashed Deen's conviction and referred it back for retrial.[26]

First, the prosecutor's fallacy relates to the legal presumption of innocence and the burden of proof on the prosecution to prove guilt beyond a reasonable doubt in criminal trials. Donnelly's arguments suggest that the expert witness's formulation presumes (or can be heard to presume) more than what the specific DNA evidence shows. Davie's statement that he is certain that Deen was the source of the semen presumes substantive knowledge about the case that should be reserved for the jury deliberation.

Second, and related to the first, is the matter of the expert witness's role in the trial court. Even though DNA evidence remained controversial in 1993, it nevertheless carried considerable weight in the courts. When a single probability figure is assigned to matching evidence, and the probability figures are very impressive (one in millions or billions), the evidence may seem to approach certainty. The *Deen* defense demonstrated, first, that various probabilities could be assigned to the match, and in some possible scenarios, the figures were not so impressive. In the absence of the sort of skilled "deconstruction" accomplished by Donnelly, Fedor, and Brinkmann, however, the 1 in 3 million figure may have stood unchallenged, of overwhelming importance when compared with other evidence.

Third, if we can be allowed to extrapolate beyond 1993 to a few years later when many nations, led by the United Kingdom, developed DNA databases, the possibility of accusing someone of a crime solely on the basis of a DNA match, with no other corroborating evidence, becomes an actual possibility. When DNA evidence from a crime scene is used as a basis for searching a large database, one or more "cold hits" may come up. Initially, all that is revealed by such a "cold hit" is that there is matching DNA evidence, along with any other information recorded on the database about the particular person's residency, age, sex, and ethnicity. In such a case, it is important to consider the difference between

---

other people in the greater Los Angeles area would be likely to match the DNA evidence used in that case. The fallacy was to imply that there was no other evidence, and that those other six were equally likely to be suspects.

26. Important for the appeal was the trial judge's summary to the jury of Davie's evidence, which suffered from similar ambiguities of meaning, and arguably committed the prosecutor's fallacy.

the probability that a given individual would match the crime sample, and the probability of finding a match in a search of a large population (presumably, a population that would be much larger than the sample recorded in the national database).

Fourth, as a theoretical matter, the defense's attack on the prosecution's DNA evidence opened up the significance of "background knowledge" about the particular case. Using the gestalt analogy, such knowledge was the tacit ground against which the probability figure was cast. With different background assumptions, one gets a different sense of the possibilities from which the probabilities are calculated. Moreover, "assumptions" is too explicit a concept to cover *pre*sumptions such as that a rapist who leaves a semen sample must have been a postpubescent male.[27] So, to take a hypothetical example, if one were to locate an adult woman or an infant whose DNA profile matched the profile developed from the crime stain, the probability that the semen sample in question originated from such a person is not 1 in 3 million, but zero. Either that, or one would have to devise a way to imagine how in the world an infant's or a woman's DNA got into the "semen" sample. Imaginable *possibilities* set conditions for measured *probabilities*.

The prosecutor's fallacy raises questions about two reciprocal aspects of the presentation of evidence: first, just how the DNA match is described in words, as well as numbers, and, second, just how a trier of fact (a judge or jury) understands the evidence. As Donnelly explained in an interview, there are various "correct" ways to formulate matching evidence. One example is "the chance that an innocent individual unrelated to the defendant would have this DNA profile." Others are "the chance that a particular individual, who is innocent and unrelated to the defendant has this profile," and "the chance that a particular, innocent individual, unrelated to the defendant, in the Caucasian population will have this profile" (Donnelly, interview 28 November 1997, 4). Donnelly emphasized that it "would be helpful to specify which population you were talking about and [to] be explicit you were talking about innocent individuals in the population." He added that the latter emphasis is "important in the definition," but that it is an open question "to what extent it matters to the jury how you word these things. . . . It's my suspi-

27. See Sudnow (1965) for an explication on "normal crimes"—informal presumptions about the possibilities included under common crime categories: burglary, auto theft, rape, etc.

cion that the jury happen to make the prosecutor's fallacy repeatedly, whether or not exactly the right wording is used" (4).

When questioned about alternative ways to present the "same" probability figures, Donnelly acknowledged that some ways may be more intelligible to jurors than others:

> Let's . . . assume that we have an agreed way of coming up with the numbers. Then, actually, I think it's really helpful. I'm very positive, and on reflection, I'm growing more positive with time [that] . . . instead of the jury having to think in terms of numbers like one in a million or one in ten million, they have to think in terms of maybe ten people in Britain having the profile, so they're thinking in terms of numbers with which they're much more comfortable. Expressing it in that form—so saying to the jury that if you believe the DNA evidence and you believe the numbers, its effect is to say that the criminal had the profile, this defendant has the profile, and there may be ten other people in Britain—maybe two or three others in London if it's a crime in London, and maybe ten in all of Britain, and maybe go on and say how many in Europe. And then . . . what the jury has to do is to decide . . . whether all the other evidence is such as to make them convinced that this guy who is in the trial of the ten in Britain is really the guilty one.[28]

Jury understanding can be described as the "black box" of the legal system, given the fact that jury deliberations are conducted in private.[29] The literature on jury understanding suggests that jurors have great difficulty understanding probability figures, but as Edmond & Mercer (1997) suggest from their review of scholarship on the matter, the jury is out, so to speak, on the question of whether, and to what extent, jurors can be led to understand such evidence, more or less adequately for the case at hand. Taroni and Aitken (1998) suggest that jurors fare no worse with

28. Donnelly acknowledged that presenting a probability estimate in terms of a concrete number of possible individuals in a region or city runs the danger of being taken too literally by a jury. A juror might interpret Donnelly's example to suggest that there were exactly ten men in Britain whose profiles would match that of the crime stain. Such a misunderstanding would become especially important in a rape trial in which a probability figure such as 1 in 40 million is given for a male population of less than 40 million. Naïvely understood, such a figure might suggest that the suspect is the only individual in that population with the profile in question. We discuss this issue further in chapter 6.

29. The Chicago Jury Project of the early 1950s ran into legal difficulty with its effort to study tape recordings of jury deliberations (Kalven & Zeisel, 1970), and the vast majority of jury studies have had to make do with mock juries. For an exceptional study of a "real" jury deliberation, see Maynard & Manzo (1993).

technical evidence than do judges or lawyers, and one key issue for all parties is just how evidence is presented. Probabilistic evidence can be presented in terms of Bayesian likelihood ratios or frequencies, and studies by Gigerenzer and Hoffrage (1995) suggest that laypersons have an easier time understanding frequencies than likelihood ratios. And, as Donnelly's example suggests, frequencies can be presented as odds, percentages, or average numbers in a population. Moreover, an expert witness can deploy effective teaching tools such as stories, analogies and well-designed visual aids. Consequently, it seems reasonable to imagine that "jury understanding" is not a fixed limit, but is instead a reciprocal aspect of just how evidence is presented. In chapter 6, we examine an unusual pedagogical aid that Donnelly used in another case to enable a jury to come to terms with a Bayesian treatment of evidence.

Conclusion

A striking feature of the *Deen* appeal was how the specific RMP of 1 in 3 million depended upon resolving ambiguities in black-and-white terms. This applied to the victim's eyewitness testimony about the perpetrator's race, the forensic scientists' testimony about visible bands on an autoradiograph, and the calculation and presentation of specific probability figures.

When we take the ambiguities associated with accounts of Deen's "race" together with those discussed earlier with the matching profiles (the "gray" bands), we can begin to appreciate that there is a range of estimates to choose from for estimating the RMP. The choice—indeed the very calculation—of a particular probability estimate depends upon what the lord chief justice called "extraneous" factors. The relevance of such factors, in turn, is set up by a story that brings into relief a set of relational identities and contingencies. Some of the identities and contingencies that arguably affect such estimates are, to borrow Fedor's expression, *unknown*. A Bayesian method of calculation enables probability figures to be generated in the face of such unknowns, but as we shall see in chapter 6, that method opens up its own set of difficulties.

Bayesians, Frequentists, and the DNA Database Search Controversy

In both the *Deen* and *Adams* cases, the defense employed Bayesian analysis to unpack and neutralize the impressive random match probabilities presented by the prosecution's experts. More recently, a controversy among statisticians broke out in the United States over the analysis of database "cold hits." We discuss this recent controversy in this interlude.

One of the more unusual aspects of the history of forensic DNA testing is that academic scientists were so prominently featured in legal controversies about techniques and interpretations of forensic identification evidence.[1] Academic scientists were involved in disputes about the admissibility of DNA evidence, and in some cases these researchers went back and forth from being experts on the witness stand to being authors of articles on the subject in the pages of *Science* and *Nature*. There were various patterns of crossover, as articles in academic journals were prominently mentioned when scientists testified in court, and some of the prominent articles in *Science* and *Nature* explicitly discussed *Castro* (1989) and other court cases (Lander, 1989; Lewontin & Hartl, 1991). Publications in scientific journals and magazines also featured collabo-

---

1. Aronson (2007) argues that some of the early cases, particularly *Castro,* introduced a novel kind of witness to the court—the academic scientist who would comment on the acceptability of forensic practices. Golan's (2004a) history of expert testimony in English and U.S. courts shows that, to the contrary, academics often testified in court about scientific and metascientific matters, so that Aronson's argument applies in a more restricted way to forensic (or, specifically, individual identification) evidence in the U.S. courts in recent decades.

rations between lawyers, forensic scientists, and academic scientists (for example, Neufeld & Colman, 1990; Lander & Budowle, 1994). The most visible dispute involved arguments in the early 1990s about the combination of statistical procedures and population genetic assumptions that went into the calculation of random match probabilities (RMPs) for the allele combinations represented in a DNA profile. Several years later, well after the so-called "DNA wars" had quieted down, another debate about the calculation of RMPs crossed back and forth from court cases to academic journals. This one has been dubbed the "the DNA database search controversy" (Balding, 2002). Like the earlier controversy about population genetics and statistical calculations, this one appeared on the surface to be an arcane dispute among academic specialists. However, also like the earlier dispute, it was far from being purely of academic interest. It concerned different methods for calculating RMPs, but it also involved questions about how to present probabilistic evidence in the courtroom and, ultimately, about the legal meaning and significance of probabilistic evidence.

The DNA database search controversy had to do with different methods for calculating the RMP for a "cold hit" resulting from a "trawl" through a DNA database of presuspects or "statistical suspects"—persons whose profiles are placed on a database, and who become candidate suspects whenever a search is conducted with crime scene evidence (Cole & Lynch, 2006). A cold hit occurs whenever a search turns up a match between an individual profile and crime scene evidence, with, as yet, no corroborating evidence. With the construction of national criminal databases consisting of millions of samples, "cold hits" become commonplace. In most cases, further investigation turns up corroborating information that ties the "statistical suspect" to the crime scene, but in some cases the cold hit is the sole, or main, item of criminal evidence. (The *Adams* case discussed in chapter 6 was a cold hit resulting from a trawl of an early Forensic Science Service database. In that case, much of the other evidence appeared to contradict the DNA evidence.) The DNA database controversy is related to the prosecutor's fallacy and the Bayesian method of analysis discussed in chapters 5 and 6, but it occurred later and focused on specific recommendations in the two NRC reports (1992; 1996).

Some cold hits have turned out to be adventitious. The most famous of these was the case of Raymond Easton, which is described in chapter 7, but there have been several other such adventitious cold hits as well

(Thompson, 2006). With databases including millions of samples being searched in hundreds of thousands of cases, the chances of random
matches become more likely. The question is, how likely?

In its two reports, the NRC recognized a distinction between what
Donnelly and Friedman (1999) later called a "confirmation case" and a
"trawl case." The former is a match between crime scene and suspect evidence in a case in which the suspect already has been identified by other
means. The latter is a cold hit on a database. DNA profiles provide representations of highly variable (polymorphic) alleles at selected chromosomal loci. The size of the RMP depends upon how many loci are represented in a given profiling technique, and how frequently the specific
alleles occur in the relevant population. There were a number of contentious issues about the composition of the reference databases used for
calculating allele frequencies and the degree of independence between
alleles in a given profile. The database search controversy was about a
more restricted issue, which can be explained by analogy with a popular classroom demonstration called the "birthday problem." A confirmation case can be set up by choosing a day of the year at random, and then
asking a randomly chosen student if his or her birthday occurs on that
day. The probable frequency of a match is roughly 1 in 365, or 0.003. A
trawl case can be set up by asking if any member of the entire class has
the specified birthday. To derive that number, you multiply the number
of class members plus the instructor $(N + 1)$ by 1/365. For a class of forty
students, the result would be around 0.11. Consequently, the odds of a
match in the trawl case are much higher than in the confirmation case.[2]

According to the same logic, as a database increases in size, the
chance of an adventitious match increases proportionally. The NRC in
its two reports attempted to take account of this problem. The first report (NRC, 1992: 129) advised against presenting database trawl results
to a jury, and recommended instead that a search should be used as a
basis for further tests using markers at additional loci. The report also
recommended that, if there is a match at the additional loci, the RMP
should be calculated on the basis of those loci alone, exclusive of the
loci used for the database. The second NRC report made a simpler recommendation that when a match is found through a database trawl, the

2. The standard way to perform the exercise is to ask, "How likely is it that two people
in the class have the same birthday?" With a class size of forty, the probability that a second student would have the same birthday as the first is quite high (around 0.9). For a simu-
lation of the exercise, see http://www.mste.uiuc.edu/reese/birthday/.

RMP should be multiplied by the number of persons in the database. So, for a trawl through a database of one million, if the RMP for a matching profile is 1 in 10 million, then the adjusted RMP would be around 1 in 10.[3]

Donnelly and Friedman (1999) criticize both NRC recommendations by setting up the problem with a different question that they argue is more pertinent to the court's task at hand: "The proposition that the DNA evidence is offered to prove is not the broad one that the source of the crime sample is a person represented on the database. Rather, it is that *one particular person—the defendant in the case at hand—is the source of that sample*" (Donnelly & Friedman, 1999: 946, emphasis in original). Using the fictitious name "Matcher" for the defendant whose database profile matched the crime sample, Donnelly and Friedman (ibid.) reiterate, "It is the probability that Matcher is the source of the crime sample that is of importance to the prosecution—not the larger probability that *somebody* represented in the database is the source." To address this problem they invoke Bayes's theorem: the posterior odds of a proposition equal the prior odds times the likelihood ratio.

Bayes's theorem is named after the Reverend Thomas Bayes (1702–61), a Nonconformist English preacher with a ministry in Tunbridge Wells, Kent, who had studied logic and theology at the University of Edinburgh and developed the approach to probability from which his eponymous theorem derives. His theory of probability, published posthumously in the *Philosophical Transactions of the Royal Society of London,* shows how uncertainty about one event is contingent upon uncertainty about other events. What is today called the Bayesian approach is of much more recent vintage, representing a number of influences. One formulation of Bayes's rule is the following:

$$P(R|S) = P(S|R)P(R)/P(S)$$

P denotes the probability and the bar | denotes conditioning. Thus $P(R|S)$ is the probability that event R occurs, given that event S has occurred (Aitken, 2000). Probabilities are expressed as values between zero and one, or expressed as odds $(1 - P)$.

---

3. Donnelly and Friedman (1999: 943ff.) distinguish what they call the "frequentist" argument from the "likelihood ratio" argument, and they argue that the two arguments lead to different answers. However, they add, both approaches agree that the more profiles examined, the less probative the evidence.

The Bayesian approach currently is very influential in forensic science, though its influence in the courts has been restricted and circumscribed. Peter Donnelly (featured as an expert witness in cases discussed in chapters 5 and 6) is one of the major exponents of the Bayesian approach, as is Ian Evett, Christophe Champod, Bruce Weir, and other influential statisticians. The Bayesian approach is distinguished by a comparison between *prior odds* and *posterior odds,* and the use of *likelihood ratios* to express the probative value of evidence. The likelihood ratio is a comparison between odds, given two hypotheses: in a forensic case, the hypotheses can be stated as two questions: What is the value of a particular item of evidence, such as a DNA match, given the assumption that the suspect is the source of the crime sample? And, what is the value of the evidence, given the assumption that the suspect is not the source? The prior odds are estimated before the evidence is presented, and the posterior odds are estimated after the evidence is presented. The likelihood ratio converts prior odds to posterior odds by calculating the probability of the evidence given the hypothesis of guilt, divided by the probability of the evidence given the hypothesis of innocence (Aitken, 2000). According to the Bayesian approach, the trier of fact (judge or jury) assesses the prior and posterior odds, while the expert shows only how the prior odds change with the introduction of the evidence. The Bayesian approach attempts to comprehend all of the evidence—expert and nonexpert; quantitative and nonquantitative; objective and subjective—with one coherent chain of calculations.

Donnelly and Friedman (1999) note that for a suspect identified through a cold hit, the prior odds—the odds that the particular individual was source of the crime sample—are extremely low. Presumably, the hit greatly raises the posterior odds. Assuming that there are no other hits in the database—that none of the other profiles in the database matches the crime sample—it might seem that the probative value of the matching evidence increases as the database grows larger and no other matches occur. It should seem relevant for determining the probative value that other (very many other) profiles have been tested and found not to match the crime sample.[4] In a hypothetical case in which

4. Donnelly & Friedman (1999: 949) point out that not all suspects in a suspect population are equally suspicious. The police often single out a number of suspects based on prior evidence and subject them to DNA testing. If their profiles fail to match the crime sample, then their elimination from the suspect pool is, in principle, more significant than the elimination of a randomly chosen member of that pool.

the database includes the entire suspect population (and leaving aside the possibility of laboratory error), then a single match with the criminal evidence would be definitive.[5] Actual DNA databases are, of course, relatively small compared with the relevant populations. Even the U.K. national database, with its more than 4 million samples, contains less than 10 percent of the male population. But, as such databases grow larger and approach the limit case of the entire suspect pool, one should expect the probative value of matching evidence to increase, not decrease (especially when a search turns up only one matching profile).

Donnelly and Friedman (1999) provoked an argument that remains unsettled. One of their critics (Devlin, 2000: 1276) remarked that the very fact that there was disagreement among the statisticians was likely to confuse people in the legal community, who are "largely not schooled in statistical theory." This uncertainty also was featured *as an uncertainty,* and resolved in an interesting way, in a recent case, *United States v. Jenkins* (2005), in which the D.C. Court of Appeals overruled a successful challenge by a defendant to database evidence. This murder case initially involved a suspect, Stephen Watson, who possessed the victim's credit cards and items taken from the victim's home. Watson, who in the pre-DNA era would likely have been convicted of the crime, was exonerated when DNA testing of a couple of drops of blood, which were presumed to have come from the attacker, did not match his sample.

Watson, the conventionally generated suspect, was then replaced by a "statistical suspect," Raymond Jenkins, whose link to the crime consisted solely of a eight-locus cold hit when the blood evidence was searched against the Virginia state database. Jenkins's defense pointed to the discrepant statistical approaches to argue that the scientific community had not reached consensus on how to treat database trawl results. The court agreed, but argued that the scientific community was not disagreeing over *science:* "These competing schools of thought do not question or challenge the validity of the computation and mathematics relied upon by others." Rather, they were disagreeing about how the probative value of the evidence should be conveyed, which, the court felt, fell into the legal, not the scientific, domain, and, therefore, the court did not need to defer to the consensus of the scientific community: "What is and is not relevant is not appropriately decided by scientists and statisticians." In

5. In principle, the likelihood of laboratory error also can be included in the Bayesian approach.

other words the court engaged in boundary-work (Gieryn, 1983) by separating the conceptual questions about the meaning and probative value
of the evidence from statistical calculations of probability. This ruling
contrasts with the Bayesian insistence that the conceptual framing of the
evidence is at least as important as the probative odds that result from
the calculations.

Another imposition of boundaries occurs in the *Adams* appeal cases
discussed in chapter 6, but in that case the justices demarcate the scien-
tific/quantitative evidence from the commonsense/nonquantitative evidence, ignoring the defense's insistence that the two must be weighed in
terms of the same Bayesian formula. Superficially understood, Bayesian
treatments of evidence are reductionist, as they weigh different kinds of
evidence on a single quantitative scale. They convert imaginable *possibilities* about how a crime may have occurred into quantified *probabilities*. But in the contexts we are examining, Bayesian treatments potentially open up a creative repackaging of evidence that disrupts settled
legal and vernacular distinctions between "expert" and "commonsense"
knowledge. In *Jenkins*, and as we shall see in *Adams*, the courts insis-
tently imposed those distinctions in the face of the Bayesian challenge.

CHAPTER SIX

Science, Common Sense, and DNA Evidence

It is commonly said about courtroom testimony that judges and jurors piece together a story of the events in question, and that the story must be meaningful and plausible in terms of their prior experience of the world.[1] This "story model," as it is sometimes called, supports the traditional role of the jury in the trial court, because it suggests that background knowledge gained from living in a community provides an overall framework for hearing testimony, making judgments about credibility, and developing an overall picture of the case. Such background knowledge is, for the most part, ordinary; largely consigned to "common sense." In Melvin Pollner's (1974; 1987) terms it involves "mundane reasoning"—an indefinite, partially articulated, largely tacit sense of what is sensible, normal, real, and right. Such knowledge can, of course, go wrong. Indeed, to call it "knowledge" is itself dubious, if knowledge is held to be more certain than mere belief. Expert evidence, and DNA evidence in particular, is often invested with the hope of transcending "subjective" (often prejudicial) judgments with "objective" facts, so that court decisions can rest upon a secure, scientific foundation. However, gathering from the previous chapter, we can begin to appreciate that impressively quantified "scientific" evidence is intelligible on the basis of "background knowledge" that involves less impressive, more ordinary presumptions. Paradoxically, this lesson was driven home though an

1. See Gluckman (1963); Emerson (1969: 192–201); Bennett (1978); and Lynch & Bogen (1996) for accounts of the organization and significance of stories in testimony. Thompson (1996) applies what he calls the "story model" of testimony to jurors' understanding of the meaning and significance of DNA evidence in criminal trials.

intervention by an expert statistician taking a Bayesian approach to the evidence.

The lesson from the *Deen* case is that the story model of testimony applies to probability estimates no less than it does to discursive testimony. Indeed, as the previous chapter illustrates, DNA evidence is meaningful only when it is embedded in stories that mention other evidence, possible suspects, and how the evidence itself was handled and interpreted. This story model conflicts with a view of DNA evidence as an exceptional kind of evidence; a source of evidence so powerful that it overrides the credibility of all other evidence. According to the latter view, story elements shift around the intrinsic meaning and truth of the scientific evidence. "DNA" becomes the bedrock, while all other story elements give way, shifting into a configuration that conforms to the bedrock truth.[2]

A decade after the *Deen* case, it became common (especially in the United States) to read news stories of DNA tests being used to exonerate defendants who had been serving long prison terms, sometimes on death row. Take, for example, a story in the *New York Times* (3 August 2005), "DNA Tests Come to Prisoner's Rescue," by Abby Goodnough and Terry Aguayo. The story describes a recent effort to vacate the conviction of Luis Diaz, who was convicted of rape in 1980 based on eyewitness testimony of eight victims. The story quotes the trial judge as having said at the time of the original conviction, "I've never seen a case where I was more convinced of a man's guilt." Twenty-five years later, Diaz was still in prison despite the fact that two of the victims recanted their testimony a decade after his conviction. According to the *New York Times* article, DNA tests (which, of course, were not available in 1980) were recently used to analyze evidence from two of the rapes (the only two for which usable evidence was still available), and the results revealed that the two victims were raped by the same perpetrator, whose DNA profile differed from that of Diaz. The prosecutors agreed to ask a judge to vacate all of the charges, even though they continued to suspect that Diaz might have perpetrated some of the other rapes for which he was convicted.

The article quoted attorney Barry Scheck as saying that of 160 exonerations that had occurred in recent years due to DNA retesting, 120 of the original convictions were primarily based on eyewitness identifications. In many of those cases, victims and prosecutors continue to express certainty even after DNA tests contradict the eyewitness identifications.

---

2. The imagery in this sentence alludes to Wittgenstein (1969: sec. 99).

As we shall elaborate in this chapter, when DNA and other forms of evidence clash, "science"—or "DNA"—tends to win the contest. Eyewitness testimony had previously been criticized as untrustworthy, subjective, and subject to bias and police suggestion (Loftus, 1975; Loftus & Doyle, 1997),[3] but until the advent of DNA testing, such criticisms had limited influence on the legal system's strong investment in such testimony.[4] As stated in the article "Lawyers for Mr. Diaz, now 67, say that his case is the best evidence yet that witnesses can make devastating mistakes, and that such testimony, however earnest and convincing, cannot be trusted" (1).

Although we laud efforts to exonerate innocent defendants, in the present chapter we raise questions about the exceptional status now assigned to DNA evidence (also see Aronson, 2005b). To shed further light on these questions, we turn to another historic British case *R. v. Adams,* which was appealed twice, in 1996 and 1997. This case included some of the elements of the *Deen* case discussed in the previous chapter, but it was particularly significant for the way the appeal court justices deployed a sharp disjuncture between "scientific" and "common sense" evidence. Like the recent case of Luis Diaz, the *Adams* case also involved an apparent contradiction between DNA evidence and eyewitness testimony (as well as other forms of "commonsense" evidence). In Adams's case, however, the DNA evidence was positive—showing a match between

3. See Golan (2004a) about early twentieth-century challenges from psychologists to eyewitness testimony.

4. In *New York v. Anthony Lee* (2001), the New York Court of Appeals ruled that a judge has discretion to admit expert testimony on the reliability of eyewitness testimony. The defendant had moved to introduce testimony by a "psychological expert" who would explain to the jury that an eyewitness's level of confidence has little correlation with the accuracy of an identification. The original trial judge ruled against admitting this expert. The appeal court affirmed the trial judge's discretion to exclude or include such expert testimony, but stated, "Despite the fact that jurors may be familiar from their own experience with factors relevant to the reliability of eyewitness observation and identification, it cannot be said that psychological studies regarding the accuracy of an identification are within the ken of the typical juror." In a commentary on that case, Dorf (2001) noted that the decision contradicted "conventional wisdom" that eyewitness testimony is stronger than circumstantial evidence. He mentioned that DNA evidence is one form of circumstantial evidence that is "extremely reliable"—much more so than eyewitness testimony—and so testimony by psychologists about the fallibility of eyewitness testimony can counteract the presumed credibility of such evidence. In other words, the extraordinary credibility assigned to DNA evidence has indirectly aided the credibility of psychologists when acting as expert witnesses to counteract commonplace belief in the accuracy of eyewitness evidence.

the defendant's DNA profile and that of the semen recovered from the victim—while the eyewitness evidence was negative. Unlike what happened in *the Diaz* case, the DNA evidence did not simply trump the eyewitness testimony; instead, the *Adams* defense effectively attacked the credibility of the DNA evidence and developed a highly unusual procedure for weighing it together with the eyewitness testimony. The appeal court eventually disallowed the defense's strategy, but the judgment highlighted some significant questions about the relationship between "scientific" and "commonsense" evidence.

The Case

On 6 April 1991, a young woman identified in court documents as Miss M was walking home in the early morning hours in a town north of London.[5] A stranger approached her and asked for the time, and as she glanced at her watch he grabbed her from behind, overpowered her, and forcibly raped her. She saw the man's face only for a few seconds.[6] Miss M reported the attack to the police and described her assailant as a white, clean-shaven man with a local accent, who was between twenty and twenty-five years old. Miss M underwent a physical examination during which the police took a vaginal swab. Forensic analysis revealed traces of semen on the swab, and a DNA profile was developed. At the time, the Metropolitan Police had compiled a database consisting of a few thousand DNA profiles from unsolved crimes and convicted criminals. The profile developed from the semen did not match any of the profiles on the database and was stored for future reference.

Two years later, Denis John Adams was arrested in connection with

5. This narrative is reconstructed from the judicial summaries of the appeal hearings: *Regina v. Adams* (Denis John), 1st Appeal (1996); 2nd Appeal (1997); and Ronald Thwaites Q.C., and Marc Brittain, "Perfected Grounds of Appeal & Skeletal Argument," *Denis John Adams v. Regina,* in The Court of Appeal, Criminal Division, 25 September 1997. Ruth McNally also attended the second appeal, and McNally and Lynch interviewed Peter Donnelly (24 November 1997), an expert witness for the defense, and Ian Evett (18 December 1998), an expert for the prosecution.

6. The judicial summary stated that the victim saw her assailant only briefly. Donnelly (interview, 24 November 1997), provided a different version of the witness's testimony, and stated that the day after the rape the victim told the police that she had a good view of her attacker and was confident that she would be able to recognize him again. In fact, according to Donnelly, she did claim to see him walking down the street a couple of years later, but the police were unable to locate the man at the time.

another sexual offence. The police took his blood sample, and forensic scientists developed a DNA profile from the sample, using the single-locus probe (SLP) technique. Forensic investigators ran Adams's DNA profile against the unsolved crime samples on the police database, and it matched the evidence recovered during the investigation of Miss M's rape. Adams was charged with rape, and according to the judicial summary of the case, the matching DNA profiles provided the only substantial evidence against him.

The *R. v. Deen* case discussed in the previous chapter involved the multilocus probe (MLP) technique. As described in interlude A, this was the first DNA profiling technique used for purposes of criminal investigation in the United Kingdom. By the early 1990s the MLP technique was being phased out and replaced by the single-locus probe (SLP) technique—the one used in the *Adams* case. Unlike the MLP technique, each probe used in the SLP technique was designed to mark a single polymorphic locus, resulting in one band for homozygous and two for heterozygous alleles. Frequency estimates were given for each of the alleles found in reference samples of population groups and subgroups. Estimates of allele frequency varied with the loci and alleles in question, ranging from a fraction of 1 percent for the least common alleles in a given population to upward of 50 percent for the most common. Several probes (usually four or more) were thus used to improve the discriminatory power of the SLP technique. On the (contested) assumption of allele independence, the probabilities assigned to each allele in a sample were multiplied together, often with a correction factor being used to compensate for nonindependence. The resulting probability figure varied with the number of matching alleles and the estimated frequency of the specific alleles in the relevant population group or subgroup.

As noted earlier, in the *Adams* case the DNA profile of the semen sample recovered from the rape victim was stored on a computer database. The database was a relatively small one that was used in the early 1990s.[7] This early database was held at the Metropolitan Police Forensic Science Laboratory in Lambeth, South London, and consisted of a computerized index of SLP profiles developed from a few thousand persons convicted of crimes. It also included a smaller number of body fluid sam-

7. In 1994 the Home Office announced an initiative to construct a national DNA profile database. This national database uses STR, a more recently developed technique that was less costly and time-consuming than the SLP system. The first profiles were entered in April 1995.

ples collected during investigations of unsolved crimes. The vast majority of these crime stains were semen samples associated with sexual assaults.[8] Forensic scientists were able to check profiles from new suspects against the criminal and crime stain profiles in the database. This was how forensic employees at the Metropolitan police laboratory were able to find a match between the DNA profile developed from Denis John Adams's blood sample and the profile from the semen stain collected during the earlier investigation of Miss M's rape.[9] This database match was thus an early (perhaps the first) instance of a cold hit; a phenomenon that became commonplace a few years later (Cole & Lynch, 2006).

Dr. Harris, a forensic scientist, analyzed the matching profiles, and his statement for the prosecution included an estimate of 1 chance in 297 million, rounded off to a more "conservative" estimate of 1 in 200 million, that a match would be found between the crime stain and a randomly chosen unrelated man from the European Caucasian population.[10] Dr. Harris and other forensic scientists testifying for the prosecution stated that this was very strong evidence that Denis John Adams had "shed the semen" recovered from the victim.

As mentioned earlier, other evidence in the case, particularly the eyewitness testimony, did not corroborate the DNA evidence. The case summary described Denis John Adams as a Caucasian man, who was thirty-seven years old at the time of his arrest. Miss M described her assailant as a Caucasian man between twenty and twenty-five years old. After Adams was arrested, he was placed in an identification parade (a

---

8. See Allard (1992) for a description of the database.

9. As noted earlier, our narrative of the case depends upon case documents, including judicial summaries. When we speak of "crime stain" and "Miss M's rape" we are borrowing presumptive language from those documents. A more cumbersome, but "correct" narrative would include constant reminders that the "rape" in question was not a simple fact, but was contingent upon the language of a complaint and criminal charge, and that the identity and significance of the "crime stain" depended upon acceptance of the prosecution's tentative determination that a crime of rape had occurred. The identification of the substance recovered from the examination of Miss M as semen from the perpetrator also relied upon the victim's testimony and the prosecution's charge. To our knowledge, Adams did not contest the fact of the rape, and did not put forward a consent defense, and so the presumptions in question had the status of "agreed facts of the case" (see Brannigan & Lynch, 1987).

10. In other chapters we use the common designation RMP (random match probability) as a shorthand description of such figures, but in this case the court used different terminology (including a unique expression that was used at the time in the British courts—"random occurrence ratio"), and the question of just how to describe the figure was one of the questions under appeal.

police lineup), but Miss M failed to identify him or anyone else in the lineup as her attacker. Later, when Miss M saw Adams at the committal proceedings, she said he did not look like her attacker. According to her, Adams appeared to be much older than the attacker. The prosecution proceeded with the case in spite of the victim's failure to identify him. At the trial, Adams also claimed that he spent the entire night with his girlfriend on the date of the attack, and his girlfriend corroborated this alibi. However, the DNA evidence proved more persuasive to the jury and Adams was convicted in January 1995. In 1996 he successfully appealed the conviction and was retried. Again he was convicted, and again he appealed. The second appeal was heard in October 1997, and this time the court rejected the appeal.

The grounds of both *Adams* appeals had to do with the way the trial judge instructed the jury on how to weigh DNA and non-DNA evidence in a case that involved an apparent disjuncture between the two forms of evidence: most, if not all, of the non-DNA evidence supported Adams's not-guilty plea, whereas the DNA evidence provided almost the entire basis for the prosecution's case. In both trials and appeal hearings, the defense mobilized some of the strategies we discussed in the previous chapter to attack the probability estimate assigned to the DNA match. The defense also made use of Professor Peter Donnelly as an expert witness, and in this case he used an innovative strategy to weigh the DNA and other evidence on the same scale. The judicial summary deployed a distinction between "scientific" and "common sense" evidence when reviewing this innovative strategy.

We place quotation marks around the words "scientific" and "common sense" for two reasons: first, because the distinction marked by those terms is problematic, and second, because the terms explicitly appear in the appeal court summary of the case, and how the court used them is significant for the way the case was decided. Consequently, while we would be hard-pressed to defend the court's distinction between science and common sense, we shall treat the *performative* use of that distinction as a "datum."[11] What interests us in this chapter is the way the

---

11. J. L. Austin (1962) devised the category of "performative utterance" to describe linguistic expressions that, when uttered in the appropriate circumstances, *institute* real-world events. So, for example, under the appropriate ceremonial conditions, an official can christen a ship or pronounce a couple to be married. Such declarations do not describe pre-existent conditions; instead, they serve to bring about the state of affairs they "describe." In the present case, the appeal court does more than *discuss* the relationship be-

court deployed the distinction to mark out domains of competence and legitimate authority. In other words, we are interested in how the court practically used a conceptual distinction—a distinction with a long and contentious philosophical pedigree—to resolve the case at hand.[12] The salience of the distinction for the court's "boundary work" is independent of the question of whether the distinction accurately "maps" the kinds of evidence deployed in the case.[13]

The *Adams* case was interesting because virtually all of the evidence other than the DNA profile match supported the defense. According to the court of appeal, *R. v. Adams* was "the first case in which the Crown had relied exclusively upon DNA evidence."[14] According to the court's summary, the "scientific" evidence supported the prosecution whereas all of the "common sense" evidence supported the defense. Because of this, the *Adams* defense argued, the trial judge should have excluded the DNA evidence for being "inconclusive in itself and inadequate to found

---

tween "scientific" and "common sense" evidence; it stipulates how that relationship should be organized in the case at hand and other cases like it. What the judges say, therefore, serves to *effectuate*, within specific institutional circumstances, a particular set of procedures related to the general theme of "public understanding of science." A further aspect of the judicial summary that is important to keep in mind is that it was written in the face of contested claims about the relationship between "scientific" evidence and public (juror) understanding of such evidence. Within the circumstances of the case, the judicial performance effectuated a resolution of a controversy about how the public can, and should, understand and use expert evidence. For a generalized use of the concept of "performative" in social studies of science, see Pickering (1995).

12. To make a very long story short, the relation between science and common sense is inconsistently and ambivalently treated, both by scientists and philosophers of science. Common sense is sometimes denigrated in the fashion of Bacon, as a source of "idols"— myths, superstitions, interested biases—or, in the fashion of Galileo, as surface appearances that fail to grasp the underlying mathematical laws revealed by mathematical analysis and demonstrated through experiment. Alternatively, and even at the same time, common sense is valorized as the tacit ground of scientific knowledge, an unexplicated intuitive foundation that supports the practical discovery of novel, sometimes counterintuitive, phenomena.

13. The term "boundary work" is used by Thomas Gieryn (1983; 1995; 1999) to respecify the science/nonscience distinction as a social phenomenon. Unlike the philosophers who have attempted to formulate principled demarcation criteria, Gieryn recommends a social-historical perspective on authoritative efforts within authoritative bodies to lend or deny scientific status to candidate practices and phenomena like phrenology, creation science, and cold fusion. Steve Woolgar, who originally may have coined the term "boundary work," uses a related expression, "ontological gerrymandering" to critically examine the way social science writers identified with constructionist perspectives tacitly grant or withdraw the accent of reality in discussions of social problems (Woolgar & Pawluch, 1985).

14. *Adams*, First Appeal (1996), 2.

the prosecution case."[15] Citing a provision in the Police and Criminal Evidence Act of 1984, the defense submitted that the evidence should be allowed only if confirmed by other evidence. The appeal court did not accept this ground for appeal, arguing that the trial judge recognized that the DNA evidence stood alone and properly exercised discretion to allow the jury to consider it along with the other evidence: "There is no principle of law that DNA evidence is in itself incapable of establishing guilt."[16]

To say that the DNA evidence was the sole prosecution evidence is something of an overstatement, because some of the circumstantial evidence supported (or, at least, did not conflict with) the prosecution's case. Adams lived in the local area and he was Caucasian. And, while he differed from the victim's description, it is possible to discount the discrepancies. Eyewitness testimony is notoriously fallible, and in this case the victim was asked to identify an assailant she had seen only briefly during a traumatic encounter two years earlier. Had Adams been, say, an eighty-year-old man of Asian extraction, the discrepancies would have been far more difficult for the prosecution's case. It also is not too difficult to discount Adams's alibi and the girlfriend's corroboration of it. Nevertheless, the *Adams* trial was construed by the court as a kind of test case in which a single powerful item of "scientific" evidence was weighed against various forms of "common sense" evidence.

For our purposes, the case exhibited a special kind of "reality disjuncture": a dramatic confrontation between incommensurable accounts of "the same" event. In many rape cases there is a disjuncture between the accused man's and the victim's accounts of the alleged event and the circumstances leading to it.[17] For example, when pursuing a "consent" defense the accused rapist argues that the victim explicitly or implicitly consented to the sexual act. In the absence of corroborating evidence, the jury is faced with choosing between conflicting narratives. In the *Adams* case, the evidence supporting the accused rapist's guilt not

15. Ibid.
16. Ibid.
17. Pollner (1975) introduced the idea of "reality disjunctures" to describe incommensurable accounts of (arguably) the same experience. Examples include patient reports of experiences that psychiatrists attribute to hallucinations, and contrary testimony about the speed of a particular automobile by defendant and arresting officer in traffic court. Also see Coulter (1975) for a related discussion of "perceptual accounts and interpretive asymmetries." Drew (1992) examines testimony at a rape case in which the victim and defendant give different versions of the circumstances leading to the alleged crime.

only conflicted with the evidence supporting his innocence, it was a systematically different *kind* of evidence (or, at least, it was viewed that way by the appeal court). As the court's summary mentions, the prosecution relied upon expert evidence, whereas the defense relied upon ordinary forms of identification and description.[18] This is only one case, but it is interesting for analytical purposes because of the unusually clear way it juxtaposed the credibility and weight of expert evidence, and specifically of DNA profile evidence, with that of other, more ordinary, forms of evidence (though, as we shall see, the defense used an expert to interpret the ordinary as well as scientific evidence with a Bayesian scheme). The court summary aligned the two sides of the case with a set of epistemic distinctions.[19]

Prosecution	Defense
Expert judgments	Common sense judgments
Probabilistic evidence	Nonprobabilistic evidence

Deconstructing the Evidence

Because of the way the DNA evidence stood alone, as the primary if not the sole item of prosecutorial evidence, the *Adams* case provided a test of the relative credibility and weight of DNA evidence, as compared with the other evidence. Adams was represented by Ronald Thwaites, QC, a prominent barrister with strong civil libertarian leanings. During the first trial, and throughout the subsequent trials and appeals, the defense used a two-pronged attack.[20] As a first prong of attack, Thwaites and his main expert witness Donnelly challenged the impressive, seemingly unassailable, probability figures the prosecution assigned to the DNA evidence. This form of challenge was not unusual, and in many respects it resembled the arguments by Donnelly and the other expert witnesses in the *Deen* appeal. The second prong involved an unusual, if not unique,

18. Risinger (2000a) distinguishes ordinary fact testimony from various categories of expert evidence. The former requires no special training and is limited by the hearsay prohibition, whereas the latter is based on specialized training.

19. Judgment, 1st appeal, 9.

20. For simplicity and ease of presentation, we present a single summary of arguments and statements that appeared in the two trials and appeal hearings, and we draw out differences between the separate trials and appeals only when they are relevant.

attempt to employ a Bayesian approach to quantifying the weight of the "common sense" as well as the "expert" evidence. The two prongs of attack reinforced each other, and arguably reduced the totality of evidence against Adams to the point of reasonable doubt.

(1) The First Prong: Attacking the Probability Figure

The match between the DNA profiles developed from the vaginal swab and from Adams's blood sample was translated into probabilistic language by the forensic scientists who summarized the evidence for the court. The scientists used different forms of probabilistic statement at different times, when the evidence was analyzed, challenged, and reanalyzed during the trials and appeals. Some evidence statements used nonnumerical terms to describe the DNA match.[21] In a written statement dated 10 December 1993, Dr. Harris, a forensic scientist who analyzed the DNA evidence for the prosecution, stated, "There is very strong evidence that Denis John Adams . . . has shed this semen." Another forensic scientist, Jane Lygo, in a statement dated 17 November 1994 made a similar, though more formal, claim: "The scientific evidence provides strong support for the allegation that the semen staining on the high vaginal swab (ZH/11) originated from Denis John Adams." A third forensic scientist, James Lambert, on 21 December 1994 also stated a nonquantified assessment, while referring to the results of two different methods of analysis: "[T]he DNA SLP and STR profiling results, when considered together, provide very strong support for the [prosecution's] view."[22]

The defense did not take issue with these nonquantified statements about the probative value of the DNA evidence. The main points of contention concerned the use of quantitative probability estimates that lent "very strong support" for the prosecution's case. Perhaps because of the way the Crown's case against Adams rested on the matching DNA profiles, the forensic scientists who analyzed the evidence sought to maximize its weight by combining a series of probes and quantitative estimates. According to testimony of Harris, an SLP technique with four

---

21. Quotations of witness statements by Harris, Lygo and Lambert were taken from Thwaites and Brittain, "Perfected Grounds of Appeal & Skeletal Argument," p. 2, sec. 3.1.
   22. Ibid.

probes was used, and seven matching bands were visible in a comparison between the DNA profile of Adams's blood sample and the profile from the vaginal swab.[23] In his testimony, Harris acknowledged that when he examined the autoradiograph, one of the matching bands was so faint that he highlighted it with a marking pen in order to make it more clearly visible. He included the match in his estimate of the frequency of the profile's combination of alleles in the white European population, and he came up with a figure of 1 in 2 million. Harris also said that he later added a fifth probe, which produced two additional matching bands. The revised probability estimate based on the nine matching bands became 1 in 200 million. This number was rounded down from a more precise figure of 1 in 297 million, which Harris estimated was the chance that a randomly chosen unrelated man in the white European population would match the DNA profile developed from the crime stain. He justified the rounded estimate in the interest of "conservatism": to avoid exaggerating the weight of the evidence against the defendant and to make the figures easier for the jury to understand.

The defense attacked the probability estimate by suggesting that Dr. Harris had engaged in an effort at "improving the numbers" to the point that "the probes became weaker and the technology was exhausted."[24] In other words, the defense accused the prosecution of stretching the technique beyond its limit of resolution in order to generate frequency estimates that would persuade the jury to convict Adams on the DNA evidence alone.[25] Thwaites and his expert witnesses argued that analysis with the fifth probe was not performed under the same conditions as the earlier analysis, thus invalidating the enhancement of the probability figure from 1 in 2 million to 1 in 200 million. They also challenged Harris's procedure of manually enhancing the evidence. They argued that Harris's manual mark was illegitimate, and that the particular allele should be excluded from the probability estimate. According

23. The reason that four probes would result in seven bands is that three of the probes identified heterozygous bands in the particular samples (two each for three probes) and one identified a homozygous band.

24. Thwaites & Brittain, "Perfected Grounds of Appeal & Skeletal Argument," p. 3, sec. 3.5.

25. This is an instance in which the features attributed to "pathological science" (Langmuir, 1989 [1968]) are used as argumentative resources for discounting (in this case, reducing the weight) of particular evidence. For a study of particularistic uses of arguments to support and discount evidence, see Gilbert & Mulkay (1984).

to the multiplication rule and the frequency assigned to the particular allele, this would alter the estimate by a factor of ten, changing the estimate from 1 in 2 million to 1 in 200,000.

In addition to using different versions of the SLP technique, the forensic scientists also used STR, a more recently developed technique. This is a DNA profiling method that employs the polymerase chain reaction (PCR) together with other laboratory techniques and instruments to analyze "short tandem repeat" (STR) sequences of DNA. The STR technique was adopted by forensic services in the United Kingdom and elsewhere and became the basis for the national DNA database, which was developed by the U.K. Forensic Science Service (the first profiles were placed on the STR database in 1995). In the first *Adams* trial, the judge decided to exclude the STR evidence. Had the evidence been presented to the jury, and had the resulting estimate been numerically combined with the SLP estimate (a procedure that the defense was prepared to criticize), the resulting probability figures would have been even more impressive. The decision not to present the STR evidence to the jury had to do with avoiding potential confusion arising from the combination of different techniques.

To complicate matters further, the defense contended ("somewhat faintly," according to a judicial summary) that the analysis should take account of the fact that Adams had a brother.[26] The DNA experts acknowledged that the brother "was a complicating factor whose existence reduced the probability of a match to 1 in 220."[27] The court chose to disregard this possible complication, because the brother's blood was not profiled and the defense did not suggest that he might have been involved in the crime.[28]

Without going more deeply into the matter, it should be clear for some of the same reasons we discussed in chapter 5 that the question of just how to formulate the match probability was contentious. Even

26. *Adams* (1996), Judgment, 3.

27. *Adams* (1997), Judgment, 3. The statement is confusing, because a match probability of 1 in 220 is much higher (and not lower, as the statement implies) than a probability like 1 in 200 million. Perhaps the statement meant to say that including the brother would reduce the odds of a profile from someone other than Denis John Adams matching the crime stain.

28. See Redmayne (1995) for an illuminating discussion of how burden of proof for including particular persons in the suspect pool is shifted from prosecution to defense in DNA profiling cases.

though the more precisely quantifiable SLP technique was used in this case, a whole range of alternative probability calculations could have been made. Probability estimates varied from 1 in 200 million to 1 in 200,000, depending upon the number of probes used, and upon which visual results were counted as adequate. If the brother was included in the narrative, the number became 1 in 220. Given these discrepancies, the rounding off of 297 million so that it became 200 million seemed a minor adjustment (although this too was challenged by the defense for having no clear biological or statistical rationale).

In spite of the variety of numerical and non-numerical formulations that were given, the figure of 1 in 200 million was used throughout the *Adams* trials and appeals. The defense continued to contest this figure, but also used it as a starting point.

(2) The Bayesian Questionnaire

A distinctive strategy was used during the *Adams* trials and appeals to weigh the non-DNA evidence. This strategy involved an effort to instruct the jury on how to use a Bayesian procedure for translating all of the evidence into comparable probability estimates. The defense appealed to "elementary fairness" when arguing that all of the evidence should be given an equivalent kind of analytical treatment. According to the argument, the "common sense" evidence presented by the defense should be weighed on the same scale as the "scientific" evidence presented by the prosecution, and both should be analyzed with the Bayesian formula: posterior odds = prior odds × likelihood ratio. The defense argued:

> Unless the defence-oriented evidence could be successfully translated into numerical form and combined by the jury with the prosecution statistical estimate, the defence were prejudiced because of the tendency of the single large statistic to overwhelm the other evidence. The jury were in effect given only one half of the equation: the prosecution reasoning being how unlikely he is to be innocent, given the finding in relation to the DNA. But the defence were attempting to say how unlikely he is to be guilty, given the rest of the evidence, if it could be put to the same form to enable to jury to complete the equation.[29]

29. Thwaites & Brittain, "Perfected Grounds of Appeal & Skeletal Argument," 4.

The defense emphasized that Bayes's theorem already was incorporated into the "practices and procedures used by the FSS [Forensic Science Service] in their out-of-court preparation for the presentation of DNA evidence in court."[30] According to this argument, the prosecution had already opened the door to the Bayesian approach. The only question was whether the door should be left open for the defense evidence.

It might seem at this point that there is a clear distinction between the quantitative DNA evidence presented by the prosecution and the qualitative "common sense" evidence presented by the defense, but this is exactly the distinction that the defense challenged. Thwaites and his expert witnesses argued that forensic scientists use empirical data from studies of reference populations when making estimates of the probability of a random match, but that the "single large statistic" they develop for a specific case depends upon what the scientists deem relevant to the case at hand. As noted earlier, in the *Adams* case the estimate of 1 in 200 million developed by the forensic analysts was based on reference figures for the European Caucasian population, and it did not take account of the possibility that Adams's brother might be included in the suspect pool. The figure also incorporated contested judgments about which lines on an autoradiograph were clear enough to declare as matching evidence, and how much rounding of numbers was appropriate.

Recalling the *Deen* case, we also can get a less formalized sense of the Bayesian strategy for *leveling* the evidence. In the *Deen* appeal hearing, the three defense witnesses insisted that probability estimates should be assigned to nonmatching as well as matching bands, and concerning the victim's testimony on the assailant's color, Donnelly also insisted that "we would want to try and assess the chance that she might be wrong." For a Bayesian like Donnelly, such an assessment should be quantified, along with the bands that were counted as matching. A single "scientific" probability figure (of 1 in 3 million in the *Deen* case) did not stand alone, and was argued to be part of a chain of judgments, only some of which entered into the prosecution's calculation.

Similarly, according to the defense's argument in the *Adams* case, the derivation and preservation of the "single large statistic" of 1 in 200 million depended upon a series of decisions about relevant evidence, adequate procedure, and reasonable estimation. In brief, the defense argued that the prosecution's quantitative evidence was based upon a series of

30. Ibid., 5.

judgments about the case at hand; judgments that, within the specialized domain of forensic analysis, were not essentially different from the
"qualitative common sense" judgments that entered into the weighing
of the defense evidence. Consequently, the defense proposed to extend
quantitative methods of estimation to cover the relevance and weight
of the eyewitness misidentification, the alibi, and other evidence in
the case.

In his testimony for the defense, Professor Donnelly argued that
Bayes's theorem was "the only logical and consistent way" for the jury to
consider the DNA evidence together with the other evidence in the case.
His intervention was more than a matter of registering an expert opinion. He also developed a questionnaire to be used by the members of the
jury. The prosecution did not object to the procedure, and a prominent
forensic scientist with the Forensic Science Service, Ian Evett, reviewed
the questionnaire before it was introduced into the trial. During both
Adams trials, the judge ruled that the questionnaire would be acceptable for individual jurors to use on a voluntary basis.

The questionnaire listed seven questions, asking individual jurors to
make probability estimates about four items of evidence: (1) that a local man would have committed the offense, (2) that the victim would not
have identified the defendant, (3) that Adams would have given the evidence he did in favor of his innocence, and (4) that Adams would have
been able to call alibi evidence on his behalf. With the exception of the
first item, which Donnelly said was a "starting point," there were two
questions for each item—one asking for an estimate of the likelihood of
the evidence given the assumption of guilt, and the other asking for an
estimate given the assumption of innocence. Donnelly did not tell the jurors which figures to use for their estimates, but he did supply some illustrative figures when he demonstrated the procedure. After working
through each question and developing ratios between "subjective" estimates of the likelihood of each item of evidence, under the assumption
of innocence versus the assumption of guilt, he arrived at the conclusion
that there was 1 chance in 3.6 million that Adams was guilty on the basis of the non-DNA evidence. This figure represented a hypothetical juror's estimate of the likelihood that Adams was guilty rather than innocent, leaving aside the DNA evidence and knowing nothing about the
case other than that he had been picked out of the population of Caucasian men in the greater Manchester area, that the victim failed to identify him, that he had an alibi, and so forth.

Having developed this estimate for the non-DNA evidence, Donnelly then combined it with the prosecution's 1 in 200 million figure for the DNA evidence. According to Donnelly's hypothetical calculation, if Adams was the actual rapist, the chance of a match between his blood sample and the crime scene evidence would be 1:1 (100 percent), because both samples derived from the same source.[31] If, however, he was not the rapist, the chance of his DNA profile matching the crime stain would be 1 in 200 million (for the sake of the illustration, this figure accepts the prosecution's estimate, though as noted earlier the defense challenged that figure).

When Donnelly's ratio of 1 chance in 3.6 million that Adams is guilty, based on the non-DNA evidence, is weighed against the prosecution's figure of 1 chance in 200 million that Adams is innocent, the resulting estimate is 1:200/3.6, or 1 chance in approximately 55 that Adams is innocent. In other words, according to Donnelly's illustrative exercise, taking all of the evidence into account, it is 55 times more likely that Adams is guilty rather than innocent.[32] Note that a successful challenge to the prosecution's estimate of the probability of the DNA match also affects the result. As mentioned earlier, Donnelly challenged the visual match between Adams's sample and the crime scene evidence by stating that the indistinct autoradiograph band that Dr. Harris enhanced with his marking pen should not be counted in the multiplication of allele frequencies in the profile. If this band is excluded, then the DNA match estimate is reduced by a factor of ten, becoming 1 in 20 million. Consequently, if we divide 1 in 20 million by 1 in 3.6 million, we get an estimate that it is 5.5 times more likely that Adams is guilty rather than innocent.

31. The 100 percent figure ignores the chance of a false negative (for example, a non-match between Adams's blood profile and the crime stain due to an error in the collection or analysis of the samples).

32. This calculation must assume that the series of estimates covers *all* of the relevant evidence. Otherwise, it would be inappropriate to make a guilty or not-guilty judgment on the basis of the series of estimates. It is obvious that the four items of defense evidence can be subdivided. One could, for example, treat the witness's misidentification of Adams in the police lineup as a separate item from the mismatch between Miss M's description of the perpetrator's age and Adams's age. However, according to Donnelly (interview, 24 November 1997), as long as all of the evidence is represented in the series of estimates, the overall calculation should, in principle, not be affected by the way it is parsed into separate items. Others, including a Mr. Lambert, who testified for the prosecution in the second trial, expressed misgivings on this point, arguing that the questionnaire that Donnelly prepared did not cover all of the relevant evidence, or all of the evidence that particular jurors might consider relevant (*Adams,* 1997, second appeal summary, 7).

If the defense had managed to knock down the DNA estimate by another factor of ten, to 1 in 2 million, the result would drop accordingly: 1
in 2 million divided by 1 in 3.6 million, or roughly one chance in two that
Adams is guilty.[33] And, finally, if Adams's brother had been included in
the suspect pool, according to the Bayesian procedure, the balance of
evidence favors his innocence, as there is now a chance of only 1 in 18
that he is guilty. In a criminal trial in the United Kingdom (and in many
other systems), jurors are instructed to register a not-guilty verdict if they
entertain "reasonable doubt" about the totality of evidence against the
accused person. There is no single point in a continuum of probabilities
at which "reasonable doubt" begins, but it should seem clear that, while
a probability of guilt of 1 in 290,000,000 (a number several times the size
of the U.K. population) might seem unassailable, probabilities like 1 in
55, and surely 1 in 5.5, can much more easily provide a basis for doubt.

Although the jury was allowed to use the questionnaire during the
first trial, it was unclear whether it was used effectively, and the verdict
indicated that the jury believed that the evidence was sufficiently strong
to prove Adams's guilt. Adams appealed the conviction, on the grounds
that the DNA evidence presented by the prosecution was insufficient by
itself to establish guilt, and that the judge had inadequately instructed
the jury to apply Bayes's theorem to the facts of the case. In its judgment,
the court of appeal announced that it accepted the defense's position
that the trial court had not prepared the jury properly, and it referred the
case for retrial. However, the court also stated that it is the jury's proper
business to decide whether the DNA evidence is sufficient for conviction in the absence of corroborating evidence. Importantly, for our purposes, the court of appeal also commented on the appropriateness of using Bayes's theorem to summarize the non-DNA as well as the DNA
evidence: "[W]e have very grave doubt as to whether that evidence [using
Bayes's theorem] was properly admissible, because it trespasses on an
area peculiarly and exclusively within the province of the jury, namely
the way in which they evaluate the relationship between one piece of
evidence and another."[34] The "grave doubt" in this case had to do with
the principle of "jury usurpation": an encroachment upon the jury's role
as trier of fact in criminal cases. In the English common law tradition,

33. Donnelly also criticized the forensic scientists for not adding a correction factor for
sampling error. This had a relatively minor effect (1 percent) on the resulting estimate.

34. *Adams* (1996), Judgment, first appeal (9 May).

expert witnesses are recognized as persons with specialized knowledge
or experience that lay members of the court are unlikely to share. Although expert testimony about relevant facts can be authoritative (especially when such testimony is not challenged by other expert witnesses),
it is also restricted. In principle, experts inform and advise the court
about particular matters relevant to the case; they are not supposed to
pronounce upon the ultimate guilt or innocence of the defendant.[35] The
judicial summary went on to say,

> The Bayes theorem may be an appropriate and useful tool for statisticians
> and other experts seeking to establish a mathematical assessment of prob-
> ability. Even then, however, as the extracts from Professor Donnelly's evi-
> dence cited above demonstrate, the theorem can only operate by giving to
> each separate piece of evidence a numerical percentage representing the ra-
> tio between probability of circumstance A and the probability of B granted
> the existence of that evidence. The percentages chosen are matters of judg-
> ment: that is inevitable. But the apparently objective numerical figures used
> in the theorem may conceal the element of judgment on which it entirely
> depends.[36]

The court's criticism was both legal and methodological. Much in the
fashion of a skeptical science studies analyst, the court reasoned that the
quantification procedure would create only the *appearance* of objective
judgment. More seriously, the court suggested that the procedure would
substitute a limited calculus for the jury's legal mandate to examine all
of the evidence in the singular case. The court argued that

> the theorem's methodology requires . . . that items of evidence be assessed
> separately according to their bearing on the accused's guilt, before being

35. Rule 704(a) of the U.S. Federal Rules of Evidence is less restrictive on the matter,
stating that within certain limits, an expert's "testimony in the form of an opinion or inference otherwise admissible is not objectionable because it embraces an ultimate issue to
be decided by the trier of fact." Mnookin (2001: 30) observes that early in the history of
fingerprinting, there was some dispute in U.S. courts about whether a fingerprint examiner's use of the language of "fact" rather than "opinion" when declaring matching evidence usurped the jury's prerogative to determine "ultimate facts" in the case. In the view
of most courts, according to Mnookin, declaring a match did not encroach upon the "ultimate issue" of guilt or innocence, because that determination would depend upon other
evidence. However, as with DNA evidence, when a match is equated with guilt, the distinction collapses.

36. *Adams* (1996), First appeal, op. cit., 15.

combined in the overall formula. That in our view is far too rigid an approach to evidence of the type that a jury characteristically has to assess, where the cogency of (for instance) identification evidence may have to be assessed, at least in part, in the light of the strength of the chain of evidence in which it forms a part.[37]

The court's rejection of the Bayesian procedure drew a strict demarcation between a formulaic mode of individual reasoning and a collective form of common sense deliberation:

> More fundamentally, however, the mathematical formula, applied to each separate piece of evidence, is simply inappropriate to the jury's task. Jurors evaluate evidence and reach a conclusion not only by means of a formula, mathematical or otherwise, but by the joint application of their individual common sense and knowledge of the world to the evidence before them.[38]

The court's judgment clearly rejected the rationale Donnelly had earlier given for introducing the Bayesian procedure. As he put it, Bayes's theorem was "the only logical way to assess all the evidence." He did not distinguish the procedure from common sense, nor did he associate it with "objective" outcomes. Instead, he recommended it as a tool that would enable each juror to represent and combine the weights of different evidential judgments. He did not tell the jurors which probabilities to associate with specific items of evidence, such as the alibi and the victim's failure to identify Adams. Far from being a source of "objective" judgments, for Donnelly and Thwaites the procedure was a way to counteract the *apparently* objective numerical figures assigned to the prosecution's DNA evidence; figures which they argued (in what we called the first prong of attack) depended on a chain of practices and judgments. The defense thus justified the use of the Bayesian method as a democratic procedure in the numero-politics of a criminal trial; a procedure that enabled "common sense" evidence to be weighed on the same scale as "scientific" evidence.[39]

---

37. Ibid.
38. Ibid.
39. Ian Hacking (1995) coined the term "memoro-politics" to describe disputes over the evidential status of "recovered" memories of childhood abuse. "Numero-politics" is our reference to disputes about particular counts, estimates, and measures, and the methods for generating them.

Because none of the parties to the first appeal hearing had argued against the admissibility of the Bayesian questionnaire, the court of appeal did not give an authoritative ruling on the matter. However, the court did express the opinion *per incuriam* that Bayes's theorem was not appropriate for aiding jurors to weigh all of the evidence in the case.

In the retrial, the defense again invited the jury to use the questionnaire to calculate probabilities for all of the evidence. The judge expressed reservations about the procedure, but permitted individual jurors to use it on a voluntary basis. The judge instructed the jury, "If you feel able to use the questionnaire to operate Bayes's theorem and you find it almost as easy as kiss your hand [*sic*] to give answers, then you have the opportunity to do it, having not only your own copies but you will have when you go out an extra blank one to fill in your collective view if you want to." The judge added, however, that the jurors' duty was to "consider your verdict amongst yourselves, all of you together and not with one huddled in a corner with his calculator."[40]

The arguments by the prosecution and defense were much the same as in the first trial, and once again the jury convicted Adams. The defense again appealed. In a written summary, Thwaites argued that it was only fair that the defense should be allowed to use Bayes's theorem to represent the non-DNA evidence in a quantitative way, equivalent to the prosecution's representation of the DNA evidence.[41] He added that the Bayesian approach was sound, logical, and approved by expert opinion. Thwaites also submitted that the trial judge's instructions to the jury gave insufficient (and even "flippant") instructions about using the Bayesian approach. He added that when the trial judge contrasted the Bayesian approach to the jurors' "common sense" reasoning, he misled the jury by ignoring the way Donnelly's questionnaire provided a means for making "scientific" and "common sense" evidence explicit and comparable.

The court of appeal considered the second appeal, but rejected each of the defense's grounds. The court also used the opportunity to give further authority to its earlier misgivings about the Bayesian procedure. Once again, the court drew a firm distinction between the Bayesian approach and "a more conventional application of judgment." The ruling stated that quantification was appropriate for the prosecution's DNA

40. Summing up of His Honour Judge Pownall, quoted in *Adams* (1997) 2nd appeal, 8.
41. A summary of Thwaites's arguments was presented in the lord chief justice's summary, *Adams* (1997), 2nd appeal, note 4, pp. 5–8.

evidence, insofar as that evidence was "based . . . on empirical statistical data, the data and the deductions drawn from it being available for the defence to criticise and challenge."[42] The court characterized the defense's evidence as "non-scientific, non-DNA evidence" and rejected the argument that such evidence should be represented in statistical form. To support this decision, the court cited the judgment on the first *Adams* appeal, which as noted earlier, stated misgivings about whether the Bayesian procedure should be allowed. The court also cited a more recent appeal court ruling in the case *R. v Doheny and [Gary Andrew] Adams,* which endorsed the first (Denis John) *Adams* appeal court's objection that the Bayesian procedure "plunges the jury into inappropriate and unnecessary realms of theory and complexity deflecting them from their proper task."[43] This chain of citations performed a kind of a bootstrapping operation on the "misgivings" initially stated in the judgment on the first *Adams* appeal. Now, a year later, the second *Adams* appeal ruling could cite the earlier judgment together with the *Doheny & Adams* judgment as "previous rulings" on the matter in question. The court concurred that "we regard the reliance on evidence of this kind in such cases as a recipe for confusion, misunderstanding, and misjudgment, possibly even among counsel, but very probably among judges and, as we conclude, almost certainly among jurors."[44] The court affirmed the "conventional" manner in which juries assess the evidence,[45] and concluded by saying, "We are very clearly of opinion that in cases such as this, lacking special features absent here, expert evidence should not be admitted to induce juries to attach mathematical values to probabilities arising from nonscientific evidence adduced at the trial."[46] The court dismissed the appeal, and at that point, the *Adams* case was closed. Adams was about to come up for parole, so the outcome of the appeal was no longer crucial to his fate. The ruling about the Bayesian procedure may have more lasting significance, however.

42. Ibid. (16 October), p. 10.

43. *Regina v. Adams* (1996) 2 Court of Appeal, R 467, cited in *Regina v. Doheny and Adams* (1997) 1 Court of Appeal R 369, 374 G, which in turn is cited in *Adams* (1997), 2nd appeal (16 October)., p. 11.

44. *Adams* (1997), 2nd appeal, note 4, p. 12.

45. Interestingly, the court exemplified the "conventional" approach by describing how a juror might work through the various elements of the *Adams* case. This description included a sequence of probabilistic judgments, much as a Bayesian might describe them, except that the judgments would not be numerical.

46. *Adams* (1997), 2nd appeal, note 4, p. 13.

The Ultimate Issue

The *Adams* case raised interesting issues that are not limited to the particular case or even to criminal law. As noted in the previous chapter, statistical, probabilistic reasoning has become increasingly prevalent in many areas of professional life and public decision making (e.g., Porter, 1995). Where the word "probability" once had no association with numbers, and non-numerical evaluations were deemed sufficient for almost all decisions, the precise calculation of risk is now said to be an emblem of late-modern society (Beck, 1992). Many of the scholars who characterize this global trend express concern about the intimidating sense of objectivity associated with statistical treatments of matters that were once consigned to "common sense," and they raise questions about the ability of the numerically challenged masses to participate in decisions that are increasingly consigned to experts (Smith & Wynne, 1989; Irwin & Wynne, 1996). This is exactly what troubled the court of appeal. The court openly worried about the possibility that a technical method of calculative decision-making would usurp the jury's (or, in a trial in which there is no jury, the judge's) responsibility to decide the "ultimate issue" of guilt or innocence.

Both *Adams* appeal judgments affirmed the role of "common sense" for making a holistic assessment of the totality of evidence presented in the singular case. As the courts recognized, such assessments not only involve discrete questions of fact; they also take into account the demeanor of witnesses and the credibility of testimony. They involve elements of trust which are fallible, difficult to justify, and impossible to quantify (Shapin, 1994; 1995). The court also stated that the jurors' common sense rests upon knowledge of the world, which cannot be reduced to an expert system.[47] The court emphasized the social, as opposed to individual, basis of jurors' judgments and affirmed the necessity for jurors collectively to deliberate about the ultimate issue. In sum, the two *Adams* appeal court decisions characterized the Bayesian approach as an individualistic, reductive calculus that creates a misleading or potentially confusing *appearance of objectivity* when applied to "nonscientific" evidence.[48]

47. This distinction resonates with criticisms of artificial intelligence and expert systems (Dreyfus, 1979; Collins, 1990).
48. In many respects the court's objections to the defense's use of the Bayesian protocol for organizing jurors' judgments parallel science and technology studies arguments

However, the *Adams* appeals did not simply result in an across-the-board rejection of a probabilistic calculus, nor did they simply uphold common sense against a technocratic procedure that threatened to subjugate (or "usurp") it. The court did not discount the appropriateness of the Bayesian method for assisting judgments about *scientific* evidence, and it did not go along with the defense's argument for an equivalent quantitative framing of the DNA and non-DNA evidence.[49] Donnelly and Thwaites argued that the courts should give the same quantitative treatment to all evidence in the case, regardless of whether it is associated with empirical science. The court rejected this argument and continued to demarcate "scientific" DNA evidence from "common sense" non-DNA evidence, and it ruled that numerical estimation was appropriate for the former type of evidence but not the latter. The court of appeal affirmed the jury's jurisdiction over the "ultimate issue," but at the same time it affirmed the "scientific" legitimacy of probability figures. The court's use of the science/nonscience distinction produced an interesting twist on the theme of boundary work (Gieryn, 1983). The court did not simply assign special authority to "expert" or "scientific" testimony. Instead, it stipulated limits to "scientific" testimony and ascribed global authority to the jury's collective "common sense" deliberations.

Like other forms of graphic and statistical evidence, DNA profiles and probability estimates can confer such impressive weight on evidence that it may seem pointless to dispute them. A concern voiced soon after DNA profiling was introduced into criminal prosecutions was that technically unprepared judges and jurors would simply cave in to DNA evidence, so that forensic reports of DNA matches would function, in

---

about reductive protocols used in, for example, evidence-based medicine (Berg & Timmermans, 2000).

49. Readers familiar with the sociology of scientific knowledge may be reminded of the "equivalence postulate" advocated by Barry Barnes and David Bloor (1982) in their proposals for a "Strong Programme" in the sociology of scientific knowledge. Barnes and Bloor argue that sociologists of knowledge should give the same form of explanation to all instances of knowledge or belief, regardless of whether those beliefs presently enjoy the status of scientific truth. A more familiar formulation of the equivalence postulate is contained in Bloor's (1976) methodological proposals for "symmetry" and "impartiality." A series of parallel themes between the sociology of scientific knowledge (SSK) and the arguments by the defense in the O. J. Simpson trial is worked out in Lynch (1998). Note that in the *Adams* case, it is possible selectively to associate aspects of the defense and the appeals court judgments with SSK arguments.

effect, as verdicts.[50] Accordingly, the immense odds against the possibility that an innocent defendant's DNA profile would match the criminal evidence threatened to overwhelm (and thus "usurp") the jurors' assessment of all the evidence. The illustration of the Bayesian approach that Donnelly gave during his testimony showed that the reported odds against a defendant, even when as high as 200 million to 1, do not preclude a judgment of reasonable doubt about the totality of evidence. The juries in the two *Adams* trials apparently were not swayed by the demonstration, but the defense argued that the jury was inadequately prepared by the judge to use the method. From the defense's point of view, the Bayesian procedure, far from being a reification of expert evidence at the expense of common sense, was designed to give common sense a fighting chance when faced with the scientific authority and impressive probability figures associated with DNA evidence.

The Thwaites-Donnelly equivalence strategy was a matter of fighting fire with fire. Another equivalence strategy also can be imagined, even if one agrees with the court's pronouncement that "the apparently objective numerical figures used in the [Bayesian] theorem may conceal the element of judgment on which it entirely depends." To imagine this strategy, it is necessary to consider that the court's objection applies no less strongly to the prosecution's DNA evidence than it does to the defense's non-DNA evidence. As the *Adams* defense's first prong of attack suggested, the probability figure assigned to the DNA match glossed over a chain of judgments; different judgments yielded different probability figures, and the basis for the judgments was not always transparent. But rather than trying to develop a more complete and comprehensive quantitative measure, one could conclude that *any* probabilistic exercise is likely to be deceptive under the circumstances. This is what population geneticist Richard Lewontin concludes. Lewontin argues that, for at least two reasons, probability figures should be barred from jury trials involving DNA evidence. One reason is that it is impossible within the time constraints of a trial to sufficiently educate a jury to critically evalu-

50. Thompson & Ford (1988), Neufeld & Colman (1990) and Neufeld & Scheck (1989) sounded the alarm about the way DNA evidence was accepted without challenge in early cases. For a sociological discussion of procedural and statistical problems associated with early versions of DNA profiling, see Derksen (2000). A key point in criticisms of court treatments of DNA profile evidence is that such evidence should be treated as "reports" by particular agents and agencies rather than simple scientific facts (Koehler, 1996: 868).

ate probabilistic evidence. (According to Lewontin, it is difficult enough
for undergraduate students at an elite university to grasp the relevant
aspects of probabilistic reasoning after several months of study.)[51] The
other is that someone who *is* capable of critically evaluating such evidence, will appreciate the folly of trying to reduce complex judgments
about criminal evidence to a single probability figure.[52]

Dispensing with numerical estimates altogether presents its own hazards, however. Latent print examiners thus far have resisted giving numerical estimates when declaring matches between latent and rolled
prints. Instead, their declarations presume certainty, and when challenged about the lack of a quantified error rate for fingerprint comparisons, they insist that the error rate for the practice (if not for the practitioners) is zero. A quantified error rate (presumably above zero) would
at least give the jury some indication that the practice is not infallible.
Practitioners of other forms of traditional comparison evidence, such as
microscopic hair and thread comparisons, also provide the court with
non-numerical probabilistic judgments, but even though these are not
presented as absolutely certain judgments, they present problems of
their own.

For example, press coverage of the Larry Peterson case—a man
who was convicted of a brutal rape and murder that occurred in 1987—
contrasted the forensic hair comparison that helped convict him with the
DNA test that was used to exonerate him after he had spent more than
seventeen years in prison. Unlike the DNA evidence, the hair comparison was presented in the 1989 trial without any numerical estimates. Instead, Gail Tighe, a forensic scientist for the State of New Jersey, stated
in court that Peterson's hair "compared" with hairs collected with the
criminal evidence—which meant that they "fit the same criteria, fit into
the [same] range."[53] The hair evidence was retested with mitochondrial
DNA analysis in 2005, and the results indicated that the hairs recovered

51. Richard Lewontin, Harvard University, interviewed by Kathleen Jordan, 7 April
1998. Lewontin (1993) makes a similar point.

52. See our discussion in chapter 5, of Lewontin's (1992; 1994a) arguments about the
contingency of random match probabilities on assumptions about suspect populations.

53. This quotation is taken from the first part a two-part series, "The Exoneration of
Larry Peterson," presented by Robert Siegel and prepared by Julia Buckley and aired on
National Public Radio on 12 June 2007 (the second part, "Rebuilding a Life, Seeking Redress" was aired on 13 June 13 2007). The quotations from Gail Tighe (including the brack-
eted portion) are from the online version of the program, available at http://www.npr.org/

from the crime scene belonged to the victim rather than to Peterson. In addition, DNA testing of semen recovered from the victim's body did not match Peterson's evidence.[54] Although not presented as uniquely identifying, the assertion that microscopic examination showed that the suspect's hair and the criminal hair evidence "compared" was framed by the trappings of science and expertise, and apparently was convincing to the jury.

Somewhat weaker, but no less ambiguous, declarations are used when microscopic thread comparisons are presented in court. Toobin (2007: 31) quotes a forensic scientist working for Queens County, New York: "'We never use the word "match,"' [Lisa] Faber, a thirty-eight-year-old Harvard graduate, told me. 'The terminology is very important. On TV, they always like to say words like "match," but we say "similar," or "could have come from" or "is associated with."' '''Such qualified expressions do express differences in confidence, in the sense of the degree of certainty in *judgments*. The word "match," with its association with fingerprinting, signals bottom-line certainty, and was the focus of a highly publicized, albeit short-lived, ruling by Judge Lewis Pollak in 2002 (*United States v. Llera Plaza I*) that declaring a fingerprint "match" is prejudicial because of the connotation of absolute certainty. Qualifications such as "similar" and "could have come from" concede indefinite degrees of uncertainty in judgment and source attribution, but they are not arrayed along a single, stable formal scale.[55] Currently, such judgmental formulations

---

templates/story/story.php?storyId=10961075. Also see the summary of the case on the Innocence Project Web site: http://www.innocenceproject.org/Content/148.php .

54. According to the NPR story (ibid.), the original conviction was based on testimony (some of it later retracted) and circumstantial evidence in addition to the hair comparison. While acknowledging that the forensic evidence had been weakened, the prosecution resisted the defense's effort to seek a new trial, and even after a judge vacated the original conviction the victim's family remained convinced that Peterson was the perpetrator and the prosecutor sought to retry him. When vacating the conviction, the judge did not rule out the possibility of a retrial based on testimony and other evidence, and the prosecution immediately pursued such a trial while Peterson remained in jail. He was released on bail in August 2005, and the prosecutor dropped the case in May 2006. A timeline that chronicles the remarkable history of this case is available on the NPR Web site for the story (ibid.).

55. See Latour & Woolgar (1986 [1979]) for an attempt to relate the inclusion or deletion of "modalities" (qualifying words or phrases, and syntactic forms that express more or less certainty about the object of reference) to the historical trajectory in which a microbiological "fact" is constructed or deconstructed in scientists' discourse. Pinch (1985) deploys a similar analysis of linguistic expressions to describe formulations and negotiations over the degree of "externality" attributed to evidence of solar neutrinos. Lynch (1985a)

are contrasted with the precise, effectively certain, and objective status of DNA evidence. Indeed, cases of exoneration, such as the Larry Peterson case, are now used to calculate the error rates for hair comparison and other forms of forensic and nonforensic evidence. Unlike qualified verbal declarations, with their contingent relation to background circumstances, declarations of DNA matches and exclusions are treated as baseline facts. However, as our analysis of the *Adams* appeal cases suggests, the contrast between "DNA" and other forms of evidence comparison may not be so stark.

Both the Bayesian argument to quantify all the evidence in a case and Lewontin's argument to avoid quantifying probative value altogether challenge the distinction between "scientific" evidence which is legitimately stated in quantified form, and "nonscientific" evidence which should be left to the conventional, "common sense" judgments and deliberations of the jury. Although the *Adams* appeal court raised concerns about jury usurpation, its categorical distinction between scientific and nonscientific evidence obscured the possibility that its objections to the quantitative reduction of non-DNA evidence also applied to the DNA evidence.

To recall the story model mentioned at the start of this chapter, consider two intertwined types of story. One type places particular evidence items in what the court characterized as a "chain of evidence." The court rules that the jury is the appropriate authority for piecing together and assessing such stories, as the justices invest trust in the traditional deliberative process to work out what might have happened in the world outside the courtroom. Questions such as "How could the victim not recognize Adams as her assailant?" would be answered with accounts such as, "Well, it was two years later, after all, and she never did get a good look at the perpetrator." Evidence assimilated into this kind of story is largely a product of police statements, victim's testimony, and the accused's defense (as presented by his legal representatives), and it is framed by the jurors' knowledge of a common world. We could say that this genealogy of the evidence is produced through a crime story. Another kind of story is composed from expert witness statements. In the case at hand, this story reconstructed the chain of practices and judgments that

---

presents a more localized analysis of "objects and objections" in laboratory discourse that problematizes any effort to trace a clear epistemic trajectory based on linguistic forms, because of the shifting and contingent interactional circumstances in which formulations of fact and artifact are enunciated.

resulted in the declaration of a match and the associated probability estimates. It included testimony about how the evidence was collected, analyzed, and interpreted in material and circumstantial detail. Questions such as "Was it okay to use a marking pen to enhance the visibility of a faint band?" would be answered with accounts such as, "No—that allele should not have been included in the figure!" The Bayesian procedure attempted to integrate the two stories into one calculative chain, but the opposite movement seems at least as plausible: to de-compose the probabilistic calculations into narrative genealogies of evidence—both of possible actions in the world that produced the criminal evidence, and of possible actions in the crime lab that produced the DNA match and the probability figures associated with it.

Conclusion

On 26 June 2003, British home secretary David Blunkett announced that the 2 millionth DNA profile was about to be listed in the national DNA database. An article in the *Guardian* pointed out that the date of Blunkett's announcement happened to coincide with the one hundredth anniversary of George Orwell's birth. Blunkett was quoted as proclaiming, "Every week our national DNA database matches over 1,000 DNA profiles taken from crime scenes with names on the database. Around 42 percent of those matches are turned into detections within an average of 14 days. That is a huge achievement."[56]

In 2003 the NDNAD deployed different techniques, and was far larger, than the database that had been used by the Metropolitan Police more than a decade earlier to ensnare Denis John Adams. Nevertheless, the *Adams* case points to an issue that is likely to become ever more salient as criminal databases expand: "cold hits," or matches that turn up in database searches without corroborating evidence. Given the size of the database, it is expected that a proportion of these hits will be "adventitious matches" (coincidence matches between crime scene evidence and an innocent person on the database).

56. See Travis (2003). The article also reported that Blunkett gave a figure of 5.5 million sets of fingerprints in police files, adding that the development of a digital fingerprint registry allowed the police to scan fingerprints recovered from crime scenes and find out within minutes if they matched prints on file. The British Home Office prided itself with being at the international forefront of these innovations.

The appeal court's pronouncement in the first *Adams* appeal that "[t]here is no principle of law that DNA evidence is in itself incapable of establishing guilt," suggests that a judge can admit an uncorroborated DNA match, and a jury can convict on that evidence alone. In a later case (*R. v. Lashley,* 2000) the court of appeal clarified this point by ruling that DNA evidence alone is insufficient to convict a suspect unless supported by other evidence. A closer examination of the *Adams* case should indicate, however, that it would be virtually impossible to introduce a DNA match into the story of a case without bringing into play inferences about suspect characteristics, other possible suspects, locality of the crime, possible modus operandi, and many other possibilities. Information bearing upon guilt and innocence that is not *given* in the form of evidence, may be *given off* by the case description and the suspect's appearance and demeanor. In both the *Deen* and *Adams* cases, a Bayesian method was put forward by the defense to formalize and mathematize the way matching DNA evidence becomes significant, meaningful, and even calculable, in light of other evidence. Even if we do not go along with the particular method of formalization, we may appreciate the value of the lesson on the difference between matching DNA evidence and the establishment of guilt.

CHAPTER SEVEN

Fixing Controversy, Performing Closure

For a number of reasons, the DNA fingerprinting affair is an interesting case for studying the *closure* of a controversy.[1] This particular controversy was simultaneously legal and scientific, and its closure involved demonstrations of scientific *consensus* refracted through legal institutions, combined with the distinctive form of *settlement* imposed by judicial decisions. In other words, while the authority of science was crucial, legal deliberations were the primary forum for opening up questions about DNA and eventually closing the controversy. In this chapter, we look into the question of how closure was achieved. A default explanation is that the techniques were improved and their implementation became highly reliable. Consequently, while it remained possible for mistakes and misuses to occur in particular cases, the techniques passed the test of scientific acceptability—that is, scientific acceptability, viewed through the lens of the law. Without discounting the role of technique, we shall emphasize that closure was largely a bureaucratic or administra-

---

1. The term "closure" is used as a technical term in S&TS studies of controversy. The term also is used in the field of conversation analysis (CA). The two uses of the word are independent of one another. This chapter discusses closure in the context of S&TS, though some metaphoric purchase can be had by deploying the CA theme of "opening up closure" (Schegloff & Sacks, 1973). In CA the theme describes how participants work toward ending an ongoing conversation. This involves complex coordination, rather than simple signaling that one or both parties wants to end the conversation, and it involves conventional moves or steps that approach or set up the possibility of mutual closing. Although remote from closure of technoscientific controversies, the theme of opening up closing can be appropriated to describe the work that anticipates, and seeks to effectuate (or, as the case may be, resist), closure.

tive achievement. To say this does not demean the reality of DNA testing or the reliability of its results. Instead, it provides a specific locus for the credibility of DNA evidence that tends to be obscured when commentators speak of it as unassailable; as truth derived from science.

Standard histories of DNA fingerprinting describe the first few years after its introduction in the mid-1980s as uncontroversial until the *Castro* admissibility hearing in 1989.[2] That case touched off the "DNA wars" of the early 1990s, which culminated in the televised spectacle of the O. J. Simpson trial in 1994–95. Then, according to such histories, the controversy rapidly diminished with the publication of the second National Research Council report (NRC, 1996). Closer study of legal cases, publications in law journals, and other sources of information provide a more ambiguous picture. Problems with DNA typing were evident prior to *Castro,* and DNA evidence was contested (albeit unsuccessfully) in admissibility hearings as early as 1987. Moreover, criticisms continued to be published well after the NRC released its second report. However, *Castro* was the first successful challenge in U.S. courts to the admissibility of DNA evidence, and it had dramatic impact on public and legal evaluations of DNA testing. The second NRC report and its recommendations were frequently cited as grounds for ending debate about the reliability of DNA typing and the probabilities assigned to matching DNA profiles.

Many other things besides the NRC report were involved in the gradual shutting down of controversy in the late 1990s, and we will discuss some of them in this chapter. We shall treat closure as a fact, in the sense of a fait accompli that would take an immense effort to undo. That it is a fact, in our view, is uncontroversial, even if explanations of, and reactions to, that fact remain contentious. By the late 1990s, DNA profiling was widely regarded as having "passed the test"—and, in fact, it had passed tests of admissibility in many national and state court systems. Moreover, the controversy entered what we are calling a "postclosure" phase in which DNA profiling replaced fingerprinting as the gold

2. By standard histories, we are not referring to scholarly histories such as Aronson (2007), but to popular sources such as Levy (1996: 62ff.), science magazine editorials and essays such as Brown (1998) and Dawkins (1998), and official reports such as the NRC (1996) report itself. See Lynch and Bogen (1996: 58ff.) on "conventional histories": narratives that are constructed and recited by official investigative bodies and popular media sources. They include, but are not limited to, official histories, and sometimes they are bound up in struggles between narratives and counternarratives.

standard and became a basis for questioning the reliability of all other forms of criminal evidence, including fingerprinting.

Assuming that closure is firm and the controversy is now relegated to history, it is all too easy to portray DNA profiling in triumphant terms. An investigation like the present one, which began before the most intensive period of controversy and continued through its closure phase, has advantages over a retrospective analysis of an already closed controversy. Retrospective analysis can, of course, reconstruct what participants in a controversy knew or believed before the dust of history settled upon their dispute, but such reconstruction is likely to be less vivid than a study of a living, and as yet unsettled, episode. Not only can such a perspective counteract the retrospective illusion ("Whig history") of using present-day understandings as incorrigible grounds for interpreting the past, it can inform us about how a controversy is not just an event in time. As we shall emphasize, talk of controversy and consensus about DNA fingerprinting was *internal* to the prosecution of controversy and closure—the very idea that there was controversy was itself a subject of controversy.[3] Moreover, projections that the controversy would soon become closed were featured prominently in the practical and discursive production of closure.[4] This fact about controversy and closure was for-

3. An ever-more-familiar feature of public controversies in which scientific evidence plays a key role is that active efforts are made to *produce* and *promote* the existence of controversy, or, alternatively, to deny that there is (or ever was) controversy. So, for example, controversies over global climate change often take the form of "meta-controversies" over the very "fact" of controversy. It is not a simple debate among some scientists who say that global climate change is predominantly due to human burning of fossil fuels, and others who attribute the evidence to natural events and normal long-term fluctuations. Instead, it is a debate about whether or not there is consensus among the relevant experts about the existence and causes of climate change (see, for example, Naomi Oreskes' [2004] measure of the consensus on anthropogenic climate change, an intervention that itself proved controversial). One side claims that there is massive consensus, while the other argues that the experts have yet to resolve some crucial issues. Moreover, supporters of the "massive consensus" view argue that corporate-sponsored "think tanks" are the main instigators of the (apparent) lack of consensus, not only attributing vested interests to those who promote controversy, but also arguing that there has been a long-standing, well-funded, and concerted effort to obfuscate the results of disinterested scientific research by interest groups that fly the banner of "sound science" (for an entertaining chronicle of such efforts, see Rampton & Stauber, 2001).

4. Social phenomenologist Alfred Schutz (1964b) identified a property of social time that was simultaneously prospective and retrospective. To use our example, the theme applies to grammatical formulations of the form "the DNA fingerprinting controversy *will soon be* closed"—that it will undergo challenge for a time, experience bugs in the system, but eventually settle down to become a reliable and widely trusted method. Fingerprint-

malized in admissibility hearings, but it also appeared in media coverage, advisory panel reports, and more casual discussions.

In the sections that follow, we discuss the concept of closure in relation to the legal and scientific disputes about DNA typing, before outlining some of the technical, legal, and administrative "fixes" that operated effectively to end that dispute.

The Concept of Closure

The term "closure" has currency in science and technology studies (S&TS).[5] "Closure" is not a technical term, in the sense of being an unfamiliar word or an ordinary word with a specialized meaning. Participants in controversies easily grasp what "closure" means, and they deploy the concept or synonyms of it. For example, as we discuss later in this chapter, Eric Lander and Bruce Budowle used the headline "DNA Fingerprinting Dispute Laid to Rest" in their 1994 article in *Nature*. The article drew criticism for seeking to preempt further dispute with a premature declaration of victory, but there was little confusion about what the authors meant by the words "laid to rest." They meant that the controversy was over, finished, closed, though they acknowledged that the trial phase of the O. J. Simpson case (which was just about to get underway) would very likely give the public a misleading tutorial on problems and uncertainties associated with DNA evidence.

As we noted in chapter 2, the core set of individuals and relevant social groups that had a direct part in the prosecution and resolution of the controversy included, but was not limited to, technical experts from

---

ing in the early twentieth century was sometimes cited as a precedent for such projections. Pragmatically, such projections are more than simple forecasts, as they enjoin others to be confident about the outcome, and to "invest" in the future of the technique.

5. As discussed in chapter 2, H. M. Collins (1985) hitched an empirical program for studying controversies to philosophical arguments that Barnes et al. (1996) gloss as "finitist"—emphasizing the skeptical problems that arise from the inevitably "finite" amount of data supporting empirical generalizations. Accordingly, closure of dispute about such generalizations can never be entirely a function of data, and Collins argued that social agreements and exclusions are necessarily involved (also see Bijker et al., 1987, for an extension into technology studies). This conception of controversy was itself controversial. Select quotations from Collins became a focus of objections to his "empirical relativism." The most widely known sources are Gross & Levitt (1994), Sokal & Bricmont (1998), Gross, Levitt & Lewis (1996). For a more open, less dogmatic, exchange, see Labinger & Collins (2001).

several fields of scientific research and forensic practice. A broader set also included judges and lawyers, science advisory panels, criminal justice administrators, and science media writers and their audiences.[6] It was made up of a diverse assemblage of professional parties and institutional agencies. Consequently, the social organization of closure was not simply a matter of social and cultural influences impinging upon the technical choices made by a restricted set of scientists or engineers. Instead, legal precedents and rules of evidence furnished a basis for identifying controversy, consensus, and closure, and admissibility hearings, trial courts, and courts of appeal provided a locus of dispute in which experts were recognized *as* experts with relevant knowledge about whether DNA typing was "generally accepted."[7]

Judges, legal scholars, science advisory panels and expert witnesses articulated technical and administrative criteria under which closure was (or could be) achieved, and court decisions marked the historical fact of closure.[8] Popular, and also legally specific, conceptions of controversy and closure were thus embedded in a dispute that played itself out in the courts, the media, and in various science advisory groups. Key words in studies of scientific controversy—"science," "controversy," "consensus," and "closure"—were articulated and mediated by reference to statutes and legal precedents, and actions in court provided a locus for applying those terms in specific cases. Judicial rulings imposed a distinctive form of closure by settling disputed matters for later cases as well as the current trial. Consequently, a sociological attempt to study such a controversy encounters institutional discourse that is already infused with tendentious uses of "controversy," "closure," and related terms.[9] Moreover, any attempt to explain sociologically how closure occurred is secondary to the various tendentious explanations (including lay-sociological explanations) that featured prominently in the spectacle.[10]

6. This set included what Edmond (2001) calls "the law set," as well as the science media.

7. Many of the key decisions in U.S. state and federal courts occurred under the *Frye* "general acceptance" standard.

8. See Jordan (1997), Derksen (2003), and Aronson (2004) for dissertations addressed to the role of these advisory panels and working groups in developing technical standards for forensic DNA typing.

9. See Lynch & Bogen (1996: chap. 2) for a more elaborate discussion of how public media spectacles include historical and social "analyses" as constituent features.

10. Garfinkel (2002: 197) sometimes professes to be writing "tendentiously"—using a term at the start of an argument, the meaning of which will only become clear later. Our use of "tendentious discourse" is consistent with the more commonplace sense of partisan

Closure as Tendentious Discourse

Closure was more than a phase in an actual history of controversy, as it was part of the discourse that constituted the controversy and its phases. As might be expected, such discourse was tendentious: it was a partisan expression of hope, and it actively pursued resolution in accordance with specific objectives and interests. Even when the DNA wars were most heated, some commentators confidently asserted that the current problems would be resolved.

Predictions about the acceptance of DNA typing eventually were vindicated, though dissenters could argue, and did argue, that such predictions were partisan statements, and their vindication was less a product of independent confirmation of inherent truths than of a political settlement.[11] Instead of marking an end to controversy, announcements that the controversy was over can be viewed as rhetorical "phase work," a temporal variant of Thomas Gieryn's (1983) geographical metaphor of "boundary work." The most famous such announcement was the aforementioned 1994 article in *Nature,* by Eric Lander of the MIT Whitehead Institute, who had been one of the most prominent critics of DNA fingerprinting, and Bruce Budowle, the FBI's director of forensic laboratories and a proponent of the technique. The alliance expressed by the coauthorship was itself a conspicuous display of closure. The headline "DNA Fingerprinting Dispute Laid to Rest" differed from earlier forecasts that the dispute *would be* closed. Whereas confident forecasts projected an inevitable future, Lander & Budowle used the present tense to suggest that the dispute already was closed. Letters to *Nature* vigorously objected that their headline was tendentiously designed to invite acceptance of the "fact" it announced, thereby helping to effectuate that "fact."

---

or promotional talk, but like Garfinkel we also want to call attention to the way a "theoretical" term (in this case, "closure") is not simply reference with a meaning, but is a polemical instrument that is reflexive to a discursive situation. The *reflexive* place of sociological analysis and sociological explanation within the production of controversy and closure was vividly brought home when one of the coauthors of this book (Simon Cole) was subjected to an admissibility hearing in the case *New York v. Hyatt* (2001). See Lynch & Cole (2005).

11. Bruno Latour's (1999; 2004a) distinctive "take" on the politics of science makes punning use of the terms "constitution" and "settlement" to dramatize the connections between the scientific constitution of nature and the use of evidence to settle disputes and the familiar political senses of constitution as a foundational document and settlement as a compromise. (No doubt, Latour would be able to make creative use of the common expression "settlements" to describe Israeli-occupied enclaves in Palestine, as such settlements stand in the way of any compromise that would resolve a protracted boundary dispute.)

A crucial point of reference for the headlined announcement was a preliminary draft of the second NRC report (eventually published in 1996). Lander and Budowle cited recommendations in the draft report and announced that their implementation would assure the courts that persistent problems with laboratory error and probabilistic estimation were sufficiently controlled to justify the admissibility of DNA forensic evidence. Lander and Budowle (1994: 735) also predicted that the nationally televised O. J. Simpson double-murder trial, which was about to begin in a Los Angeles courtroom, would "feature the most detailed course in molecular genetics ever taught to the US people." Their mention of this public tutorial was ironic, and they added that it would be unfortunate if the attempts by the *Simpson* defense team to undermine the prosecution's DNA evidence were to encourage the public to think that such evidence was inherently problematic. Letters objecting to Lander & Budowle's article were published in later issues of *Nature,* and other letters were submitted but not published (Derksen, 2000). The letter writers included population geneticists Richard Lewontin (1994b) and Daniel Hartl (1994), who were well-known critics of DNA profiling. Lewontin (1994b) stated that "Lander and Budowle, in declaring the end of the controversy over the forensic application of DNA technology have presented a piece of propaganda that completely distorts the current situation in a very difficult matter at the nexus of science and law." And, as Lander and Budowle forecast, the *Simpson* trial featured extensive criticisms of DNA evidence, and while the defense made ineffectual efforts to challenge the "scientific" basis for the techniques, it effectively cast doubt about the Los Angeles Police Department's implementation of those techniques. Repercussions of the trial spread widely in the international forensics community, and police departments and forensic services within and beyond the United States made efforts to demonstrate that the Keystone Cops version of LAPD forensic practice dramatized by the *Simpson* defense did not apply to their districts. Nevertheless, with the 1996 publication of the second NRC report, and the continual adoption of DNA profiling methods and the development of DNA databases, talk of controversy quickly subsided.

By 1998, even Lewontin was willing to admit that the controversy was over. However, unlike Lander, who had conspicuously joined Budowle in preemptively announcing the end of the controversy, Lewontin did not relent from his earlier criticisms. In an interview in April 1998, he was asked, "Do you still believe that the controversy has not ended?"

RL. Well wait a minute. I still believe what I said there [in the 1994 *Nature* letter], that they [Lander and Budowle] present a piece of propaganda which doesn't confront the real issues, but do I believe as a practical political matter the controversy is over. My answer is yes, the controversy is over.

INTERVIEWER. Can you say more about that it's— it's over in terms of—

RL. The NRC rep— the second NRC report was designed to end the controversy.

INTERVIEWER. And it has.

RL. Oh sure. It has ended the controversy because the establishment has spoken.

INTERVIEWER. Right.

RL. I mean, look, anybody who is engaged over a long time in such struggles knows that usually you lose.

INTERVIEWER. emm hmm

RL. I mean the power of the state has been marshaled uh, to end the affair.[12]

Lewontin acknowledged the end of controversy, but only "as a practical political matter," and continued to maintain that the NRC evaded the "real issues" that he and others had raised. His stated position is close to that of social constructionists who explain the closure of scientific controversies by reference to politics and rhetoric, rather than the intrinsic nature of the objects in dispute (indeed, Lewontin is well-versed in political sociology and social studies of science). In the above excerpt, he discounts the intrinsic reliability of the way DNA profiling techniques are implemented, and emphasizes the role of political interests and official rhetoric in closing the dispute.[13]

In debates about DNA fingerprinting Lewontin was often cast as the most persistent and extreme critic among the prominent scientists who

12. The interview took place at Harvard University on 7 April 1998. Kathleen Jordan was the interviewer.

13. Unlike proponents of SSK, Lewontin does not attempt to give a "symmetrical" explanation of closure. That is, his emphasis on the interests and politics involved, and the failure to address the real issues, suggests that closure was both real and illegitimately achieved. Bloor's (1976) "symmetry" and "impartiality" postulates would have us treat politics and interests as "causes" of closure that do not necessarily imply falsehood or illegitimacy. An alternative approach propounded by Gilbert & Mulkay (1984) is to treat talk of interests and politics as a discursive "repertoire," alternative to a repertoire that emphasizes the "real issues"; with both repertoires being featured in the furtherance of controversy. Viewed in their terms, Lewontin's expression of resignation both prosecutes the controversy and acknowledges that it is dead.

took part in the dispute.[14] His acknowledgment that the controversy had
ended thus marks a strong point of consensus. However, unlike Lander,
who also had been a prominent critic, Lewontin makes clear in his acknowledgment that he maintains deep unease about the reasons for closure. While Lander and Budowle cite what they consider good scientific and organizational reasons for accepting forensic DNA evidence,
Lewontin suggests that power politics and propaganda forced closure in
the face of unanswered criticisms. His account of closure is consistent
with his earlier participation in the controversy. He acknowledges that
the controversy is over, but he does not abandon the arguments through
which he and others helped fuel the controversy. Nevertheless, like
Lander and Budowle, Lewontin acknowledges that closure is a historical
fact. The question that remains open is just what kind of "fact" is it?

   In the remainder of this chapter we address the "fact" of closure. Like
Lewontin, we construe closure as a practical and political matter, and
not an inevitable result of technical improvements. However, we also believe that changes in technique and the development of technical standards were crucial for ending what, in many respects, was a *legal* controversy. In our view, closure was not a direct result of technical or political
factors. Instead, technical "fixes" made it difficult, if not impossible, to
sustain many of the courtroom arguments that made trouble for forensic
DNA evidence in the early 1990s, and together with authoritative efforts
to set quality assurance standards for forensic science, technical changes
helped to persuade the courts that forensic DNA profiling was scientific
and thus admissible.

Technical, Legal and Administrative Fixes

Like many controversies, the DNA typing controversy did not reach discrete resolution. There was no crucial test, definitive legal judgment,
or moment of truth. The controversy concerned a whole nest of problems and uncertainties that were addressed in different ways at different
times. DNA profiling techniques changed throughout the period of controversy, and forensic organizations and consortia reacted to criticisms
by initiating reorganization schemes and making efforts to set techni-

14. William Thompson, whose background is in law and psychology, continues to publish articles on problems with DNA profile evidence. See chapter 8.

cal standards. Questions about the admissibility of forensic DNA profiling were prominent in the United States, but different state courts adhered to different standards and not all states moved in concert. The issues raised in cases such as *New York v. Castro* and *California v. Simpson* had international influence, partly because of the global circulation of articles in *Science, Nature,* and other publications, and the massive amount of national and international media coverage, especially of the *Simpson* case. While the NRC reports were directed to the problems in the United States, the reports and their recommendations had broader influence. Global influence did not always fan outward from the United States, however. DNA fingerprinting was first developed and implemented in the United Kingdom, and the British Home Office and Forensic Science Service were ahead of the curve in adopting the STR system, adjusting civil liberties to facilitate its use, and constructing a national database. The national database, and some of the legislation and policy that facilitated its development, provided a model for other nations to follow. Efforts were made to "harmonize" DNA profile and database standards in Europe, and other international efforts were made to coordinate forensic methods in order to track suspected "mobile" criminals and terrorists across national borders.[15] These international efforts to implement and standardize DNA profiling contributed to closure in particular localities by demonstrating that the technique was widely used and had become a fixture of many criminal justice systems.

Just as there was no single event that marked a point of closure, there was no single type of cause that was responsible for it. As we interpret the historical record, three interrelated "fixes" came into play: technical fixes, legal fixes, and administrative fixes. It may seem odd to use the vernacular term "fix" to describe what might otherwise be called "factors," "reasons," or "causes" responsible for shutting down a public controversy, but we have specific reasons for preferring this term. The colloquial expression "technical fix" refers to a mechanical innovation or adjustment that is offered as a solution to a problem that has broader political or personal significance. S&TS scholars tend to be skeptical of technical fixes and suspicious of the technocratic rationality implied by the idea that disinterested mechanical interventions can solve deeper

15. We are grateful to Peter Martin for hours of conversation in 1999 on the topic of European efforts to harmonize forensic standards, and for other information about the development of DNA profiling in the United Kingdom.

social and personal problems.[16] The term "technical fix" sometimes is used with ironic connotations,[17] implying a deceptive bait-and-switch strategy in which a technical adjustment is put forward as a solution to an interpersonal or political difficulty. The ironic use implies that the fix is an obvious or tangible surface action that masks a less obvious, more persistent, or perhaps inadmissible, reality. By adopting and broadening the idea of fixes we do not intend to endorse the idea that particular technical adjustments, legal stipulations, or administrative programs directly caused problems to be fixed and controversies to be settled. Instead, we suggest that these fixes frequently were mentioned in accounts of closure, and that frequent and felicitous mention of these fixes was part of the performance of closure. However, the fixes were not simply figures of speech or figments of a technocratic imagination. Without question, there have been numerous substantive changes in technique, procedure, and accounting practice. They are fixes in the way they circumvent, bypass, bury, displace, and effectively extinguish recurrent sources of controversy.

Technical Fixes

As noted in previous chapters, DNA profiling is a family of techniques. The earliest and most commonly used were the multilocus probe (MLP) and single-locus probe (SLP) techniques (often glossed under the RFLP, for restriction fragment length polymorphism, acronym). Both MLP and SLP were developed in the mid-to-late 1980s. The polymerase chain reaction (PCR) also was developed and commercialized in the mid-to-late 1980s. One of the earliest DNA profiling techniques to make use of PCR was the DQ-alpha system implemented by the FBI in the late 1980s. By

16. A partial exception to this tendency is Latour's (1992) treatment of "mundane artifacts" such as automatic seat belts, speed bumps ("sleeping policemen") and other devices that replace or reinforce rules, speed limits, and human enforcement of order. Latour treats such nonhuman devices as effective agents in systems of action, which do indeed fix specific problems with rule compliance and enforcement. Latour recognizes that such devices can be bypassed or disabled, but he credits them with an agency that is on the same plane as that of the rule-governed human actor or rule enforcer. Contrary to a reductionist model that explains systems of social action in terms of mechanical causes, Latour's actor-network theory deploys idioms of social action to "irreduce" systems in which mechanical agents are embedded.

17. The term "fix" also has colloquial uses that associate it with disreputable activities such as drug addiction (a "fix" of heroin) or an illegal deal.

the mid-1990s, the British Forensic Science Service had adopted the short tandem repeat (STR) system, which makes use of PCR as one of its constituent procedures. In addition, there was a gradual increase in the number of markers deployed in the SLP and STR systems, along with many other changes that enhanced the speed, lowered the cost, and increased the automation of the constituent procedures. By the end of the 1990s, forensic laboratories in the United States, United Kingdom, and many other nations had converged on the STR system. The number and selection of loci continued to vary, but cross-national efforts were made to establish a subset of standard loci.

As noted in earlier chapters, legal disputes developed around some of the distinctive features of MLP and SLP techniques, and the STR system circumvented some of the main points of uncertainty and dispute.

1. MLP: Sir Alec Jeffreys, the inventor of the MLP technique, gave it the name "DNA fingerprinting," thus drawing an explicit analogy with fingerprinting and individual identification. The markers used in this technique visualized a large, but indefinite, number of "bands" in an autoradiograph. The technique did not lend itself to precise statistical estimation, and the degree of individual identification was difficult to assess. The large number of loci marked by the probes resulted in a highly complex pattern. Ideally, the complexity of the DNA "fingerprint" enhanced the power of a match, but it also raised the likelihood of finding mismatched bands. As we saw in the *Deen* case (chapter 5), in order to "save" the evidence it was necessary to explain away the (alleged) artifacts, which provided openings for adversary attack. Assigning probabilities to matching bands also was problematic, and the procedure used in *Deen* appeared arbitrary.

2. SLP: This technique had the advantage of relative simplicity, and also (compared with MLP) more refined methods of probability estimation. With the addition of further markers, the technique became very powerful. However, problems that emerged in courts and science journals focused on methods of visual examination and statistical estimation. Criteria for assessing when bands in an autoradiograph matched or did not match and estimating the likelihood of matching (or nonmatching) bands proved contentious, as we saw in the *Adams* case (chapter 6). In addition, compared with PCR methods, SLP required larger amounts of bodily sample containing DNA, and the small amounts of crime sample often were used up during the analysis, precluding repetition or checking.

3. PCR DQ-alpha: This technique had the advantage of requiring very small samples, but because it employed a single locus it had limited power compared with SLP. Usually, it was used as a supplementary method or when the amount of crime sample was too small for SLP.

4. STR: This technique was said to have distinct advantages over the earlier methods. It was PCR based, and thus usable with small samples. This facilitated its use for criminal databases, because cheek swabs could be taken instead of more intrusive blood samples. In addition, profiles were digital rather than analog, and storable on electronic databases. Partly because of the scale of the U.K. database project, the STR system made use of automated procedures, such as ABI Prism Genetic analyzers, which used laser reading of capillary gels (fig. 7.1). Graphic output, which was said to give an exact measure of the number of nucleotide bases in specific STR sequences, replaced the older method of visually matching the bands (fig. 7.2). Moreover, the various

FIGURE 7.1. ABI analyzer. This device is a key component of the STR system. It uses a laser mechanism to "read" molecular samples run through gels or capillary tubes, thus replacing the procedure of visually assessing gels—a step in the procedure that was often criticized in court cases and critical articles in the science press. Photograph taken by Michael Lynch in 2002, with permission, at the Forensic Science Services Laboratory, Lambeth, London, U.K.

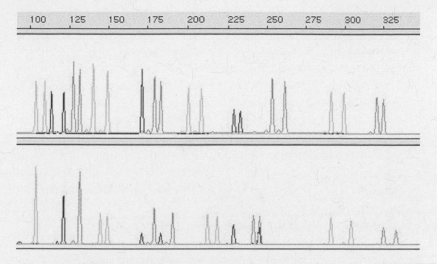

FIGURE 7.2. STR output. Unlike earlier systems, the STR system produces color-coded graphic peaks that are said to provide an exact measure of the number of base pairs in the alleles at each locus. Source: http:///www.fbr.org.

STR sites were selected so that each was spaced at some distance from the others in the array of human chromosomes (reducing the likelihood of linkage artifacts), and the number of sites was such as to greatly reduce the theoretical likelihood that more than one individual (with the exception of very closely related persons) in the entire human population on earth would have the same DNA profile (figs. 7.3 and 7.4).

There are different ways to view the technical fixes. One view, which is quite often expressed by proponents of DNA profiling, is that sources of uncertainty have been eliminated by a whole series of technical improvements. Statistical measures and reference databases have been refined, adequate precautions have been taken against erroneous inferences, and sources of human error and "subjectivity" have been eliminated through automation and standardization. Another view is that sources of uncertainty have been made more obscure, and thus more difficult and expensive to deconstruct in criminal trials. It is difficult to choose between these two views, but it seems clear that technical and administrative changes have worked effectively to foreclose many of the challenges to DNA evidence that had been raised in admissibility hearings, trials, and appeal cases. It also is clear that the technical fixes were framed by legal precedents and standards of admissibility.

   The STR system seems to eliminate the entire set of problems associated with visual matching of bands on autoradiographs. Arguments
about the relative alignment of bands across lanes, the use of arbitrary
correction factors for band-shifting, and suspicions about "subjective" visual inspection no longer seem salient when judgments are programmed,
and molecular weights of STR sequences are read automatically and visualized as discrete, color-coded, graphic peaks. While it very well may
be the case that personal judgment enters into the preparation of samples, the design of computer programs, and the calibration of instrumentation, such judgments no longer are transparently available for review
and challenge by adversary attorneys. Moreover, proponents of the evidence (usually the prosecutors) can appeal to the credibility of "mechanical objectivity" (Porter, 1995; Daston & Galison, 1992; 2007, chap. 3). It
is not impossible to contest such evidence, but the costs in effort, technical education, and money have become prohibitive for defense attorneys in routine cases. The ABI analyzers adopted by the British Forensic

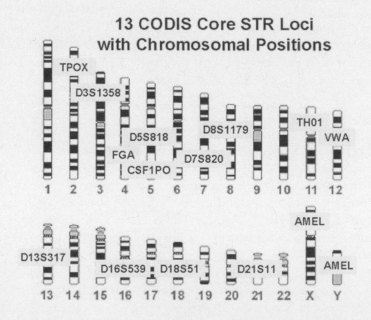

FIGURE 7.3. FBI CODIS STR loci. The FBI uses a system with thirteen STR loci, called
Combined DNA Index System (CODIS). The loci are distributed across the chromosomes
of the human genome in order to be distinguishable from one another, and are selected for
high heterozygosity, regular repeat units, and ease of PCR amplification. Source: http:///
www.fbr.org.

STR Marker	#Alleles	Random match probability (FBI Caucasian)
CSF1PO	11	0.112
FGA	19	0.036
TH01	7	0.081
TPOX	7	0.195
VWA	10	0.062
D3S1358	10	0.075
D5S818	10	0.158
D7S820	11	0.065
D8S1179	10	0.067
D13S317	8	0.085
D16S539	8	0.089
D18S51	15	0.028
D21S11	20	0.039
	Product	0.000000000000001683
	One in	594,059,679,247,540
		1 in 594 trillion

FIGURE 7.4. Probability table, CODIS. A sample calculation of random match probability of 1 in 594 trillion for a profile with thirteen CODIS STR markers. Source: http:///www.fbr.org.

Science Service (FSS) in the mid-1990s to work in concert with the STR system cost in excess of $200,000 each. In principle, defense attorneys had a right to use the services of the FSS to reanalyze samples, but this option was rarely if ever taken. Because of the expense, few independent services were available for testing and retesting criminal evidence, and so in the vast majority of cases "DNA" was a tool of the prosecution.[18] (It has, of course, become a widely publicized tool for the Innocence Project in the United States, but the use of DNA evidence by defendants is still rare compared with its massive use for criminal prosecutions.)

Technical fixes are deeply intertwined with legal decisions and administrative programs. For example, one change that is envisaged, and lobbied for, by the FBI has to do with the epistemic status of DNA matches and the protocols for reporting upon such evidence in court. Currently, courts in the United States, United Kingdom, and elsewhere require forensic scientists to give random match probabilities (RMPs)

18. Interview, Paul Debenham (London, 10 July 1996).

when presenting evidence of matching suspect and crime scene profiles. The calculation of RMPs and the protocols for presenting them in court have been sources of legal dispute and subjects of judicial review. Some proponents of forensic DNA evidence advocate a wholesale solution to those problems. Citing the current power of the technique, they suggest that a forensic expert can avoid quibbles about numbers and simply declare that, like the traditional fingerprint but for different reasons, a DNA profile is unique to an individual.

Budowle et al. (2000), for example, cite the power of the current CODIS system of STR profiling used by the FBI. This method uses a system of thirteen polymorphic loci (fig. 7.3). Budowle et al. argue that, except in cases in which an immediate family member of the suspect also is included in the pool of suspects, the probability of an adventitious match is so miniscule that it is virtually zero. For example, the combination of alleles listed in figure 7.4, when multiplied together in accordance with the product rule, delivers a probability of 1 in 594 trillion, a number many orders larger than the total number of humans that ever inhabited earth. Budowle recommends that DNA profiling has thus passed a threshold of individualization, so that there is no longer any need to confuse jurors with debates about statistics and population genetics. Were the courts to accept Budowle's recommendation, the whole set of debates associated with probability and population genetics would seem to be ruled out of play.[19]

Another wholesale fix has to do with the hypothetical expansion of DNA databases. Currently, databases include (or are supposed to include) samples from suspects and convicted criminals. The national database in Britain currently holds DNA profiles of samples taken from more than 4 million individuals. Although this is by far the highest proportion of a national population represented in such a database, it is still a relatively small proportion of a resident population of more than 60 million. As noted in interlude C, some judges and police officials in the United Kingdom have proposed expanding the national database to in-

19. Other forensic scientists do not agree with Budowle's recommendation. In an interview (March 2002), Ian Evett objected strenuously to the idea that profile matches could be formulated with a language of practical certainty. Evett also recommended that fingerprinting should be formulated probabilistically, because in his view as a proponent of Bayesianism, all evidence is probabilistic. However, he also mentioned that the British FSS had recently developed the policy not to declare RMPs lower than one in a billion. With the current number of matches being used, that would mean that the FSS will use the "one in a billion" figure in most cases, which would tend to avoid the statistical disputes.

clude all residents and visitors to the nation. It also has been proposed
that all newborn infants should be profiled.[20] Thus far, governments have
not endorsed such proposals, but if they were implemented, many of the
problems with statistical estimation would seem to disappear. A crime
scene profile would simply be run against the database, and if one and
only one match were found (and the database were assumed to be in-
clusive), then there would be no apparent need to calculate the probability of an adventitious match. Nor would it be necessary for the police
to engage in mass screening of local populations near a crime scene: the
population would already have been screened. Civil libertarian concerns
about mass screens, and the jeopardy risked by former criminals whose
profiles remain in a database, would no longer apply, though there would
be concern about the rights of citizens to opt out of a national database.

In addition to broader changes in technique, there have been countless
small changes in the design of more mundane tools and procedures for
handling, tracking, and analyzing samples. They include such devices as
bar-code stickers placed on evidence samples, and hand-held and door-
mounted bar-code readers (figs. 7.5, 7.6, 7.7); tamper-evident containers for evidence (fig. 7.8); and color-coded pipettes with disposable tips
(fig. 7.9). These quotidian tools grade into what we are calling "administrative fixes," because often they have to do with maintaining the integrity of chains of custody and implementing quality assurance/quality
control regimes. Many of them (along with architectural barriers, work
assignments, and organizational rules) are designed to protect against
the possibility of cross-contamination. Successful strategies of interrogation, such as those employed by Scheck, Neufeld, and other members
of the O. J. Simpson "Dream Team" of defense attorneys who worked so
effectively to embarrass the Los Angeles Police Department's criminalists and forensic analysts, provided vivid lessons on what to avoid for forensic labs worldwide:[21] Simpson's lawyers made much of the LAPD's
lack of a manual of standard procedures for handling DNA evidence, the

20. Alec Jeffreys has made such a proposal in the interest of civil liberties. See Jha,
2004. Jeffreys was quoted as saying that DNA criminal databases in certain parts of England would lead to over-representation of certain minority groups. Two possible solutions
to the problem were to delete records from suspects as soon as they are cleared and to extend the database to everyone in England.

21. An administrator with the British Forensic Science Service told us that the organizational and practical problems that were aired during the *Simpson* case were closely followed by members of his service (Chris Hadkiss, FSS, London, interviewed by Michael
Lynch & Ruth McNally, 10 February 1995).

FIGURE 7.5. Bar code sticker on evidence tube. Source: Forensic Science Service, U.K.

FIGURE 7.6. Bar code reader, above door. Photograph taken by Michael Lynch in 2002, with permission, at the Forensic Science Services Laboratory, Lambeth, London, U.K.

FIGURE 7.7. Hand-held bar code reader. Photograph taken by Michael Lynch in 2002, with permission, at the Forensic Science Services Laboratory, Lambeth, London, U.K.

FIGURE 7.8. Evidence bags. Photograph taken by Michael Lynch in 2002, with permission, at the Forensic Science Services Laboratory, Lambeth, London, U.K.

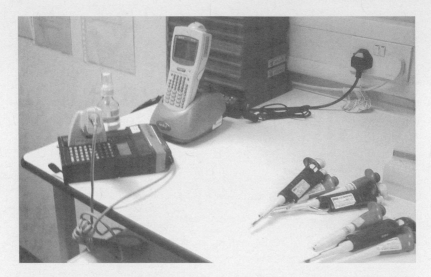

FIGURE 7.9. Color-coded pipettes with disposable tips. Photograph taken by Michael Lynch in 2002, with permission, at the Forensic Science Services Laboratory, Lambeth, London, U.K.

apparent irregularities in the way evidence was collected, stored, and handled, and the "gaps" between mundane protocols (for changing disposable gloves with each sample handled; writing the criminalist's initials on each "bindle" of evidence collected, and so forth) and documented practices.

Administrative Fixes

Administrative fixes are continuous with technical fixes. The difficulty of making a clear distinction between the two indicates the extent to which idealized experimental protocols for controlling scientific inferences overlap with the rational procedures attributed to modern bureaucracy. Indeed, Robert K. Merton's (1973) well-known and much-criticized scheme of the "ethos of science"—consisting of four institutional norms, communalism, universalism, disinterestedness, and organized skepticism—draws heavily on Max Weber's (1968) well-known and much-criticized ideal type of bureaucracy. At a more practical level, advisory groups such as the National Research Council and the FBI's Technical Working Group on DNA Analysis Methods (TWGDAM) addressed many of the technical problems with forensic DNA analysis by recommending administrative procedures. These were grouped under "quality control" (QC) measures designed to ensure that DNA typing and interpretation meets specified standards, and "quality assurance" (QA) regimes taken by laboratories to "monitor, verify, and document" their performance (NRC, 1996: 24). One important recommendation in connection with QA was proficiency testing (both "blind" and "open") of laboratories and specific personnel. One can think of QA/QC as both an administrative regime and communicative protocol: as an administrative regime it covers the architecture of laboratory space, the design of equipment, the training and testing of personnel, the implementation of a dense network of procedural rules and electronic monitors; as a communicative protocol it covers internal communications within staff hierarchies and, at least as important, the "assurances" of reliability and "best practice" that are passed to the courts along with analytical results.

Perhaps more obviously than technical fixes, administrative fixes specifically address arguments that have been made in court. "Standards" is one of the broadest categories of administrative fix, and one type of standard—technical standards—is both technical and administrative. Aronson (2005a) describes early efforts by the FBI TWGDAM group to develop standard protocols for DNA typing that would guide local forensic laboratories. These included specifications of the set of restriction enzymes, probes, and DNA sites featured in the single-locus probe (SLP) technique deployed in the late 1980s and early 1990s. Other advisory bodies such as the now-defunct Office of Technology Assessment (OTA) and the NRC also weighed in with standards for assigning probabilities to DNA matches. The first NRC report (1992) presented a "ceiling principle"—a way to develop supposedly conservative estimates of the likelihood of a coincidence match between a suspect's and perpetrator's profiles.[22] Many of the features of the STR system discussed above

22. As noted in the second NRC report, the ceiling principle proved to be the most controversial recommendation in the earlier report: "The ceiling principle calls for sampling 100 persons from each of 15–20 genetically homogenous populations spanning the racial and ethnic diversity of groups represented in the United States. For each allele, the highest frequency among the groups sampled, or 5%, whichever is larger, would be used. Then the product rule would be applied to those values to determine profile frequency" (NRC, 1996: 35). The 1992 report also recommended an interim ceiling principle to be used before the ceiling principle was established. This was a matter of calculating "the 95% upper confidence limit of the frequency of each allele . . . for separate US 'racial' groups and the highest of these values or 10% (whichever is larger) should be used. Data on at least

under the rubric of technical fixes incorporated "standard" reagents, probes, tools, representational formats, and so forth, that were built into the technique itself, and which were designed to enhance the credibility of the results. Black boxing techniques for detecting and displaying matching alleles, selecting alleles so as to reduce linkage artifacts, and multiplying the number of probes used in a multiplex system all worked together to enforce standardization and to bypass apparent sources of "subjective" variation.

With the rapid increase in cases involving DNA evidence, and especially with the development of national databases, DNA profiling was conducted on an industrial scale. In comparison with the "cottage industry" style of work in a small university biology laboratory, the larger forensic laboratories handled large batches of samples, and converged upon standard equipment, reagents, and protocols. Often obscured by invidious comparisons between "real" science and criminal forensics was the fact that the forensic laboratories deployed a relatively stable complex of equipment, reagents, and protocols for routine casework. Moreover, because errors in forensic analyses (and also in diagnostic testing) are associated with severe costs and liabilities, monitoring regimes are much more prominent in such laboratories than in a university lab, where it is often assumed that mistakes will be ironed out in a trial-and-error process (see Jordan, 1997; Jordan & Lynch 1992; 1993).

Standardization applied to the credentials and training of practitioners, as well as to their practices. The FBI's TWGDAM developed and disseminated standards of practice and practitioner training. In addition, the work environments where forensic DNA analysis was performed were designed and outfitted with a dense network of devices to protect the techniques and their results against established criticisms about sources of contamination, sample mix-up, and mislabeling. Walls were built and labeled with signs to segregate reference samples, crime samples, and suspect samples (fig. 7.10). Rules were affixed to signs posted on walls, and reminder labels were affixed to equipment. Laboratory clothing and tools such as pipette tips were color coded and dedicated to particular types of samples (fig. 7.11). Disposable tips for swabs and pipettes were designed to discourage handling and reuse, thus reducing contamination (figs. 7.12, 7.13), and evidence bags were designed to be tamper-evident.

_____

three major 'races' (e.g., Caucasians, blacks, Hispanics, east Asians, and American Indians) should be analyzed" (NRC, 1992; quoted in NRC, 1996: 35).

FIGURE 7.10. Segregated area for "reference" samples. Photograph taken by Michael Lynch in 2002, with permission, at the Forensic Science Services Laboratory, Lambeth, London, U.K.

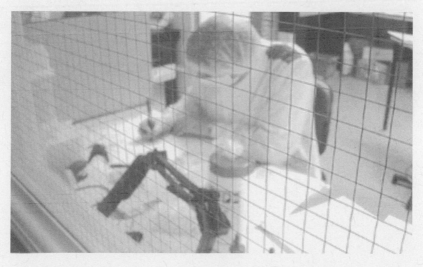

FIGURE 7.11. Color-coded lab coats and equipment. Photograph taken by Michael Lynch in 2002, with permission, at the Forensic Science Services Laboratory, Lambeth, London, U.K.

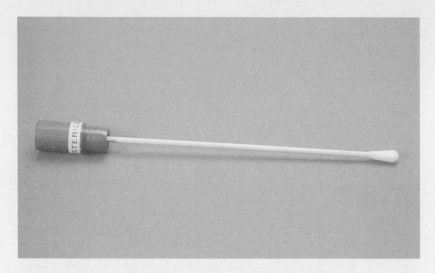

FIGURE 7.12. Evidence swab. Source: Forensic Science Service, U.K.

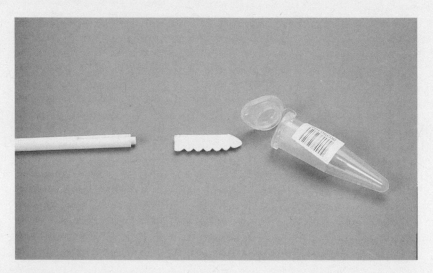

FIGURE 7.13. Evidence swab with detachable tip, and container. Source: Forensic Science Service, U.K.

Steps in chains of custody were witnessed in accordance with older bureaucratic and newer automatic mechanisms: clerical signatures and the
ubiquitous bar-code recordings that stood proxy for them.

It would be an exaggeration to say that these various mundane devices
and practices were responsible for closing the controversy about DNA
profiling. Perhaps it would be more accurate to say that they worked together, and with other legal, technological, and procedural changes, to
*fore*close further controversy. They acted as ostensible evidence that
quality control/quality assurance regimes were in place, and while they
did not offer infallible guarantees against contamination, sloppy work, or
forensic malpractice, they were not mere signs or even rules designed to
assure the courts that all was well with DNA typing. Sequestered space,
color-coded equipment, tamper-evident bags, bar-code scanners, and so
forth, were material things integrated with the production of DNA evidence. There is nothing intrinsically scientific about a bar-code sticker or
a color-coded lab coat—these are organizational things that have widespread use in a variety of commercial and government activities. They
are "rational" in the Weberian sense of rational bureaucratic administration: normalized, hierarchically supervised, standardized, iterated,
coded, recorded, and subject to review. The relevant sense of "objectivity" associated with these things also is administrative, bureaucratic.
From histories of Western concepts and practices of objectivity, we learn
of a trend toward divesting truth and trust from the sovereign individual
and emplacing it in material and bureaucratic mechanisms. "Mechanical objectivity" trumps personal judgment, and "quality assurance"
mechanisms overshadow trust and personal testimony. It is easy to exaggerate the extent to which personal trust and testimony are overridden
by bureaucratic mechanisms of assurance. It is not so much that personal
and interpersonal acts of judgment, witnessing, testifying, entrusting,
and so forth, completely leave the picture; instead they become compart-
mentalized in a detailed division of labor, and distributed within and between the interstices of the paper trails. A signature is not only a sign, it
is something *signed* by somebody who can be called to testify. A chain
of custody is thick with trust, and with potential avenues of evasion, even
though most agents are rarely summoned to testify before a court.

Administrative objectivity is not identical with mechanical objectivity—the consignment of judgment and trust to automated devices
(for example, replacing the competency and honesty of a bank teller
with the reliable operations of an ATM; replacing hand-signed and

manually counted election ballots with touch-screen machines that automatically record votes). It includes many such mechanisms, but adds a layer of "quality assurance" and administratively produced "verification" or "validation." The slogan used by critics of touch-screen voting machines—that they should include a paper receipt as a "voter-verified audit trail"[23]—expresses a demand for this administrative "layer" that stands guard over the operations (as well as the malfunctioning and systematic misuse) of the mechanisms.[24]

Although they were important for closing controversy about DNA profiling, administrative fixes offered their own potential for creating ever more trouble to fix. A recent newspaper story underlines this point, announcing that "[t]housands of people could be accused of a crime they did not commit as a result of errors in records on the national DNA database" (Cockcroft, 2007). The article quoted a spokesperson for the National Police Improvement Agency (NPIA), an agency responsible for managing the database, saying that the "Data Quality and Integrity Team" discovered "1,450 demographic discrepancies" between January and November 2007. Many of these were mundane administrative and clerical mistakes, such as spelling errors and wrongly recorded dates and police codes. Such mundane mistakes are not necessarily represented in proficiency tests or other audits that focus on laboratory procedures, and they are not the sort of errors that preoccupy population geneticists and statisticians. The NPIA claims to have "rectified" these mistakes, but acknowledged that further errors could still be made.

Legal Fixes

Legal decisions were highly significant for opening, closing, and foreclosing controversy about forensic DNA profiling. The most obvious, and perhaps the most important, legal fix was the gradual acceptance of DNA profile evidence by federal and state courts in the United States, and courts throughout Europe and many other parts of the world. Especially in the United States, in the wake of publicity about *Castro*, con-

23. See Help America Vote Act, Public Law no. 107–35 (29 October 2002, 116 Statutes, codified at 42 U.S.C. §15301 et seq., at 42 U.S.C. 15481. For a discussion of voting machinery, see Lynch, Hilgartner & Berkowitz (2005).

24. Cambrosio et al. (2006) coined a term, "regulatory objectivity," that also identifies an infrastructure of bureaucratic rules and procedures that rationalizes local practice without tracing back to a coherent, cognitive rationality.

tested admissibility hearings were portrayed as a battleground for legally "testing" DNA typing. Many of the scientists who debated the issues in *Science* and *Nature* also appeared as expert witnesses or consultants in some of the early admissibility challenges. Admissibility hearings were requested in federal cases, in each state, and often repeatedly in a given jurisdiction whenever new DNA profiling techniques were introduced. They also were requested to challenge the particular evidence in a case.[25] The accumulation of favorable decisions (and the vast majority were favorable) provided a basis for further decisions in favor of admissibility. The fact that DNA had "passed the test" in previous trials enabled it to pass further tests in the courts, eventually to be accepted in many cases without significant challenge. Admissibility continues to be challenged in occasional cases, but the accumulation of favorable decisions tends to narrow the avenues of attack to the particular evidence used in the case rather than to the generic type of evidence.

Acceptance in the courts went hand in glove with *entrenchment* in criminal justice systems. The expansion of databases from regional, to national, and to international levels of scale was facilitated by partial standardization of technique, and also put pressure on local forensic authorities to harmonize with the emergent system. By the mid-1990s DNA evidence became, as they say, a fact on the ground. This combination of legal and technological momentum made it increasingly difficult to argue that such evidence lacked general acceptance or was unreliable. The massive use of DNA profile evidence, and the enlargement of databases to include millions of samples, provided a growing mound of statistics that were cited to support and justify the continued use of evidence and to identify technical problems to be overcome. Even well-publicized cases of error could be used to support the progressive growth and development of the technique.

For example, a case in the United Kingdom involved a man named Raymond Easton, who lived in Swindon, England. In 1997 his DNA profile was placed on the national criminal database following a "police caution" in connection with a domestic dispute. In 1999 he was charged with breaking and entering a house in Bolton, nearly two hundred miles north of where he lived. The sole evidence was a cold hit between his

25. Accounts of the narrowing of avenues for challenge were given to Michael Lynch and Simon Cole, in interviews with Peter D'Eustachio (22 May 2002); and Barry Scheck & Peter Neufeld (21 May 2002).

DNA profile and blood evidence left at the scene, apparently by the burglar. The forensic analysis identified six matching STR loci, and declared an RMP of 1 in 37 million. The police charged Easton with burglary, despite some striking anomalies: Easton suffered from severe Parkinson tremors and could not drive, dress himself, or walk for more than a short distance; and he had an alibi that other family members could corroborate. Although, according to his solicitor, he had never traveled north of Birmingham, the police tried to discount his evidence and to imagine scenarios in which he secretly traveled to Bolton, and perhaps with the help of an accomplice, managed to hoist himself through a kitchen window. The case went on for months, until he was exonerated after the prosecution retested the samples with a technique using 10 STR loci.[26] Although this story illustrates how DNA matches can misidentify a suspect, it also illustrates how the progressive improvement of the technique can overcome the limitations of earlier variants. It is a story of progress as well as error.

At a more specific level, court decisions have both facilitated the implementation of DNA profiling and "fixed" actual or potential windows of controversy. A small sample of legal facilitations and fixes include the following examples. The 1994 Criminal Justice and Public Order Act was a broad effort by Parliament to revise the criminal code. It included a rather interesting revision that was geared to the implementation of DNA databases. This revision had to do with what sociologist Erving Goffman (1971) once called "territories of self," or in this case the more concrete territories of the body and the authorization needed for the police to encroach on those territories. The mouth had previously been defined as an "intimate" area, which meant that the police needed to get written authorization from a magistrate before they could enter a person's mouth and collect evidence without that person's permission. Borrowing a provision from earlier legislation applying to terrorism investigations in Northern Ireland, the 1994 act redesignated the mouth as "nonintimate" for purposes of evidence collection. This change happened to coincide with the British government's plan to build a national DNA database. The plan was to use the STR system to build searchable databases from convicted criminals, suspects, and crime scenes for

26. See Mnookin (2001: 50). This case received heavy press coverage. See, for examples, Willing, 2000; Chapman & Moult, 2000. Also see British Broadcasting Company, *File on 4*, BBC Radio 4 (4 July 2000).

a broad array of recordable offenses. Unlike earlier techniques, which required a relatively large sample, the STR system was said to be capable of analyzing small traces of DNA. Specifically, a cheek swab yielded an adequate sample, and so it was unnecessary to draw blood from the person being tested. In many respects, it was an easier system for police investigations.. This legal revision thus responded to the development of the STR technique, while at the same time it facilitated the wholesale use of that technique by the police.

Another example, or rather a set of examples, has to do with statute-of-limitation laws in the United States. This ability to recover precise identity information from old and degraded bodily traces has been crucial for the ongoing effort of the Innocence Project to overturn criminal convictions (Scheck et al., 2000). Its advantages also have not been lost on prosecutors, and changes in statutes of limitations have facilitated the effort to keep criminal investigations alive for longer periods than had been permitted under many state laws in the United States. Legislation to modify statutes of limitation, and prosecutorial practices such as "John Doe" warrants (warrants for the arrest of the unknown source of a DNA profile), are not legal fixes of sources of controversy about DNA evidence, as much as they are legal reactions to the technical repute lately assigned to such evidence.[27] Both the statutory exceptions made for DNA evidence, and the "John Doe" strategy for working around existing statutes of limitations point to the way laws and legal practices anticipate and help to facilitate novel technological potentials.[28] Like the British reclassification of the mouth as a nonintimate bodily region, the reworking of statutes of limitation in several U.S. states is an example of a reflexive historical process (a "co-production," as Sheila Jasanoff prefers to call it)[29] in which law and technology are worked and

27. According to standard accounts of PCR, it employs enzymes and reagents involved in cellular reproduction to "amplify" the initial amount of DNA within specified regions of the genome. In principle, PCR can work with an initial sample containing only a few molecules of DNA. It also can work with partially degraded DNA samples (thus having considerable value for archaeological, and even paleontological, analysis). Consequently, stored physical evidence, such as blood samples or semen-stained clothing from decades-old criminal investigations—evidence that would not furnish a sufficient quantity or quality of analyzable material for older forms of DNA typing—can be analyzed using the STR system.

28. See Alsdorf, 1999.

29. See Jasanoff (2004). Latour (1987) first coined the term "coproduction," and the term "coconstruction" also is common in S&TS circles. Either term has an advantage over the term "coevolution," because they are terms of action. The reciprocal relations between

reworked, retrospectively and prospectively, in reciprocal relation to each other. In addition, the aura of science is in the picture, because courts and legislators often seem inclined to treat DNA evidence not simply as an effective criminological tool, among other such tools, but also as an exceptional source of objective truth that transcends all other forms of evidence, which by contrast are deemed subjective, traditional, and merely credible.

Further legal fixes resulted from judicial stipulations. For example, in the *Adams* case discussed in chapter 6, the court of appeal stipulated that it was appropriate to present "scientific evidence" but not "common sense evidence" in a quantified probabilistic form. This decision reinforced the exceptional status granted to DNA evidence, and protected the impressive probability figures assigned to such evidence from the leveling strategy offered by a Bayesian procedure. The British and American courts also have consistently ruled that match probability figures should be presented separately from estimates of practical error rates. Jonathan Koehler (1996) and William Thompson (1996) have argued against such segregation of probability figures, by pointing out that error rates are likely to be much higher than currently estimated rates of finding a coincidence match for a DNA sample. Accordingly, the higher rate should be used as the effective measure of the likelihood that an innocent suspect's DNA would be alleged to match the crime sample. In the courtroom and press, however, the lower rate (the estimate of how likely a randomly chosen, unrelated individual's DNA profile would match a given profile, under the assumption that there was no error or fraud in the collection or analysis of the evidence) is invariably given as a "scientific" estimate based on population statistics and genetic assumptions. Results of proficiency tests and other measures of practical reliability are the stuff of social science; subject to variations in test conditions, vulnerable to the experimenter effect (subjects reacting to test conditions), and dependent on vernacular definitions of terms such as "error." It seems clear that if error rates were used as the effective measure for assessing DNA matches, they would appear a great deal less impressive and more variable. Legal stipulations thus have helped protect the presenta-

---

novel technology and law do not simply happen; they involve deliberate efforts in technological design to take account of (and sometimes to bypass) laws and regulations, and, in turn, the latter are frequently reworked in reference to, and in anticipation of, technological change. For vivid examples in the domain of electronic recording devices and copyright law, see Gillespie (2007).

tion of DNA evidence in terms of idealized probabilities. This does not mean that measures of error are rendered irrelevant in a criminal case, but that impressive, "scientific" probabilities are given a privileged place in the courtroom, and the burden shifts to the defense to demonstrate that error may have occurred.

A related form of judicial boundary work was mentioned by William Thompson in an interview. During the interview, Thompson reminded us of an early article in *Nature* by Eric Lander (1989), in which Lander distinguished between two broad types of problems with DNA typing: one involved laboratory practices and interpretations and the other involved estimates of the frequency of specific profiles in the population. Lander treated the second (statistical/population genetics) type of problem as more challenging, and predicted that the procedural/interpretive problems would be solved earlier through the promulgation of standards and improvements in technique. When reminded of this in the interview, Thompson exclaimed, referring to the problem of laboratory error,

wt. Oh it's still not resolved. I mean, I don't think it's ever been addressed, ever. I mean people *pretend* that it's been addressed. But, but what happened is that . . . in the process of appellate litigation, the appellate courts decided early on— there was some opposition, I mean there's some people that argued that all these issues just went to the weight of the evidence and were not admissibility issues. So you know the (. . .) camp thought everything was an issue of weight for the jury.

ml. Yeah.

wt. Uh right. Until it got to the jury trial and then anything having to do with fundamental issues was you know no longer addressable there either (laughs). But, but I mean, there's some really interesting papers put up by prosecutors on how one evaluates science and sort of at the admissibility stage, you can't, you can't, . . . you know— at the admissibility stage you can't raise any of these issues because they're weight issues, but when you get to the jury you can't raise any of those issues because they're admissibility issues, so you're never able to raise them.

ml. Oh, that's very nice (laughs).

wt. So but, uh, so there were some— there were some who argued from the beginning that all of those issues are not—, should not be of concern to the appellate courts, the issue should just go to the jury. But the appellate courts decided early on, by and large, that the issues having to do with population genetics were admissibility issues and therefore, uh, those

became, uh, threshold issues that had to be resolved before the DNA
evidence could even be used as admissible evidence. Whereas the issues
having to do with the adequacy of the inferential procedures were issues
of weight that could be argued in front of the jury.[30]

The expression "weight issues" refers to issues that are consigned to the
jury (or judge, in a trial before a judge alone) to resolve, and not subject
to an admissibility challenge. Thompson refers to a point he made in a
publication ten years before the interview about the tendency for courts
to treat disputes about procedures for interpreting DNA matches as a
matter of weight rather than admissibility (Thompson, 1993: 49). In that
article, he mentions that, with the exception of one case, courts ruled
that disputes about procedures and criteria for determining when a suspect's profile matches that of the criminal evidence are matters particular to the case (and thus of the weight of evidence to be assessed in trial),
that do not pertain to admissibility.[31] In other words, under the *Frye*
"general acceptance" rule, alleged errors in procedure and interpretation were deemed to be exceptional matters to be addressed in trial, and
not matters pertaining to "general acceptance." However, as Thompson
also discussed in the interview, defendants were often precluded from
challenging the *general* adequacy of the prosecution's procedures for determining matches, unless they could show that specific errors may have
occurred in their specific case.

A related stipulation was made by the judge in the O. J. Simpson trial,
as well as many other trials. Partway through the trial, the *Simpson* defense team submitted a legal brief that requested a *Kelly-Frye* hearing
(an admissibility hearing under the *Frye* "general acceptance" standard,
supplemented with a California case, *People v. Kelly*). Usually such hearings are held prior to the trial, and this procedural irregularity was one
reason the judge gave for refusing the request. The request was a lengthy
document that spelled out numerous problems with DNA evidence in
general, and with the particular handling of the evidence in this case
(Lynch, 1998). Among the arguments made in the document was that estimated error rates should be combined with the probability of finding
a random match. The rationale for combining them was that the probability of a false positive match depends both on the frequency of the

30. William Thompson, interview with M. Lynch and S. Cole, Irvine, CA, 5 May 2003.
31. The exceptional case was *People v. Halik* (1991).

given profile in a specific population and the probability that the match resulted from a procedural or interpretive error. In addition, procedures for calculating error rates are less well established, and subject to more obvious uncertainty. Such uncertainty would be injected into the combined figure. The court, and later the second NRC report (1996), upheld the prosecution's procedure of separating the measures.[32]

Judicial stipulations also resulted in burden of proof being shifted to the defense to show that standard probability estimates based on comparisons with broadly defined national and "racial" population samples do not apply to the case at hand. So, for example, the burden is on the defense to establish that a defendant's brother, father, or half-brother should be considered as a possible suspect, or that there are other closely related members of the defendant's extended family in the local neighborhood (Redmayne, 1995). In the absence of such arguments, the estimates use standard tables of allele frequencies for one or another broadly defined population sample, corrected "conservatively." If the burden were on the prosecution to outline a specific suspect pool for estimating allele frequencies, the whole practice of using such estimates in court would quickly become unworkable. Judicial stipulations thus fixed a potential "methodological horror" for the prosecution.[33]

Judicial boundary work thus resulted in separating domains of controversy about DNA profiling and allocating them to different phases of the case. When paired with the recommendations of the National Research Council, admissibility decisions (even in the highly contested *Castro* case) depicted forensic DNA testing as "generally accepted" in the scientific community, even while acknowledging particular areas of uncertainty and dispute about the composition of relevant reference populations, procedures for determining matches, and methods for assessing practical error rates. The problematic issues were separated from the general credibility of DNA testing. The courts and advisory panels also accepted standard procedures for estimating the probability of false-positive results, and allowed them to be presented separately from

32. Thompson, who was a member of the *Simpson* defense team, in our May 2003 interview with him stated that the second NRC report effectively killed any argument for combining random match and error probabilities, and added, "I think people were frightened by the arguments in the *O. J.* briefs."

33. "Methodological horror" is a term used by Steve Woolgar (1988) to describe a set of general, ultimately intractable, epistemological problems facing proponents of empirical evidence. Here, we use this evocative term in a more restricted way to refer to a rather specific, potentially intractable problem that is foreclosed by judicial stipulation.

error rates (if available). The burden fell on defendants to argue in terms of the particular circumstances of the case that laboratory errors might have been made, or that a closely related suspect might be at large.

Conclusion

At the start of this chapter we stated that we accept as a fact that the DNA typing controversy is closed, but that we needed to clarify just what kind of fact we are talking about. Certainly, it is a *socially constructed* fact: the very fact that earlier problems with laboratory error and statistical estimation *appear to have been solved* is crucial for closing off many avenues of legal controversy.[34] However, this fact is more than a widespread belief that guides further inference and action in a social system. It also is a "fact on the ground" in the sense of being materially entrenched in a way that resists change. But more than mere entrenchment are the many large and small innovations in technique, organizational routine, and accounting practice discussed in this chapter. These are not simply rules, ritual behaviors, or habits, as they involve architectural spaces and barriers, recording and packaging devices, protocols, and molecular biological techniques. Many of the (alleged) sources of uncertainty that arose in cases such as *Castro, Simpson, Deen,* and *Adams* were bypassed or black boxed by changes in forensic technique, administration, and law. This does not mean that they were solved in some final way, or that no avenue remains open for adversary dispute. Instead, a DNA match becomes strongly presumptive, imposing a concrete form of burden of proof on an opponent who would hope to attack its credibility. DNA matches can be attacked, and continue to be attacked in se-

34. Philosopher John Searle (1995) distinguishes "institutional facts" from what he calls "brute facts." Institutional facts are socially constructed, whereas brute facts are independent features of the world. A prototypical institutional fact is the price of a commodity, which depends upon historically and socially situated market conditions, ascriptions of value, and even collective frenzies and panics, whereas an example of a brute fact is the height of Mount Everest. As we understand it, the fact of closure is "institutional" (legally accredited, administratively managed), and the *fact* of closure is inseparable from the *belief* that the controversy is over and that a DNA profile is objective, virtually certain, evidence of individual identity. However, this fact is by no means ephemeral, as it is supported, implemented, and entrenched in a dense network of instruments, molecular biological methods, and mechanical systems of accountability.

lected cases, but it would be hopeless to argue that DNA typing results, in general, are doubtful.

The very idea that DNA evidence is generally accepted in law as well as science discourages efforts to challenge it. Technical, legal, and administrative fixes convince relevant parties that it is no longer worth attacking DNA evidence unless there are specific reasons to doubt the quality of the police or laboratory work that adduced it. It is frequently said that defense lawyers and their clients rarely challenge DNA evidence, because either they capitulate by pleading guilty or they shift the line of defense away from the forensic evidence (for example, in a rape case, by challenging the victim's allegations by insisting that the sexual act was consensual).[35]

General adherence by proponents and critics alike to the idea that the controversy is closed tends to foreclose further controversy, whether or not forensic DNA profiling is as reliable as advertised, and whether or not forensic organizations actually adhere to recommended proficiency standards. Criticisms of DNA typing continue to be pursued by defense attorneys in criminal trials, and reports of errors and ongoing problems occasionally surface. However, even in cases in which the errors are alleged to be systematic—such as a Houston forensic laboratory whose notoriously shoddy practices led to a review of evidence samples for hundreds of cases—the problems are attributed to practitioners and organizations, rather than the technique itself. When one delves into cases in depth, however, some of the issues that appear to have been fixed can be extracted from the technical black box. Nevertheless, there is no mistaking the widespread acceptance and entrenchment of DNA evidence. In fact, as we shall discuss in the next chapter, the acceptance of such evidence has become so strong, it has become a basis for doubting all other forms of criminal evidence.

35. Accounts of the first few years of DNA typing often state that defense lawyers rarely attacked such evidence, because they did not know how to do so (Aronson, 2007). By the time of the *Castro* case, defense lawyers had begun to find effective lines of attack, and they were aided in this by several prominent molecular biologists and population geneticists who took issue with forensic practices. For reasons discussed in this chapter, many of these lines of attack were closed down by the late 1990s, and defense lawyers again tend to accept DNA evidence without much struggle. Recent research on Austrian prison inmates indicates that they take a fatalistic view of DNA evidence and see little hope in challenging it (Prainsack & Kitzberger, 2007).

CHAPTER EIGHT

Postclosure

Toward the end of the previous chapter, we reviewed a constellation of technical, administrative and legal "fixes" that together helped shut down the controversy about forensic DNA evidence. The concept of "closure," and phrases like "put to rest," suggest that the end of controversy marks the beginning of a quiet and peaceful period in which consensus reigns. In this case, however, something much stronger than acceptance happened; DNA evidence wasn't just accepted by the courts as one form of evidence among others; it attained a superevidentiary status, and became a basis for invidious comparisons with all other forms of criminal evidence, including older forms of forensic comparison evidence. What's more, when DNA fingerprinting was heralded as a new "gold standard for truth telling" in forensic science (Scheck et al., 2000: 122), it deposed the reigning gold standard of dermal ridge analysis from which it had borrowed its name and credibility (Mnookin, 2003). Fingerprinting and other traditional forensic sciences were subjected to new challenges, and their scientific status was questioned.

The use of probability figures to qualify matches, which once seemed a relative weakness in comparison with fingerprinting, worked in its favor after the *Daubert* decision highlighted the importance of quantitative error rates. However, even before *Daubert,* and at a time when DNA profiling was controversial, proponents cited the controversy as a "test" that would strengthen the new technique in comparison with its predecessor. Referring to the controversy as a "rite of passage," an FBI case examiner we interviewed in 1991 noted in passing that the "absolute identifications" associated with fingerprinting might not fare so well if subjected to such challenge—a remark that proved prescient in light of challenges mounted almost a decade later: "[I]t has to be battled out in some arena, it has to go through this rite of passage. Every technique now needs to kind of pass through this. If fingerprints were going through this, I don't know that we would be able to say that fingerprints are absolute identifications."

The postclosure phase of the history of DNA profiling is of interest for at least two reasons. First, it is a continuation of our narrative history of techno-legal controversy about DNA fingerprinting. Though closure often marks the end, or resolution, of a controversy, in this case it also marks the beginning of what is perhaps a more interesting story. Second, it is of interest for what might be called a "logical" reason: the *assumption* that forensic DNA evidence (often glossed under the simple acronym "DNA") was no longer assailable became a logical foundation for questioning and evaluating other forms of evidence. In addition, some established legal institutions also began to give way in the face of the extraordinary credibility lately assigned to "DNA." Postclosure thus signals a period of time after closure was established, and also an apparent transcendence of the fallible status ascribed to other forms of legal evidence. In this chapter, we begin by discussing the extraordinary credibility currently assigned to "DNA," and then discuss some of the legal implications of that credibility. We then discuss some criminal cases that indicate that some familiar problems still attend the use of forensic DNA evidence, despite the widespread assumption of infallibility. We close with some remarks about the hazards of treating "DNA" as though it had transcended mundane criminal justice practices and the legal institutions that take into account the limitations and hazards of such practices. Challenges to latent fingerprint evidence are part of the postclosure phase of the story, but in order to give sufficient space for discussing these challenges we take them up in the next chapter.

God's Signature

In the late 1990s, the Innocence Project became headline news in the United States. This campaign was centered at the Benjamin Cardozo School of Law in New York City, and was headed by Barry Scheck and Peter Neufeld, the defense lawyers who became widely known during the *Castro* and *Simpson* trials as critics of forensic DNA evidence. The Innocence Project sought to retest available criminal evidence with current DNA typing techniques, in order to overturn wrongful convictions based on other evidence, including older methods of DNA analysis (Scheck et al., 2000).[1] In addition to making use of postconviction DNA testing, the Innocence Project made efforts to identify potential candidates for exoneration, and to conduct the legal and political work necessary to reopen their cases. By the year 2007, the Innocence Project listed more than 200 prisoners who had been exonerated, at least a dozen of whom had been on death row.[2] The capital cases had critical significance, because of the finality of the death penalty and because it had been widely assumed that elaborate precautions were built into sentencing, appeal, and review procedures to protect against the possibility of executing the innocent. Together with other efforts to review convictions, the Innocence Project was part of the most effective campaign in decades against the still-popular use of the death penalty in the United States (Turow, 2003). DNA evidence was not the only basis for such exonerations. In fact, according to the Death Penalty Information Center, 122 inmates have been released from death row with evidence of their innocence since 1973—nearly ten times as many as have been released through postconviction DNA testing.[3] However, "DNA" has provided especially strong leverage in recent years for overriding other forms of criminal evidence, particularly eyewitness testimony, older forms of forensic evidence, confessions, and the testimony of "jailhouse snitches." Reviews of the case records turned up stories that illustrated the dismal state of criminal justice for indigent defendants: overworked and incom-

1. In an interview with Lynch and Cole (21 May 2002), Scheck and Neufeld noted that their efforts to use DNA profiling for postconviction testing preceded their criticisms of prosecutorial uses of DNA evidence in *Castro* and *Simpson*. They recounted an effort starting in 1986 to use DNA technology to exonerate Marion Coakley, who was serving a fifteen-year sentence for rape (see Scheck et al., 2000: 1–5). At that time, they began to learn more about the new techniques—both how they worked in theory and how the transfer from diagnostic testing to forensic investigation was problematic. Neufeld added in the interview that he had gotten his "feet wet" in challenging forensic blood (conventional serological) evidence in a 1984 admissibility hearing. At that point, he mentioned, he became aware of a lesson that that became important for challenges to forensic DNA testing: "that there's a scientific way of doing things and then there's the way that it's done in criminal justice and the two had nothing to do with one another."

2. The 200th DNA exoneration was announced in April 2007. Using the standard shorthand attribution of agency, *USA Today* announced, "DNA to clear 200th person" (Willing, 2007). The article also mentioned that the pace had picked up in recent years, possibly due to the expansion of the project to involve more lawyers combined with the increased sensitivity of the analytical technology.

3. http://www.deathpenaltyinfo.org/article.php?did=412&scid=6 (accessed 15 December 2005). Also see Liptak, 2003a.

petent court-assigned defense lawyers who slept through trials; prosecutors pursuing convictions at all costs; and corrupt police officers and forensic scientists who finagled evidence to help maintain prosecutors' high conviction rates. When used in such exonerations, DNA evidence was identified with a higher form of certainty than other forms of criminal evidence.

Accounts of exonerations typically assign an extraordinary degree of certainty to DNA evidence.[4] In the shorthand rhetoric used by the press, "DNA" has become a primary agent of exoneration. For example, in August 2002 a prisoner named Eddie Joe Lloyd was released after serving seventeen years in prison when DNA testing showed a mismatch between his DNA profile and the profile developed from the crime scene evidence. After his release, Lloyd was widely quoted as exclaiming: "DNA—deoxyribonucleic acid—is God's signature. God's signature is never a forgery and his checks don't bounce!."[5]

The effort to retest criminal evidence sometimes ran into legal and procedural resistance, as prosecutors and courts did not immediately defer to "God's signature." For example, in May 1987, Bruce Godschalk, a twenty-six-year-old landscaper, was convicted of raping two women in separate incidents in a housing complex in King of Prussia, near Philadelphia, Pennsylvania.[6] He was sentenced to ten to twenty years in prison. Years later, after a lengthy effort by Godschalk's lawyers to gain court approval to review the evidence using DNA profile methods, laboratory tests of semen evidence indicated that the victims were raped by the same man, but that the man in question was not Godschalk.

The prosecutor who had handled the case, district attorney Bruce L. Castor, Jr., refused to release Godschalk from prison, arguing that the DNA tests were less convincing than the confession Godschalk made to the police (a confession he later recanted).[7] Castor was quoted as

4. See the discussion in chapter 6, of the exoneration of Luis Diaz.

5. Different versions of this quotation were presented in press reports and on various Web sites. We like the "And his checks don't bounce" part, which is attributed to "jive talking Eddie Lloyd" from a rather doubtful source: http://www.landmarktheatres.com/mn/afterinnocence.html. A more accurate version may be: "That's God's signature. God's signature is never a forgery" (Wilgoren, 2002).

6. This story and the quotations from Castor are taken from Rimer, 2002.

7. Another case, which also drew media attention, was the 2005 exoneration of Larry Peterson, who had spent seventeen years in jail after being convicted of rape and murder (see chapter 6 for an account of this case). After DNA tests contradicted the forensic hair comparisons the prosecution had used in the original trial, the victim's family and

saying, "I have no scientific basis, I know because I trust my detective and my tape-recorded confession. Therefore the results must be flawed until someone proves to me otherwise." Such statements from prosecutors were commonplace in early postconviction DNA cases, but the scientific authority of "DNA" has rendered such arguments increasingly untenable. Castor may have been among the last prosecutors to try to argue that a confession trumps DNA evidence. It is notable that he was criticized in places as far away as Atlanta, a locale not known for its sympathy toward accused criminals.[8] Indeed, some of the prosecutors who resisted postconviction DNA testing most fiercely have been converted to advocates of DNA testing—not just prosecutorial testing, but postconviction testing as well. Such prosecutors inevitably invoke the breathless language that has enveloped DNA testing, calling it "the greatest tool for human identification since fingerprinting."[9]

Dr. Edward Blake of Cellmark Diagnostics, the company hired by Godschalk's defense to perform the tests, expressed no such doubts about the scientific evidence. He likened the DNA profile result to a signature that "is expected to occur in no more than a few human beings who have ever lived." Another expert gave even stronger testimony to the press, saying that the matching profiles developed from the criminal evidence could come from only one person in the world. Still another expert contacted by defense lawyer Peter Neufeld stated that the chances of laboratory error in this case were "nonexistent."

In another example, the *Arizona Republic* in April 2002 used the headline "DNA frees Arizona inmate."[10] The inmate, Ray Krone, had

---

Burlington County (NJ) prosecutor Robert Bernardi refused to acknowledge that Peterson was innocent. Immediately after the judge vacated the conviction, the prosecutor rearrested Peterson in preparation for a new trial. See "The exoneration of Larry Peterson," NPR Radio (12 June 2007), available at http://www.npr.org/templates/story/story .php?storyId=10961075.

8. Tucker, 2002.

9. Topping, 2000.

10. Wagner & DeFalco, 2002. This report coincided with the Death Penalty Information Center's announcement of the 100th postconviction exoneration of a death-penalty inmate since the Supreme Court's reinstatement of that penalty. In a more recent case, DNA analysis of saliva found near a bite mark on the victim overturned the New York murder conviction fifteen years earlier of Roy Brown. According to a newspaper report, in the original trial, a bite-mark analyst had given a tortuous interpretation of the bite-mark, suggesting that Mr. Brown may have "twisted the victim's skin to fill the gaps in his missing teeth" to account for an apparent mismatch between the mark and the defendant's bite pattern. Santos, 2007.

been convicted of fatally stabbing a woman at a Phoenix lounge in 1991. He was convicted, retried after appeal, and convicted again in the second trial. After the second conviction, his defense succeeded in gaining permission to perform an STR test of a saliva sample on the victim's tank top shirt, and the results not only failed to match Krone's STR type, it matched that of another man, Kenneth Phillips, who was serving time for another, unrelated, sexual offence.

> Once labeled the "snaggletooth killer," Krone was convicted largely on circumstantial evidence, particularly expert testimony that bite marks found on the victim matched his teeth. No DNA evidence was submitted in the first trial, and genetic tracing results provided for the second trial merely failed to preclude him as the perpetrator. But all of that changed after defense attorney Alan Simpson obtained a court order, and Phoenix police produced new results.
>
> On Monday, prosecutor William Culbertson told Maricopa County Superior Court Judge Alfred Fenzel that DNA found in saliva on the victim's tank top did not come from Krone. In fact, chances are 1.3 quadrillion to one that it came from Phillips, the Florence prison inmate.

Not only is it commonplace to read in press reports that "DNA" provides unequivocal proof of innocence, more sophisticated writers also describe it as an agent of exoneration and a source of "irrefutable proof of the fallibility of eyewitness testimony" (Berger, 2004: 112), and of virtually all other forms of criminal evidence.[11] Saks and Koehler (2005: 892)

11. Steven Avery, who spent eighteen years in a Wisconsin jail on a sexual assault conviction, was exonerated when DNA testing of pubic hair found on the victim after the attack showed no match with Avery's profile and matched the profile of a man who was serving time for another sexual attack. Avery's release was celebrated by civil libertarians, until some two years later when he was arrested for the murder of a young woman. According to a news report, "Steven Avery, who was once this state's living symbol of how a system could unfairly send someone away, has left all who championed his cause facing the uncomfortable consequences of their success." However, Avery's arrest did not open up questions about the DNA test that exonerated him. A spokesman for the Wisconsin Innocence Project was quoted as saying that "the group's intent was not just to release the innocent but to find truth, and properly punish those truly responsible for crimes" (Davey, 2005). The case is in some ways a reprise of the Kerry Kotler case. Kotler was among the earliest of the Innocence Project's postconviction exonerations. Even as Kotler was successfully suing the state for wrongful conviction in a rape case, he was being prosecuted—using DNA evidence—in a second rape case. Although there were those, such as the victim of the first rape, who suggested that the second conviction was proof that the postconviction DNA testing had gotten the first rape case wrong, the general consensus now appears

present a graphic figure of "factors associated with wrongful conviction" based on Innocence Project data for eight-six exoneration cases. The most common of the factors is errors with other forms of evidence: eyewitness error (77 percent); forensic science testing errors (63 percent); and, further down the list, false confessions (17 percent). Other factors include misconduct by police, prosecutors, and forensic scientists, and incompetent defense attorneys. (The figures add up to more than 100 percent because more than one factor was found in many cases.) Although Koehler has been one of the most articulate critics of forensic practices with DNA typing, in this particular article he and Saks treat the DNA tests as an unquestioned basis for exposing errors and misconduct.

Truth, Credibility, and Stories

It is no exaggeration to say that "DNA" became a marker of factual, scientific identity, and that courts, politicians, the popular press, and other commentators ascribed a truth to DNA evidence that was more solid and objective than mere credibility. "DNA" also became a kind of superhero for public justice, exposing corrupt police practices and pervasive flaws in the system of justice. Increasingly prosecutors have yielded to such evidence, even when rape victims insist that they correctly identified the perpetrator, police officers insist that the defendant's confession was genuine, or other forensic evidence seems to point unequivocally to the defendant. Prosecutors who remain convinced of the evidence of guilt are in a bind, because DNA typing for the most part remains a prosecutorial tool. More often than not its mystique works in favor of the prosecution's efforts to identify suspects, solicit confessions, and pursue convictions.

Unqualified statements by proponents of the Innocence Project about the probative value of DNA evidence might seem startling, given the reputation of Neufeld and Scheck for attacking the reliability and integrity of forensic DNA analysis. However, they avoid contradiction by distinguishing later techniques from earlier techniques, exculpatory evidence

---

to be that Kotler was innocent of the first rape but guilty of the second. What is interesting to note is that in the narrative in which Kotler is either a rapist who happened to be falsely accused, or someone who was turned into a rapist in prison, the credibility of DNA is preserved. See Nelson, 2003.

from inculpatory evidence, and, when speaking in detail, by mentioning the necessity to use recommended quality control standards. For example, in one statement quoted in the press, Neufeld states, "It is powerful evidence both to convict and to exonerate. . . . It's kind of a truth machine. But any machine when it gets in the hands of human beings can be manipulated or abused."[12] When chronicling many of the exonerations brought about through the Innocence Project, Scheck and Neufeld discuss cases in which they overturned convictions based on DNA analysis as well as other forms of evidence. In such cases, investigators using newer, more sensitive techniques with a higher number of probes were able to analyze small and degraded samples which had been insufficient for testing at the time the case was brought to trial. In addition, retesting of samples was used to expose dubious procedures and interpretations used in earlier tests.

For example, Scheck et al. (2000) describe a case in Oklahoma in which a DNA test was used along with an eyewitness identification and forensic hair analysis to convict a man, Tim Durham, for raping a child. Despite alibi testimony of eleven eyewitnesses who placed Durham far away from the crime scene, he was convicted and sentenced to life in prison. After he had served five years in prison, the conviction was reviewed, and another DNA test performed under the auspices of the Innocence Project exonerated him. According to Scheck et al. (2000: 168), in the original trial, prosecution witness Dr. Robert Giles declared a match between one of Durham's genetic markers and the result of a lab test of patches from the victim's swimsuit containing presumed evidence of a semen stain. The criminal evidence required three attempts before a faint "dot" showed up in the test that matched Durham's evidence. Giles declared the match, and the prosecution estimated that the particular genetic marker occurred in 10 percent of the population. Giles declared the match "based on the fact I've been doing this for several years. I have been involved with gene amplification from its very beginning. . . . I have experience in knowing how to make these particular calls. There is some art involved in that, but that's why I'm trained as a scientist to do what I do" (quoted in Scheck et al., 2000: 168). The review of the case also determined that Giles observed a nonmatching marker in one of his earlier tests of the swimsuit evidence, but discounted it as an artifact. After

12. Peter Neufeld, quoted in Liptak, 2003b.

saying that Durham was released from jail on the basis of the retest, Scheck et al. observe that he was a victim "of a court system unwilling to scrutinize any evidence coated with a veneer of "science"" (170).

In this case, the earlier DNA test was retrospectively characterized as involving a fallible experiential judgment—not unlike other forms of ordinary and expert evidence that postconviction testing had over-turned. When Scheck, Neufeld, and others discuss DNA exonerations, they use language that suggests that "DNA" is an independent source of truth, separate from the other forms of ordinary and forensic evi-dence that had led to the original conviction. In addition to suggesting that the latest techniques of DNA analysis are infallible, even if earlier techniques were not, talk of truth machines glosses over a point we have emphasized throughout this book, which is that DNA evidence is inter-twined with other evidence in the story of a case. A vivid illustration of this point is provided by the first case of DNA exoneration that oc-curred in the United Kingdom: *R. v. Michael Shirley*.[13] Shirley was con-victed in 1988 for the brutal rape and murder of Linda Cook. The pros-ecution's serological experts declared a match between Shirley's blood protein group and that of the semen recovered from the victim's body, and stated that 23.3 percent of the British adult population shared that group. The prosecution also cited other forensic evidence linking Shirley to the victim. DNA analysis was not performed at the time of the trial, because the quantity of analyzable material was deemed insufficient for the techniques available at the time. In 2001, using a newer PCR-based technique, the Forensic Science Service reanalyzed a remaining portion of the semen sample, and the result did not match Shirley's profile. The case went before the appeal court in 2003, and the court quashed the conviction.

In an insightful analysis of the appeal case, Paul Johnson and Robin Williams (2004) point out that the DNA evidence did not automatically exonerate Shirley. The Crown prosecutor accepted that the DNA ev-idence showed that Shirley was not the source of the semen, but con-tended that other evidence was sufficient to show that he committed the murder. The prosecutor suggested that the victim had consensual sex with an unidentified man earlier on the day that she was raped and murdered. The defense countered this argument with eyewitness tes-timony about the victim's movements during the day, and expert testi-

13. *Regina v. Shirley* (2003).

mony about the condition of her underwear, which apparently had been removed during the rape. Defense witnesses testified that no semen was found on the underwear, and that semen from sexual intercourse earlier in the day would very likely have drained onto the garment if the prosecutor's account had been correct. Consequently, the appeal court concluded:

> The truth is that, taking the scientific evidence together with such knowledge as we have of Miss Cook's movements in the hours before she was attacked, the overwhelming probability is that *all* the semen found in the intimate swabs was deposited by one man on one occasion, that is to say by her killer when he raped her (*R. v. Shirley*, 2003; quoted in Johnson & Williams, 2004: 80).

Johnson and Williams argue that language that treats DNA technology as a source of certain "truth" and an agent of exoneration ignores the necessity to "contextualize the significance of any particular instance of DNA profiling" (80). Although the availability of DNA technology was crucial for reopening the case, and all parties to the case accepted that the analytical result showed that Shirley was not the source of the semen, the appeal turned on a consideration of a larger story compiled from circumstantial evidence and several formal and informal probability judgments.[14]

When DNA evidence from crime scenes fails to match defendants' evidence, this would seem to aid the defense, but prosecutors can be adept at contextualizing and recontextualizing DNA evidence in stories. Perhaps the most notorious instance of this was the infamous Central Park jogger case in New York City, which went through two phases, the first of which was in 1989–90, when an alleged gang rape and near-fatal assault occurred on a twenty-eight-year-old Wellesley-educated

---

14. A similar case in the United States is described in a *New York Times* article on difficulties released prisoners have had in getting compensation from the state for having served time in prison based on a wrongful conviction: "In one case, Vincent H. Jenkins, who served 17 years in prison on a rape conviction, was released in 2000 because DNA tests of a semen sample showed he was not the rapist. But the prosecutor argued that Mr. Jenkins was guilty anyway—that he was present when others committed the rape. . . . Mr. Jenkins . . . had to argue his case for two years before winning $2 million. He died last month, at the age of 66" (Purnick, 2005).

As in the Shirley case, the prosecution discounted the significance of the nonmatching evidence by using other evidence to recompose the story.

investment broker who had been jogging in the park, and five African American youths were tried and convicted during an intense media frenzy; and the second of which occurred in 2002 when new evidence led authorities to vacate their convictions.

The first phase is described by former Manhattan prosecutor Harlan Levy (1997: 70–103) in a popular book that for the most part heralds the triumph of DNA testing. The jogger case is one in which the DNA evidence did not corroborate the case against the defendants, which largely rested on confessions that the defendants retracted soon after making them. Semen recovered from examination of the victim and her clothing was analyzed by the FBI using the single-locus probe technique, but (according to Levy, 1997: 71), only a single band showed up in analyses of samples from the vaginal swab and the victim's sock. The prosecution initially declared this to be "inconclusive," although the FBI had determined that the profiles—which matched each other—did not match profiles taken from any of the suspects or the victim's boyfriend. Levy discusses the dilemma faced by the prosecution:

> So when the DNA results came in, [lead prosecutor Elizabeth] Lederer and I began to discuss how we could address this problem. We dismissed out of hand the possibility of attacking the reliability of DNA testing. We saw such testing as a development that our office should, and would, embrace as part of a consistent posture. We knew that a district attorney's office cannot argue for the reliability of DNA testing in some instances and denigrate it in others. (94–95)

When presenting the evidence to the jury, Lederer emphasized the conditions under which the evidence was collected: because of the hospital's intensive efforts to save the victim—Levy (97) mentions that she lost 75 percent of her blood, and had severe fractures to the skull—it wasn't until early the next morning that the semen sample was taken. Lederer also mentioned other contingencies, such as the possibility that some of the defendants participated in the rape but did not ejaculate.

> In short, Lederer sought to convince the jury that the DNA evidence was consistent with the other evidence, and that it was consistent with guilt. In this scenario, far from representing a problem, these results were exactly what one would expect in the circumstances. It was a reasonable, credible argument, but whether it would be sufficient to convince the jury remained to be seen. (97)

Inconsistencies in the confessions also where handled with a familiar interpretative strategy: they indicate that the confessions were not concocted (for example, as a result of police suggestion). The prosecution's arguments apparently were convincing to the jury, as they obtained convictions for the five defendants, who were sentenced to long prison terms.

The second phase began thirteen years later, after the defendants had completed their sentences, when a long-term convict Matias Reyes, claiming to have been motivated by religious conversion, sought out authorities and confessed that he alone had committed the rape and assault. When the DNA evidence was retested using contemporary techniques, Reyes's profile matched the swab and sock evidence. The police resisted the single-perpetrator story, but the convictions were vacated and dismissed by justice Charles Tejada in December 2002.[15]

With hindsight (and as the defendants alleged during the original trial), the police elicitation of the confessions seemed coercive and dubious. Levy (1997), without implying that there was anything wrong with such tactics, described how police interrogators cleverly used parents as intermediaries to shame the kids into admitting the "whole story" after they initially mentioned nothing about the rape. He also recounts how NYPD detective Thomas McKenna elicited a confession from Yusef Salaam, a key fifteen-year-old defendant, by claiming, "we have fingerprints on the jogger's pants or her running shorts, they're satin, they're a very smooth surface." After McKenna proposed that he was going to compare Salaam's prints to those recovered from the pants, the young man supposedly blurted out a confession that also implicated the other defendants. As McKenna freely admitted when testifying in the trial, this was a ruse: no fingerprints had been recovered, and the pants were not satin (Schanberg, 2002). However, even after the DNA evidence was retested in 2002, the confessions and other evidence of youth gangs running amok in the park that night assaulting strangers (the term "wilding" was coined in connection with the story) provided the police and prosecution with a basis for incorporating the DNA evidence into a story that implicated the defendants in the rape and/or assault on the victim.

Neufeld and Scheck use the term "unindicted co-ejaculator" to allude to a type of argument that is sometimes used by prosecutors in

15. See Executive Summary: Central Park Jogger Case Panel Report, available at http://www.nyc.gov/html/nypd/html/dcpi/executivesumm_cpjc.html.

rape-murder cases when postconviction DNA testing excludes the convict. In such instances, the prosecution accepts that the nonmatching evidence exonerates the convicted man of rape, but asserts that it does not necessarily exonerate him of murder. Although the Central Park jogger barely survived the assault, a similar logic applied. The victim had no recollection of the assault, and there were no eyewitnesses. The fact that the DNA evidence excluded the defendants did not exonerate them in the view of the prosecution. Instead, confessions and circumstantial evidence were used to implicate them in the assault and possibly the rape as, so to speak, "indicted nonejaculators." In several other notorious cases, prosecutors have changed their theory of the crime postconviction, transforming single-attacker cases into gang rapes. Thus, prosecutors can explain the appearance of alleles in the semen sample that are not matched by those in the suspect's sample by citing the possibility of a second, previously unmentioned man who would have had sexual intercourse with the victim. And the relevance of stories extends beyond the narrative of the crime to include the local histories of the evidence samples (what we called in chapter 4 the "careers" of the samples).[16]

One particularly interesting example of such a case was the prosecution of an African American man named Wilbert Thomas for a rape that occurred in Huntington, West Virginia, in 1987. The assailant entered the house, removed light bulbs, and attacked and robbed the victim. The victim stated that there was no penetration or ejaculation during the attack. Thomas's lawyers claimed that the victim denied having any consensual sexual partners around the time of the attack, but the government contended that she testified only "that she had not dated any African-American men at or near the time of the rape" (Response Brief, *Thomas v. Trent,* 1999).

Crucial physical evidence against Thomas was testimony identifying his fingers as the source of latent prints recovered from two light bulbs removed from inside the house, and one removed from a neighbor's porch. In addition, a "Negroid" pubic hair was recovered from the crime scene. Thomas later shaved all his body hair in jail, which the prosecutor

16. "Local history" refers to the singular course of action in which a sample is collected, contained, stored, analyzed, and so forth, on the way to becoming criminal evidence. For an account of the local history of observations in astronomy, see Garfinkel et al. (1981). Terms of art such as "continuity of evidence" and "chain of custody" refer to formal requirements for assuring the courts that the movements of evidence were continuously monitored and that proper analytic procedures were adhered to.

contended was an indication of guilt. Finally, serologist Fred Zain testified that Thomas could be the source of a semen stain found on the victim's nightgown (the very presence of the stain contradicted the victim's
account; this was before DNA profiling was commonplace). Zain would
later be discredited for falsifying results (Scheck et al., 2000: 112ff.).
Thomas maintained his innocence, claiming that he had been home with
his family, and his wife supported his alibi.

Thomas was actually tried three times. In the first two trials, he was
convicted of burglary but not of rape. In the third trial, in 1990, he was
convicted of rape. Around 1995, the court ordered postconviction DNA
testing of a vaginal swab recovered at the time of the assault. The testing, by Marcia Eisenberg of LabCorp, used differential lysis to distinguish between different possible sources of a mixed sample and found
vaginal and sperm fractions (thus contradicting the victim's account) and
excluded Thomas as the donor of the semen. A U.S. magistrate judge
reversed Thomas's conviction, but this decision was overturned by the
U.S. District Court, which claimed that Eisenberg had not "tested for semen." Thomas, and later Eisenberg herself, responded that differential
lysis *is* a test for semen. Thomas, now represented pro bono by the prestigious Washington law firm Hogan & Hartson, demanded his release in
the ringing terms typical of postconviction DNA exoneration cases:

> Thus, with all due respect, the Court's conclusion that the DNA test results
> were "inconclusive" simply cannot be squared with the presence of someone
> else's genetic material in the sperm fraction taken from the vaginal swabs. The
> time has come to release Mr. Thomas after twelve years of unjust imprison-
> ment and return him to his family. The time has come for the state to devote
> its time and energy to catching the real perpetrator of this assault—the man
> who matches the genetic material found inside the victim and who may have
> perpetrated many more crimes since then. (Petitioner's Motion, *Thomas v.
> Trent,* 1999: 6)

However, Thomas's case did not follow the usual Innocence Project
script. The government was able to successfully trump the DNA exclusion with contextual evidence. In a 1999 response brief, the government
dismissed the exclusionary DNA evidence, concluding:

> In short, no matter what Marcia Eisenberg may say now that she has crossed
> the line from scientist to advocate, and no matter what tests she may run and

what those tests may show, it can not change the fact that *all the DNA tests in the world cannot prove the Petitioner's innocence in this case and the non-DNA evidence conclusively proves him guilty.* (Response Brief, *Thomas v. Trent,* 1999: 9; emphasis in original)

Crucial to the failure of DNA evidence to win Thomas's exoneration was the existence of a factual scenario that explained both the exonerating DNA evidence and the incriminatory non-DNA evidence: the possibility that the victim had been in a sexual relationship that she preferred not to acknowledge. Particularly interesting is the failure of "all the DNA tests in the world" to overcome the latent print evidence, which would be in marked contrast to the later *Cowans* case (see chapter 9). Thomas served out his sentence before the legal battles fully ran their course, so the ultimate "truth"—legal or factual—of the case is unlikely to ever be fully resolved.

The Innocence Project, and related events such as former Illinois governor George Ryan's moratorium on executions due to a lack of confidence in the fairness and accuracy of the verdicts, provided the most effective movement against the practice of state execution since 1976, when the Supreme Court ruled in favor of granting courts the discretion to give death sentences under specific conditions.[17] Ryan, a Republican governor facing corruption charges who did not intend to run for reelection, became convinced that many Illinois inmates were on death row for capricious and illegitimate reasons. Just before leaving office, Ryan pardoned four inmates on death row and commuted the sentences of 167 others. His actions were inspired not only by DNA exonerations, but also by the results of a review of case records by Professor David Protess and his students at the Medill School of Journalism at Northwestern University.[18] The Medill Innocence Project documented the doubtful evidence, possibly corrupt police practices, shoddy efforts by prosecution and defense lawyers, and racial bias by juries and judges in many death penalty cases.

17. The 1976 Supreme Court ruling took up three cases: *Gregg v. Georgia* (1976), *Jurek v. Texas* (1976), and *Proffitt v. Florida* (1976), and ended a moratorium on the death penalty in the United States. In *Furman v. Georgia* (1972), the Supreme Court had ruled in a 5–4 judgment that capital punishment was contrary to the cruel and unusual punishment clause of the Eighth Amendment and also violated the due process guarantees of the Fourteenth Amendment of the U.S. Constitution.

18. See Governor George Ryan, "I will not stand for it," Speech at Northwestern University College of Law, 11 January 2003, available at http://www.worldpolicy.org/globalrights/dp/2002–0111-Ryan%20speech%20on%20capital%20punishment.html.

In an interview, Protess acknowledged the power of DNA evidence for exonerating inmates, but noted that DNA evidence was involved in only one of the death penalty exonerations arising from the Medill project.[19]

As noted earlier, in connection with the Innocence Project, "DNA" was widely heralded as a scientific tool that transcended the all-too-human limitations of the criminal justice system. Unlike the practical closure or "finality" achieved through the adversary process, DNA was frequently described as a source of proof; an exceptional form of evidence that supported judgments with an order of certainty that transcended the "tests" available through the adversary process. When used as a civil libertarian tool, "science" (namely, DNA typing) was the presumptive basis for exposing weak evidence, prejudice, and even fraud. However, this was a game that both sides could play. Predictably, such certainty was seized upon to circumvent the concerns of the civil libertarians.

The Guilty Project

In September 2003, then-governor Mitt Romney proposed to use "science" to reinstate the death penalty in the Commonwealth of Massachusetts, one of a dozen U.S. states that had not reinstated the death penalty after 1976:

> "We want a standard of proof that is incontrovertible," Romney said as he stood at a State House press conference with members of his newly formed Governor's Council on Capital Punishment. He said he wants to put "science above all else" in capital murder cases. The governor said he is directing the panel, which is made up of well-known forensic and legal specialists, to craft a narrowly defined capital punishment law that will deal with those who have committed multiple murders through acts of terrorism; killers of those in the criminal justice system, such as judges, prosecutors, and police officers; and those who commit the "most heinous violent crimes." (Phillips, 2003)

As Kara Swanson observes, Romney's statement demonstrates how certainty of innocence is readily converted into certainty of guilt.[20]

19. See Tackett, 2003.
20. Although Romney failed to persuade the state legislature to implement his ideas, he was not alone in advocating DNA testing as support for the death penalty. Senator Orrin Hatch of Utah introduced a bill in Senate entitled "Advancing Justice through DNA

The inversion of the Innocence Project to the Guilty Project attempts to constrain the moral dilemma of capital punishment, reducing the multitude of tensions generated by the exercise of an unreviewable coercive power in a democratic state into a single scientific question of guilt which can be answered through the power of modern technology, as witnessed by the jury-as-public. (Swanson, 2005: 12)

Romney appointed a governor's council, consisting of eleven experts. Their charge was to develop a new statute that set conditions for the use of the death penalty. The members included experts in forensic science, particularly DNA typing, such as Dr. Henry Lee, who gained fame as a key defense witness in the O. J. Simpson trial. The council also included experts in legal procedure and law (Swanson, 2005: 2–3).[21]

Romney's proposal was roundly criticized in the press, and polls indicated considerable public skepticism about his idea to develop an "air-tight" and "error-free" system of justice. A few years later, however, "DNA" once again was embraced publicly by proponents of capital punishment. Ironically, this occurred after opponents of capital punishment succeeded by means of a long legal struggle to commission DNA tests on the exhumed body of Roger Coleman, who had been executed in Virginia in 1992 for the 1981 rape and murder of his sister-in-law. Coleman famously proclaimed as he was strapped into the electric chair, "An innocent man is going to be murdered tonight." Had the DNA test results

---

Technology Act of 2003." This bill would provide funds for postconviction DNA testing, and also for reducing the backlog of more than 300,000 DNA samples that were awaiting analysis. In addition to giving the rationale of protecting innocent defendants against false conviction, Hatch specifically mentioned that the act would assure conviction of the guilty and improve support for the capital punishment system. See hatch.senate.gov/ index .cfm?FuseAction=PressReleases.View&PressRelease_id=900.

21. Swanson goes on to observe parallels between Romney's ambitions for a scientific death penalty verdict and those of another first-term governor a hundred years earlier— David Hill, governor of New York, who in 1885 proposed to institute the electric chair in order to eliminate the uncertainties and barbarities associated with hanging. See Trombley (1992) for examples of botched executions with the electric chair. Close descriptions of the spectacle of the electrocution written by opponents of the death penalty emphasize how it sometimes requires repeated attempts to dispatch the person being executed. But, instead of convincing authorities that the penalty is cruel and unusual, such descriptions motivated the use of more "humane" and efficient means. Swanson observes that the move by most states to lethal injection administered to a shrouded patient on a hospital gurney medicalizes the procedure, turning the condemned into a patient to whom medication is administered in a hospital-like setting. That procedure itself has lately undergone similar criticisms and calls for more humane alternatives.

been negative, his case would have been the first postconviction/post-mortem exoneration, but instead the test showed a match between his evidence and the crime scene evidence. After noting that proponents of capital punishment celebrated this outcome as "a watershed moment . . . as if the case were the ultimate proof of the universality of justice," a *New York Times* editorial advised that "the Coleman outcome is further evidence that the authorities should embrace DNA testing as a continuing tool of justice."[22]

The belief that DNA evidence delivers extraordinary, and long-lasting, proof of innocence or guilt led to some interesting shifts in legislation and legal practice. One such shift, which we briefly discussed in chapter 7, involved statutes of limitation.

Statutes of limitation require that legal proceedings for particular categories of crime must be initiated within a specified number of years after the offense is committed. The rationale for such limitations is to guarantee a speedy trial in accordance with the Sixth Amendment, and also to prevent convictions based on "stale" evidence, such as fogged eyewitness recollections, statements that can no longer be corroborated, and physical evidence that has spoiled or degraded. The reputation of PCR-based techniques, such as the STR system, to recover genetic information from old evidence goes against the presumption in favor of relatively fresh evidence.[23] After DNA evidence became a regular part of criminal investigations, many states began to devise exceptions for it. Prosecutors also have found ways to work within existing statutes of limitations by employing a very interesting strategy. When faced with unsolved crimes for which the statute of limitation is about to run out, and bodily evidence is on file (such as semen recovered from examination of a victim in states that specify a five- or six-year statute of limitation for

22. "DNA's weight as evidence," Editorial, *New York Times* (18 January 2006).

23. Interestingly, another source of pressure against statutes of limitation is one of the most notoriously "subjective" forms of criminal evidence: repressed memories of childhood sexual abuse suffered long ago but recently "recovered" through psychotherapeutic counseling. Like the recovery of legible DNA from old evidence, recovered memories supposedly operate in a way contrary to the general assumption about the advantages of recency for the collection and recollection of evidence. In *Stogner v. California* (2003), the U.S. Supreme Court ruled against legislation that used the date of "discovery" rather than original occurrence for applying statutes of limitation for repressed memory evidence in criminal trials. The ruling did not apply to civil suits brought by victims against accused perpetrators (or their employers). For a recent case, see *Powel v. Chaminade College Preparatory, Inc.* (2006). It should be noted, however, that judicial discussions of recovered memory mark it as highly controversial and express uneasiness about its "objective" basis.

rape), prosecutors have used the practice of filing a complaint against "John Doe"—the unknown person whose DNA was collected during the investigation. This use of a DNA sample as a synecdoche for an unknown person is a further instance of the contemporary trend to treat "DNA" as the essence of a person.[24] Oddly, this practice was not used for fingerprint evidence, which for decades has been held to be a unique personal identifier. Despite the much-debated probabilistic relationship between crime sample evidence and a suspect's evidence, "DNA" has been taken as a proxy for a particular individual. The strong affiliation with science appears to have impressed courts and legislatures to grant it such exceptional epistemic status. The first "John Doe" warrant that led to the arrest and conviction of a suspect was filed in California in the year 2000. The case originated in 1994, with a rape that took place in the victim's second-story apartment in Sacramento. Police believed that the same attacker was responsible for a series of rapes at around that time, all taking place in second-floor apartments. In the year 2000, the six-year statute of limitations was about to expire for that crime, and so prosecutors deployed a practice that they learned had been used in another state—filing a "John Doe" warrant. A warrant for arrest stops the clock on the statute of limitations, but unlike a standard warrant that names the suspect and mentions distinct characteristics such as date of birth, physical marks, and so forth, this warrant identified DNA profile characteristics:

The People of the State of California v. JOHN DOE, unknown male with Short Tandem Repeat (STR) Deoxyribonucleic Acid (DNA) Profile at the following Genetic Locations, using the Cofiler and Profiler Plus Polymerase Chain Reaction (PCR), amplification kits: D3S1358 (15,15), D16S539 (9,10), THO1 (7,7), TPOX (6,9), CSF1PO (10,11), D7S820 (8,11), vWA (18,19), FGA (22,24), D8S818 (8,13), D13S317 (10,11), with said Genetic Profile being unique, occurring in approximately 1 in 21 sextillion of the Caucasian population, 1 in

24. Martin (2007) discusses cases of chimerism, a condition in which an individual's body contains cellular populations having different DNA patterns. Like the upsurge in multiple-personality cases a decade or two earlier, such cases potentially threaten the one-body/one-person presumption of liberal democracy and law (Hacking, 1995). She also describes an episode in the popular American television program *CSI*, in which a suspect initially eludes identification as the perpetrator of a rape because he is a chimera whose blood cells have a different DNA profile than his sperm cells.

650 quadrillion of the African American population, 1 in 420 sextillion of the Hispanic population Defendant(s). (Starrs, 2000: 4; quoted in Imwinkelried, 2004: 95)

Note that the warrant lists a technical description of "John Doe's" features: the type of profile (STR, the current technique used by most forensic organizations); a list of chromosomal loci and alleles; and random match probability figures for three broadly defined "racial" or "ethnic" populations. These nonhuman indices stand proxy for the presumed human source of the evidence.[25] Notably, the warrants describe the particular profile as "unique," as the probability figures imply that the profile is so rare that it would be impossible to get it for any person other than the perpetrator or his identical twin. The victim had identified the assailant as African American, and so the relevant probability figure would be 1 in 650 quadrillion—a figure so small that it would seem impossible that another individual who ever lived on earth, or for that matter any other planet, would match the evidence.

Three weeks after the warrant was filed, an African American man— Paul Robinson—was identified with a cold hit on California's criminal database. Interestingly, Robinson's profile had been placed on the database by mistake—he had been arrested for a misdemeanor that the police had mistakenly believed qualified for inclusion in the database. Nevertheless, the evidence was allowed under California law as a "good faith" mistake.[26] More recent legislative efforts in many states (and also in the United Kingdom) have expanded databases to include persons convicted of relatively minor offenses, and also have relaxed restrictions against including profiles from suspects who had not been convicted.

In addition to circumventing statutes of limitations with "John Doe" warrants, many state officials aim to repeal such statutes altogether. Currently, statutes of limitations for rape prosecutions are used in roughly half of the states in the United States in order to ensure timely prosecution when "memories are still fresh."[27] The advent of DNA evidence has

25. This is a strikingly literal instance of treating a nonhuman inscription (a DNA profile) as a human "actant" or agent (Latour, 1987).

26. See Delsohn, 2001.

27. Defense attorney Michael F. Rubin, quoted in Preston, 2006. The article mentions that the New York State Senate had passed legislation to abolish the statute of limitations for rape, but that the legislation had failed to pass in the Assembly. New York cur-

put pressure on legislatures to repeal such statutes. In the view of some officials, such as John Feinblatt, justice coordinator for the New York City mayor's office, "DNA doesn't have any of the human frailties. . . . It's time for law to catch up to science."[28]

Conflating DNA test results with "justice" and treating DNA evidence as objective proof independent of the "human frailties" associated with memory and testimony, in our view, represents a disturbing trend. One needn't be a social constructionist to agree that scientific results are tentative, never absolutely certain;[29] and "scientific" results embedded in a criminal justice process are even more so. As we have seen in the preceding chapters, DNA typing results depend upon the competence and integrity of police, clerical staff, prosecutors, and many other agents in the criminal justice system. Furthermore, even if one were to accept that a DNA match provides near certain evidence of identity, a trier of fact's judgments about guilt and innocence depends upon evidence of circumstance, intent, motive, and capacity—evidence of the circumstances of the investigation and motives of the investigators, as well as of the crime and the persons implicated in it. To a large extent, the "ultimate issue" has *not* been shifted away from the arguments, assessments, and deliberations of the court and jury room and taken over by the laboratory.

Trouble in Paradise: When Things go Badly with the Gold Standard

The social constructionist line in science and technology studies emphasizes that closure of technical controversies is "social."[30] This can be a

---

rently classifies rape as a B-Felony, in the same category as burglary and grand larceny. The five-year statute of limitations for such felonies can be extended to ten years when the perpetrator's whereabouts is unknown. In addition to proposing to repeal the statute of limitations, officials in New York also have recommended reclassifying rape as a more serious A-Felony, which would not carry a statute of limitations.

28. Ibid.

29. Carl Hempel, Karl Popper, and many other philosophers of science emphasized fallibilist doctrines well before Thomas Kuhn and the sociology of scientific knowledge came on the scene.

30. Collins (1985) builds upon Kuhn's (1970 [1962]) view that revolutionary change in science occurs not because members of scientific communities are persuaded by unequivocal experimental evidence. Instead, such change involves elements of generational conflict, and it occurs through a more gradual realignment in the relevant research community around a radically new cognitive and technical nexus (paradigm). Collins extends Kuhn's

difficult point to grasp, because it does not mean that closure is *merely* social—decided prematurely because of political pressures, maneuvering in a priority race, or undue influence by powerful individuals and groups.[31] What it means is that the basis for closure is not as clear-cut as it may seem with hindsight, and that the possibility remains that issues that are closed can be reopened, at least in principle. In the case of DNA profiling, we no longer see a reprise of the kind of controversy that occurred in the early 1990s. Instead, for the most part, disputes occur in trials and appeals of particular cases rather than in admissibility hearings that address the general reliability of DNA technology, but at least some of the disputed cases hold broader implications. And, when one goes into some of these cases in sufficient detail, it is possible to find variants of some of the earlier problems with the production and interpretation of DNA evidence that seemed to have been "fixed" years ago by changes in technique and the imposition of QA/QC regimes.

When, in the previous chapter, we identified the nexus of techni-

---

idea of revolutionary science to cover less spectacular instances of controversy, and further emphasizes the role of persuasion, prior commitments, and group alignments that form and reform in a scientific dispute. Collins's picture of controversy and closure was later applied to technical change and stabilization by Bijker et al. (1987).

31. Scientists and engineers who find themselves on the losing end of a controversy commonly argue that (apparent) closure in the particular case was imposed artificially, and that the currently ascendant scientific theory or technological system is really not the best one available. At around the time that Bruno Latour published his book *Aramis* (1996), named after a failed transportation system that had been planned, and in a very limited way built, by the French government, one of the authors of the present book had an engineering colleague who had designed a similar system in the United States. That system also had been developed into a limited prototype that, like the Aramis system, involved many small vehicles hung on overhead rails, with a computerized control system that shunted the vehicles to local destinations. The engineer complained that despite comparative advantages in cost and efficiency, his system lost to a conventional light rail system in a competition for a new public transportation system in a U.S. city. The engineer blamed "politics" rather than the inherent qualities of the competing systems for the outcome. Latour's story is designed after a murder mystery, and his main character is a sociologist, taking the part of a sleuth trying to get to the bottom of the mystery of why Aramis, which once seemed such a vital possibility, was now confined to the scrap heap of history. The sociologist-hero treats accounts such as the one given by our engineering colleague as part of the mystery, rather than a correct diagnosis of it. He also refuses to settle for the more conventional forms of sociological explanation, and in the manner of a film noir detective, he delves ever more deeply into a black box in which the social and technical never quite resolve themselves into discrete "factors." He never quite solves his murder mystery, but one simple lesson we can take from the story is that closure is real, effective, and exceedingly difficult to undo (Aramis is unlikely to become Lazarus), and yet the more deeply one delves into the conditions that create and sustain closure, the murkier it gets.

cal, administrative, and legal "fixes" that helped bring about closure of the DNA typing controversy, we did not explore the murkier aspects of closure that can leave lingering questions about the extraordinary evidential power that now is commonly ascribed to forensic DNA testing. In light of the pronouncements we often read in the newspapers, and even in more technical sources, about the level of certainty associated with "DNA," it may be worth examining what, if anything, remains contestable (and is occasionally contested) about forensic DNA evidence.

A Debacle in Houston

Although the currently used techniques are broadly accepted by national and international courts and agencies, DNA evidence continues to be challenged in particular cases, sometimes successfully. According to William Thompson, serious problems remain with the way prosecutions interpret DNA evidence in many, and perhaps most, criminal trials.[32] One case that garnered widespread media attention was that of Josiah Sutton, who was convicted of rape in a Houston, Texas, court in 1999. As with the *Adams* case discussed in chapter 6, there were problems with the victim's eyewitness testimony, and the main evidence against Sutton consisted of a DNA match between his blood sample and semen recovered from examination of the rape victim.

Sutton was arrested in 1998, when a woman identified him as one of two men who had raped her. The victim claimed that two men accosted her at gunpoint near her apartment, kidnaped her in her own vehicle, and then took turns driving the vehicle and raping her in the back seat. She described her assailants as black men who were around 5 foot 7 inches tall, with one weighing around 120 pounds and the other around 135. She said that one of the assailants wore a baseball cap with the bill to the side, and the other a skullcap. A few days later, when returning to her apartment, she spotted a group of black men, including Sutton, who was wearing a baseball cap with the bill to the side, and another man, who was wearing a skullcap. She called the police, and the two men were apprehended. The police held the suspects in the back seat of a patrol car and had them wear their hats as described. While sitting within

32. Interview with Simon Cole and Michael Lynch, 18 October 2005. For a published elaboration of problems that can arise in particular cases involving STR evidence, See Thompson et al. (2003). For Thompson's on-line exhibit "DNA Testing Problems," see "Scientific testimony: An on-line journal" at http://www.scientific.org/.

another police car, the victim was asked if the two men were her assailants, and after she identified them they were arrested. Sutton, who was
sixteen years old at the time, stood 6 feet ½ inch tall and weighed 200
pounds. The police took DNA samples from him and the other man.
The crime lab's analysis excluded the other man, who was then released,
but the lab declared that Sutton's evidence was consistent with some of
the crime samples.

There were several crime samples: one was a semen sample extracted
from the back seat of the victim's car, and the others included a vaginal
swab sample, combings from the pubic hair, and a stain from the jeans of
the victim. These samples were analyzed by the Houston Police Department Crime Laboratory serology unit, which reported:

> A mixture of DNA types consistent with J. Sutton, [the victim], and at least
> one other donor was detected on the vaginal swabs, unknown sample #1, de-
> bris from the pubic hair combings, and the jeans based on PM, DQA1, D1S80
> typing results. The DNA type of J. Sutton can be expected to occur in 1 out
> of 694,000 people among the black population.[33]

During the trial in the Houston court, Sutton's defense did not obtain
independent access to the prosecution's DNA evidence or resources for
hiring expert witnesses. To say the least, Sutton's defense did not effectively contest the prosecution's analysis. In Texas, there is no system
of public defense lawyers for indigent defendants. Instead, judges appoint private defense attorneys, who are reimbursed by the court. Compared to public defenders, court-appointed defenders are more dependent on judges for a steady stream of work, and critics of such systems
argue that they are less inclined to irritate the judge by pursuing a vigorous defense. Such concerns are all the more acute in Houston, where
every single judge is a former prosecutor, and the court-appointed defense attorney is paid a relatively low fee and is not reimbursed for the
services of expert witnesses.[34] According to Thompson, Josiah Sutton's

33. Quoted in William Thompson, "Review of DNA Evidence in *State of Texas v. Josiah Sutton*" (District Court of Harris County, Case no. 800450), unpublished document
furnished by author. This was a set of notes that Thompson prepared at the request of David Raziq and Anna Werner, investigative reporters for television station KHOU.

34. Thompson (interview, 18 October 2005) informed us that Texas judges are elected,
and almost all of them had served as district attorneys. He added that no judge had ever
been elected who was opposed by the local district attorney, and so they tended to curry

attorney had the dubious distinction of having sent more of his former clients to death row than any other defense attorney in the United States. Thompson added that this lawyer had never gained an acquittal in a single case, and that when the defense attorney put Sutton on the stand, the first question he asked in direct examination was, "Isn't it true that you contracted venereal disease while you were in juvenile hall?"—thus introducing two pieces of damaging evidence to his client in a single question. The apparent rationale for asking the question was to discredit the rape allegation by suggesting that if Sutton had committed the rape, the victim would have contracted venereal disease. The prosecutor parried this suggestion by noting that a standard procedure when examining a rape victim is to give her antibiotics.[35]

The trial went quickly, and Sutton was convicted and sentenced to a twenty-five-year jail term. In 2002, after Sutton had served several years of his term, Houston television station KHOU ran a series of programs investigating the Police Department Crime Laboratory in Harris County, which had sent more people to death row than any other U.S. county. KHOU contacted several experts, including William Thompson, who reviewed trial transcripts and case records.[36] According to Thompson:

> In televised interviews I said the work of the HPD lab was some of the worst I had ever seen. The laboratory routinely failed to follow proper scientific procedures. It interpreted DNA test results and computed statistical estimates in a manner biased against the accused. Most importantly, I found several instances in which there was outright misrepresentation of scientific findings— where the lab analysts would say that two samples had the same DNA profile when the actual test results showed they did not. (Thompson, 2003: 16)

The Harris County district attorney requested an audit of the serology laboratory, and the January 2003 report was damning.[37] It described

favor with local prosecutors. Consequently, the defense attorneys they appoint tend not to be the most aggressive advocates for their clients, and the low amount of reimbursement by the state encourages a perfunctory defense.

35. Ibid. Thompson mentioned that Sutton's family eventually hired another counsel.

36. Thompson mentioned that the fact that the investigation was initiated by television journalists was a "tremendous embarrassment" to the legal system, including defense lawyers. Ibid.

37. In June 2007, an extensive *Final Report of the Independent Investigator for the Houston Police Department Crime Laboratory and Property Room* was released. The investigators reviewed 135 DNA cases (including four death penalty cases), and identified

many of the mistakes and misrepresentations that KHOU had chronicled, and stated that the lab worked under heavy political pressure
to obtain convictions. In addition, the report stated that the evidence
room had a leaky roof, which resulted in the destruction of thirty-four
evidence samples on one rainy night (Thompson, 2003: 16). Interestingly, the leaky roof serves as a concrete, visual symbol of disorder, even
though it is, for critics such as Thompson, a much less significant problem than dubious interpretations masked by impressive probability figures. Such interpretations are subtle and difficult to explain to jurors and
the media public, and the leaky roof became something of a metaphor
for the less-than-watertight evidence produced at the site. The district
attorney agreed to shut down the laboratory, pending review of problematic cases. One such case was Sutton's.

After reviewing the case evidence, Thompson objected that the
1:694,000 figure greatly exaggerated the probability that Sutton's DNA
evidence would match the crime samples. The figure was based on the
probability that a randomly drawn person would match the *defendant's*
profile, rather than that a randomly drawn person would match the criminal evidence. The problem was that the crime samples were presumed
to include DNA from two, and possibly three, different persons. One of
the analytic techniques employed was the DQ-alpha polymarker system,
which deploys PCR to "amplify" a DNA sequence (the DQ-alpha locus,
a gene associated with particular immune deficiency disorders), supplemented by markers from five other genetic loci, each of which has two
or three alleles. An early STR locus (DS 180) also was used. Given the
relatively limited array of possible alleles that show up in that type of
test, when samples from different persons are mixed, the resultant profile covers a larger array of possible markers, thus making it more likely
that a DNA profile taken from a randomly chosen person would match
a subset of those markers. Taking the mixtures into account, Thompson recalculated the probability of a coincidental match to be greater
than 1 in 15. Given that two suspects were tested, either of whom could

---

"major issues" in 43 (32 percent) of them, and the report observed that "[b]y stark contrast with the overall quality of work performed in [other] . . . sections of the Crime Lab,
the Lab's historical serology and DNA work is, as a whole, extremely troubling. We found
significant and pervasive problems with the analysis and reporting of results in a large proportion of serology and DNA cases. The Crime Lab's substandard, unreliable serology and
DNA work is all the more alarming in light of the fact that it is typically performed in the
most serious cases, such as homicides and sexual assaults" (Bromwich et al., 2007: 4).

have matched, the probability of a coincidental match rose to approximately 1 in 8.[38] Although the mixed samples made up a contingency that was not taken into account in the *Deen* and *Adams* cases we discussed in chapters 5 and 6, once again we see that adversarial reanalysis of an impressive probability statistic can greatly reduce its evidentiary weight. Thompson argued further that, contrary to what the prosecution's analysis claimed, Sutton's DNA did not match the crime sample taken from the back seat of the car. The semen apparently had come from an unknown man. That unknown man's sample was consistent with the vaginal sample. However, if the unknown "donor" was one of the two rapists, then Sutton was not the other one, because the combination of alleles in both of their profiles was inconsistent with the vaginal sample. Under the assumption that one of the two rapists produced both samples, then Sutton was excluded from being the other perpetrator.[39] The DNA samples were retested by a private firm, using a more recent STR system, and Sutton was excluded in the retest. He was released from prison in 2003, and the district attorney ordered retesting of evidence from sixty-eight convicted prisoners, seventeen of whom were on death row. Unfortunately, in many of those cases the crime laboratory had consumed all of the DNA evidence during earlier testing.

The *Sutton* debacle exemplifies many of the themes that we encountered in discussions of early preclosure cases: eyewitness testimony that, in retrospect, was highly dubious but which initially was overridden by DNA profile matches; dubious procedures for generating probability estimates; and declared matches that overlooked anomalies. To these elements were added some distinctive features of Texas criminal justice: underfunded and incompetent defense representation, and political

38. This is an estimate of the probability that a profile from a randomly chosen individual from the relevant population group would match the criminal evidence. Thompson supplied these estimates in his report to KHOU and in an interview with Cole & Lynch (15 October 2005).

39. The reason for this, according to Thompson's report to KHOU, was that the "vaginal sperm fraction" included at least one allele that was included in neither Sutton's profile nor the profile from the unknown "donor" of the semen sample recovered from the back seat. Assuming that the victim was truthful when she stated to the police that there were two rapists, and that she had not had sexual intercourse with a third man for several days before the assault, then the extra allele in the mixture was anomalous. And, since the victim's blood sample (which Thompson reports was analyzed twice by the crime lab, with inconsistent results) also did not include the allele, it apparently was not an artifact of an incomplete fractionation of the vaginal sample to separate the "sperm" fraction from the victim's DNA.

pressures to convict. We also can see that interpretations of the evidence were framed by assumptions about the "ordinary fact evidence" and assessments of the truthfulness of the victim's testimony.

One could argue that the forensic techniques used in the *Sutton* case in 1999 were precursors to the currently used STR system (which was just then being adopted in the United States), and that the reanalysis of the evidence using the more recent technology was instrumental for revealing the weakness of the earlier analysis. To address such concerns, we shall examine a more recent case (though, again, the original analysis did not deploy techniques used today), and we shall also note that popular discussion of the power of DNA tends to gloss over differences in technique that continue to work themselves through the criminal justice system.

The *Sutton* case undermines the dominant narrative in which DNA is treated as a magic bullet in an imperfect justice system. The standard narrative portrays DNA as unassailable science which exposes the flaws in less trustworthy procedures like eyewitness identification, interrogation, police investigation, lesser forms of forensic science, and, not least, the justice system itself. Crucial to this narrative is the harnessing of the authority of science to critique "nonscientific" procedures. The critique works because it plays off the perceived epistemological authority of science.

As Saks and Koehler (2005) point out, this use of science was always somewhat paradoxical, because the second leading cause of wrongful convictions exposed by postconviction DNA testing was science itself: faulty forensic science, principally serology and microscopic hair comparison. Thus, this complicates any effort to draw easy contrasts between science and law.[40] It can always be argued, of course, that science is not monolithic, and that microscopic hair comparison, if it even is science in some sense of the word, is quite different from DNA profiling. But the *Sutton* case, and at least one other like it, poses a more tortuous paradox:

40. An interesting question is to what extent Scheck, Neufeld, and other proponents of DNA exonerations are aware of this. The original breakdown of the causes of wrongful conviction in the postconviction DNA exoneration data set published in *Actual Innocence* (Scheck et al., 2000: 263), deliberately or not, downplays the role of forensic science in generating wrongful conviction by splitting "forensic science" into several categories ("hair comparison," "serology," "other"). It took Saks and Koehler's (2005) reanalysis of the same data to show that forensic science was, in fact, the second leading cause of wrongful convictions in the data set. But, see Garrett & Neufeld (2009).

a case in which someone was convicted with DNA evidence and later exonerated by DNA evidence. The inclusion of such a case among the postconviction DNA exonerations undermines the efforts of that very project to construct "DNA" as unproblematically truthful, a "truth machine." Instead, the *Sutton* case would seem to point to less spectacular culprits in his wrongful convictions: not the technology itself, but subtle (and sometimes blatant) interpretive maneuvers that can lead any technology, no matter how "intrinsically" powerful, to dubious conclusions.

Hats and Hotdogs

The *Sutton* case garnered national attention, but for a further demonstration of some of the continuing problems with DNA evidence, we now turn to a lesser-known case in which William Thompson also was involved: *People v. Robinson*. This case took place in California Superior Court, and was not reported. The details we recount were reconstructed from interviews with William Thompson in May 2003 and October 2005.

Robinson was a murder case in which two brothers—Cory and Brendale Robinson—were accused of murdering a clerk at a convenience store. Details of the crime were recovered from the store's closed-circuit video camera: the store clerk confronted two African American men who apparently were freely loading their coat pockets with items from the shelves. The clerk demanded that they return the goods they had taken, and this confrontation quickly escalated to a physical fight. While the clerk and one of the men were violently wrestling on the floor, the other man took out a gun and fatally shot the clerk from short range. The perpetrators then fled the scene. During the criminal investigation the Robinson brothers were identified on the basis of testimony by a former girlfriend, who claimed that she saw them with the stolen goods. She had been arrested on forgery charges, and the prosecution agreed to recommend no jail sentence and to cover restitution costs in exchange for her testimony against the brothers. The evidence was less than conclusive: the brothers could not be clearly identified from the videotape, because of the distance and camera angle; the ex-girlfriend's testimony was suspect because of the inducement; and there were discrepancies between the stolen goods she described and the police record of the stolen items. DNA tests also were performed on material recovered from beneath the victim's fingernails, and from the inner sweatband of a hat that been recovered at the scene. The video had shown that the shooter was wearing a white hat, and the police presumed that the hat found at the scene was the one he wore. A DQ-alpha polymarker test and an STR test with seven loci were performed, but the material under the finger-nails did not match the DNA profiles of either of the defendants. How-ever, the prosecution did manage to declare a match between the profile of one of the brothers and the hatband, though there were further com-plications with the evidence.

The forensic analysts cut out three different sections along the hat-band for the purpose of DNA testing. From their analysis of the small amounts of DNA recovered from each sample from the hatband, the crime lab concluded that more than one person had worn the hat. There were as many as four alleles at some loci, but no more than four at any single locus. Since no more than two alleles should show up at any lo-cus for the profile of one person (two being the number for heterozygous alleles), the evidence indicated that the samples included mixtures of DNA from at least two individuals. Given the pattern, it was also possi-ble, but less likely that the samples included DNA from three or more in-dividuals. To complicate things further, profiles developed from the dif-ferent samples taken along the hatband showed different patterns, using different techniques, which the prosecution's analysts inferred were due to the band's uneven contact with the heads of the different individuals who had worn the hat.[41]

Aside from the questions that might be raised about the prosecu-tion's inferences about hat wearing—that the hat was worn by two, three, or more people, and had uneven contact with the heads of the different wearers—the unknown number of hat-wearers presented an interpretive problem for assigning a probability figure to the match between Robin-son's DNA and the DNA extracted from the hat.[42]

41. Thompson explained that the forensic lab began by using the DQ-alpha test. They then decided to use the STR system, which was just beginning to go on line at that labo-ratory, and this required a different sample from the hatband. The lab then multiplied the frequency of the match on the STRs with the frequency of the match on the DQ-alpha to generate the match statistic. Thompson said (interview, October 2005) that he objected that it was not clear that the samples from different parts of the hatband came from the same person. But while, he said, "leading forensic scientists" would agree with his argu-ment, apparently the jury did not.

42. The issue of how the technical analysis depends on inferences about how many people would wear a hat brings to mind something an FBI case examiner told Kathleen Jordan during an interview in December 1992, when PCR was being developed for crimi-nal forensics: "We take cases, extortion cases where an envelope is licked. Now how many people are going to lick an envelope? . . . How many people are going to lick a stamp? How

wt. In this case . . . the government's case actually depended upon there being three different people whose DNA [was] on the hat.

ml. This is where you get into these inferences about what's plausible in these kind of ordinary ways: Would three people wear a hat? You know, that kind of thing.

wt. Well, yeah! Now, I actually think it's interesting. Part of what's fascinating about the case is that whether the DNA is strongly incriminating or totally exculpatory depends *entirely* on how many people you think wore that hat. If you think— If you think, if it's *two,* then, you know. . . . My Johnnie Cochran line that I thought of after it was over, is, you know, "If it's two, the case is through!"[43]

The hatband samples contained very small amounts of DNA, near the threshold of detection for the STR and DQ-alpha tests. According to Thompson, some of the color-coded graphic peaks shown in the output of the STR analysis were small. He explained that with small samples, and especially with older or degraded material, it is common to get peaks that are relatively low, difficult to distinguish from "noise," and it is also possible that some loci which would register with a larger or better-quality sample will not show up at all.[44]

The material under the victim's fingernails also presented an interpretive problem. Initially, it seemed sensible to assume that the clerk might have "sampled" the DNA of his assailant when grappling with him on the floor of the convenience store. Accordingly, the fact that neither of the Robinson brothers' DNA failed to match that sample would tend to exonerate them. However, the jurors used common knowledge about convenience stores to explain away the discrepancy by suggesting that the DNA derived not from another human source (presumably, the

---

many people are going to drink from a cup, how many people are going to smoke a cigarette butt? . . . In other words [we] look at things which one can reasonably assume DNA, genomic DNA . . . so consequently we do a lot of stamps, envelopes, cigarette butts, bubble gum, cups, things that one person would drink out of" (Jordan & Lynch, 1998: 792).

43. William Thompson interview, October 2005. Thompson is referring to a famous line from the closing argument of defense lawyer Johnnie Cochran in the O. J. Simpson trial, in which Cochran refers to a bloody glove presumed to have been worn by the perpetrator, but which apparently did not fit Simpson's hand: "If it doesn't fit, you must acquit."

44. With degraded material, the DNA tends to break up, especially at loci containing larger sequences, thus reducing the amount of material registered by florescent markers at the specified site, thus resulting in a relatively low peak at that site.

actual perpetrator), but possibly from a nonhuman source: hotdogs sold at the store. Ironically, according to Thompson, the specific expertise of a witness he used was recruited inferentially to support this possibility.

When assisting the defense, Thompson enlisted the help of William Shields, a long-time expert witness on DNA matters who was discussed by Halfon (1998) in an S&TS study of expert witnesses in cases involving DNA evidence. Shields was prepared to explain to the jury some of the problems with interpreting matches with mixed samples. Initially, the prosecution attempted to dismiss his testimony by pointing out that Shields was an expert on bird DNA, not human DNA.

> Q. Okay. Now, Mr. Moses asked you about some articles you've published on birds and animals. Does research on animal genetics, does that have any relevance to humans?
> A. I hope so. Animals—humans are animals. (*People v. Robinson,* 4 December 2001: 87)

This was, in fact, a common tactic for trying to neutralize Shields. For instance, in discussion of the notorious Scott Peterson "celebrity" trial, CNN legal pundit Nancy Grace reported on the television program *Larry King Live:*

> GRACE. Upon further investigation that I did on the defense expert Dr. William Shields, I found out quite a bit. One, he has never extracted mitochondrial DNA for forensic purposes. In fact, his expertise, Larry, some of his articles are "Inbreeding and the Evolution of Sex," "The Management of the Wolf." I looked up his Web site, and I found this picture of two little birds. Why is he testifying about MTDNA? Somebody explain!

When another guest on the program, Chris Pixley, attempted to interject a comment, Grace continued emphatically:

> GRACE. . . . but when your DNA expert writes articles about "Barn Swallow Mobbing: Is It Self Defense" . . .
> PIXLEY. Nancy . . .
> GRACE. A barn swallow is a bird, Chris!
> PIXLEY. I want to ask . . .
> GRACE. This is not about DNA!

And again:

> GRACE. This Dr. William Shields that they brought in, I mean, when I
> investigated him and found out that most of the publications are about the
> animal world, how they propagate, about birds and wolves and there he is
> testifying as a DNA forensic expert, in my mind, unheard of.[45]

Thompson, in his words, "rehabilitated" the witness by pointing out
the massive structural and analytical similarities between human and
bird DNA. But the close link between human and nonhuman DNA was
turned against Thompson's client when the prosecution pointed out that
the convenience store sold hotdogs, that the videotape had shown the
clerk preparing them, and that the material under his fingernails might
have come from handling hotdogs rather than from a struggle with one
of the perpetrators. In the retrial, in which Thompson did not participate, the fingernail evidence was excluded when the judge agreed with
the prosecution's argument that there was insufficient proof that the material came from one of the perpetrators.[46] Once again, the evidence
gained or lost significance based on a story of the crime, and understandings of commonplace elements at the scene of the crime. The Robinson
brothers' trial resulted in a hung jury. For reasons that are not entirely
clear, the defense did not ask Thompson to participate in the retrial, and
the second trial resulted in the conviction of both brothers.

This case vividly illustrates that possibilities of technical error and
artifact do not necessarily work to the disadvantage of the prosecution.
A kind of endogenous "sociology of error" can occur in adversary disputes in which technical sources of artifact can be used to *preserve* the
inference of a match in the face of nonmatching details.[47] Again, as oc-

45. CNN, *Larry King Live,* transcript, 4 November 2003, http://transcripts.cnn.com/
TRANSCRIPTS/0311/04/lkl.00.html.

46. One might wonder why the laboratory tests of the material taken from underneath
the victim's fingernails did not exclude the possibility of animal DNA. In fact, according
to Thompson (interview, October 2005), he had explicitly argued to the jury that the test
results were specific to humans. He noted that after two jurors were interviewed following
the trial, he realized that he had made a "tactical error" by arguing that Shields was competent to testify, because human and animal DNA are very similar. This may have led the
jurors to tune out his other point about the tests excluding a nonhuman source.

47. "Sociology of error" is a term used in the sociology of knowledge to describe the
citation of social factors in accounts that dismiss or discredit specific truth claims. Bloor
(1976) famously argued that the sociology of scientific knowledge should not limit social
explanation to circumstances in which naturalistic claims are held to be dubious. Our ref-

curred in cases we discussed in earlier chapters, nonmatching bands or peaks can be "explained" not as evidence that the samples being compared came from different sources, but rather as glitches in the particular evidence that *suppress* or *interfere with* the visibility of a more perfect match that would have been obtained with better samples analyzed under more ideal conditions. To further illustrate this possibility, Thompson described a rape and murder case in Maine, in which the key sample was collected at the coroner's office. He recalled that the defendant had been included in the pool of suspects on the basis of a DQ-alpha test, but that an STR test commissioned by the Innocence Project later excluded him. The state then argued that the mismatch was due to contamination of the sample collected by the coroner, and supported this account with forensic scientists who testified that the coroner had not been using sterile procedures at the time, so that DNA from another case may have been introduced into the sample. The Innocence Project then tested DNA samples from people who worked in the coroner's office and also tested samples from other bodies that had been examined with the same instruments in the coroner's office. Those who were tested were excluded, but it proved impossible to get samples from all of the bodies, and the Innocence Project eventually withdrew from the case.[48]

Thompson referred to other cases involving very limited samples resulting in missing alleles as well as "spurious alleles" due to either contamination or running too many amplification cycles.

ML. Of course, they can go both ways on these, can't they? I mean, they can
    discount mismatches by invoking these kinds of mistakes.
WT. Well, yeah. Exactly. Exactly. And that's why . . . I've been arguing, that
    there's a . . . lot more interpretive flexibility available to these analysts
    than— than is acknowledged. . . . Everybody views it as cut and dried,
    and in many cases it *is,* but in the cases I'm concerned about, which is a
    significant fraction, there's— there's really a *lot* of *flexibility.* . . . And it has

erence to *endogenous* sociology of error is informed by ethnomethodology (see, for example, Pollner, 1974; 1975). The idea is that the family of terms associated with mistakes, errors, misinterpretations, and, in the case of laboratory technique, artifact (Lynch, 1985a) is actively employed in social interaction to advance or discredit rival claims. Even "interpretive flexibility"—an established analytic term in science and technology studies—comes into play in participants' discourse.

48. Thompson refers to a magazine article (Thompson et al., 2003) that he coauthored about contamination issues.

a couple of implications: one is that you could end up . . . calling a match
between somebody who doesn't match, or falsely excluding someone . . .

Conclusion

In chapter 7 we attributed closure of the DNA fingerprinting controversy to three interrelated "fixes": technical fixes, administrative fixes,
and legal fixes. In this context of explanation, we prefer the term "fix"
rather than, say, "factor," because it refers more specifically to concerted
actions that address a problem, rather than a causal force or explanatory variable that impinges upon a given nexus of human actions. The
term "fix" can express the ironic connotation of a stopgap solution that
glosses over a deeper set of problems and/or a broader set of contingencies that eventually threaten it. When considering the postclosure phase
of the controversy over forensic DNA evidence, we can begin to appreciate that the fixes are challengeable, in detail, even while remaining in
force as sources of closure, in general.

Thompson and other critics of contemporary DNA profiling methods acknowledge that the STR system and the associated procedures for
preparing and analyzing samples hold distinctive advantages over earlier
techniques: they require less sample material, can recover usable information from degraded samples, replace visual judgment with automatic
laser reading of samples, and generate impressive match probabilities.
However, the cases described in this chapter exemplify problems with
reliability that had been *thought* to be out of play, but which can still
arise in particular cases. The difference—and it is a significant one—is
that a *presumption* of reliability is now a default condition for raising
challenges in particular cases. Administrative fixes are highly significant
in this context, because they provide citable resources for claiming that
prestigious scientific panels have deemed forensic DNA testing to be reliable and that forensic laboratories have adopted quality control standards to assure and enforce such reliability.[49] However, as the scandal involving the Houston Police Department Crime Laboratory exemplifies,

49. As we saw in chapter 3, in the excerpts from the O. J. Simpson trial, an adept defense attorney can use formal recommendations and standards as a resource for casting
aspersions on actual practices. However, in most cases, when there is no Peter Neufeld to
follow through with such an interrogative strategy, the presumption of reliability stands
unchallenged.

the general presumption of reliability and the formal availability of standards can cover over what happens "on the ground" (or under the leaky roof) when samples are handled and tested.[50] Legal fixes in the form of precedents on admissibility and less formal inferences about the gold standard provided by forensic DNA testing also can serve to enhance the power of presumption, especially under circumstances in which defendants have limited access to competent defense attorneys and expert witnesses. The Innocence Project and related campaigns have resulted in widespread publicity about inequities in the U.S. criminal justice system, but public discourse about the "truth machine" tends to exempt DNA testing from the critical scrutiny given to virtually all other forms of ordinary and forensic evidence (unless such scrutiny is done through the lens of an upgraded DNA technique).

It can be argued that forensic DNA testing is, in fact, highly reliable, and that cases such as the Josiah Sutton case or the Robinson brothers' case are exceptional. As Thompson noted in his accounts of some of these cases, they involved efforts by the prosecution to press the analysis of DNA evidence to the limits of its resolution. Reasons for doing so are easy to imagine: the crimes were serious and horrible, and strong incentives were present for making the most of the limited evidence available. Some of the cases involved transitional techniques (DQ-alpha polymarker, and early variants of PCR) and mixed samples. The efforts by prosecution experts to "make do" with such techniques, and to "make out" meaningful patterns in noisy fields, could be dismissed as exceptional instances of "pathological science" (Langmuir, 1989 [1968]) embedded in an otherwise reliable practice.

Such efforts are, however, undermined by continuing reports of serious errors, incompetence, and biased interpretations emerging from a wide range of DNA laboratories around the United States. A recent scandal at the Virginia State Crime Laboratory, which held itself to be among the nation's leaders in the area, undermined any claims that such problems were confined to poor-quality organizations like the one in Houston.[51]

The history of dermal ridge fingerprinting is worth recalling in connection with the possibility that DNA testing is "stretched" to, and even beyond, the threshold of resolution in some cases. As Cole (1998;

50. Aside from egregious cases such as the Houston crime lab, it remains to be seen how often, if at all, "ordinary" forensic laboratories adhere to recommended forms of proficiency testing, probabilistic calculation, and so forth.

51. See Dao, 2005.

2001a; 2005) points out, the much-heralded reliability (even infallibility) of fingerprinting is bolstered by a professional culture among fingerprint examiners that purportedly emphasizes "conservatism." That is, fingerprint examiners are enjoined not to count (declare as matching) ambiguous evidence. Moreover, the major association of fingerprint professionals has gone so far as to prohibit members from giving "probable" evidence when testifying in court: "[A]ny member, officer or certified latent print examiner who provides oral or written reports, or gives testimony of possible, probable, or likely friction ridge identification shall be deemed to be engaged in conduct unbecoming such member, officer, or certified latent print examiner."[52] Such professional discipline enhances the apparent reliability of the practice, since latent fingerprint examiners, on penalty of sanction, are supposed to declare as inconclusive any latent comparisons showing "insufficient" consistent ridge detail to eliminate all other donors. This policy enhances the credibility of examiners' declarations, but it also results in ruling out evidence that might otherwise be used for declaring probable evidence.

Given the incentives for pursuing convictions in rape and murder cases, it seems reasonable to suppose that prosecutors are unwilling to consign threshold DNA evidence to inconclusive status. Perhaps the "DNA mystique" (Nelkin and Lindee, 1995) also encourages them to play a dangerous game with such evidence. One could argue that "DNA fingerprinting" eventually will adopt conservative professional standards, just like its namesake. However, there are some indications of a reverse trend—of devising ways to utilize fingerprint evidence that is not declared because *merely* probable.[53] As we elaborate in the next chapter, the relationship between DNA typing and fingerprinting has become complicated in recent years, and it is no longer the case that the more recent technique is following a path forged by its predecessor. Indeed, the relative positions have reversed, and the status of fingerprint evidence is now being reviewed in light of the successful career of its younger namesake.

52. International Association for Identification, *Resolution V,* 30 Identification News 3 (August 1980). Quoted in Cole, 2005b: 992.

53. In an interview with Michael Lynch and Ruth McNally (19 March 2002, Forensic Science Service, London), Ian Evett pointed out that the refusal to declare "probable" fingerprint matches resulted in the loss of potentially important criminal evidence. While acknowledging the severe difficulty with quantifying the probability of such matches, Evett noted that it would enable currently undeclared evidence to be used.

Fingerprinting and Probability

In the late 1990s, with "DNA" as the model, renewed demands were made to develop probability figures for fingerprint comparisons. In this interlude we discuss why latent fingerprint comparisons have thus far proved intractable to probabilistic analysis.

As suggested by the common name "DNA fingerprinting," Alec Jeffreys and other proponents of the new technique drew a strong analogy with the established forensic practice of fingerprinting. There was, and is still, a strong basis for the analogy: both are forensic identification or forensic individualization techniques that attempt to link suspects to crime scenes by identifying similarities between crime scene samples and reference samples known to come from particular suspects. However, one difference that became prominent during the "DNA wars" in the early 1990s, and again at the end of the 1990s when the admissibility of fingerprint evidence began to be challenged, had to do with *probability*.

Although popular discourse surrounding DNA is rife with celebrations of the uniqueness of each individual's genome (monozygous twins excepted), aside from some statements made when the multilocus probe technique was first developed, forensic DNA researchers did not claim that DNA "fingerprints" were unique to individuals. As we have seen, forensic DNA researchers, partly in response to academic and legal criticism during the DNA wars, developed probability estimates by collecting samples and building databases of DNA profiles from samples drawn from various so-called racial or ethnic populations. Courts demanded that expert witnesses should quantify the "random match probability" (RMP)—the frequency of the perpetrator's DNA profile in a specific population—to assess the probative value of a DNA inclusion.

Quantitative frequency estimates had been used previously in court-room presentations of serological evidence (for example, ABO blood typing), but while these earlier techniques could exclude, for example, a rape suspect when his serological type did not match the crime sample, the probabilities of a coincidental or "innocent" match were many orders of magnitude higher than those that would later be used in DNA profiling. No such figures were given or demanded for many forms of traditional comparison evidence (for examples, fiber, hair, handwriting, bite mark, and fingerprint comparisons). The latent print examiner (LPE) profession went so far as to actually ban the use of probabilities in examiners' testimony.

Instead of characterizing the rarity of the features of the latent print statistically, LPEs adopted the convention of declaring an inclusion only when the features of the print were deemed to be so rare as to exclude all other possible donors, a condition that was known as "individualization" (Cole, 1999; Thompson and Cole, 2007). Since no one had collected data on the rarity of various friction ridge features, LPEs could only do this intuitively. There was no set number of corresponding details that would ensure that the potential donor pool had indeed been reduced to only one area of skin. Thus, LPEs restricted their testimony to only three possible conclusions: individualization, exclusion, and inconclusive (Scientific Working Group on Friction Ridge Analysis Study and Technology, 2003). A vernacular declaration of individualization was equivalent to a quantitative estimate with an RMP of zero.

The courts generally allowed LPEs "to testify about identity as if it were fact, not opinion" (Mnookin, 2001: 30). LPEs especially seemed to hold an advantage over forensic DNA analysts because of the way they made strong, unqualified claims to identity. LPEs thus enjoyed the best of both worlds: they were able to opine that all potential donors of the latent print other than the defendant had been eliminated, a claim that not even DNA analysts would make, and yet they avoided the complex arguments about probability and population genetics that dogged forensic DNA analysis during the height of the DNA wars.

When, as they often did, proponents of traditional comparison evidence made strong vernacular claims of individual identification, this put them in an advantageous epistemological position, compared with serologists, or even DNA analysts, who adduced merely probabilistic evidence. Although trace evidence lacked quantification, when presented with unqualified expressions of certainty, it seemed more probative and less complicated; it simply linked a trace to the defendant. Although DNA profiling combined quantification with very low RMPs, during the DNA wars of the early 1990s these estimates offered targets of criticism and sources of confusion.

Counting Details

Courts and practitioners assume that each individual's friction ridge skin, like each individual's genome, is unique. Indeed, even monozygous twins' friction ridge skin is not identical, whereas their genomes are assumed to be identical (Lin et al., 1982). However, it is also recognized that forensic DNA researchers do not analyze the complete genome; they examine graphic bands or peaks that denote a relatively small sample of alleles, each of which occurs with variable frequency in the population. And, as we have seen, DNA analysts have devised formulae for combining the allele frequencies in a given profile, in order to calculate the RMP for that profile. Fingerprint examiners also examine details that are assumed to vary within the human population, but despite claims dating back to Galton's famous estimate that the chance of two randomly chosen persons having the same ridge details is 1 in 64 billion, such estimates are of the probability of finding an exact replica of all the details in a complete fingertip, and they are not estimates of the rarity of the friction ridge features contained in any particular latent print. No stable, agreed-upon method for assigning probability to fingerprint matches has ever been devised. Whether they can, or even should, be devised remains an open question.

Like DNA analysts, LPEs reach conclusions by finding consistencies between crime scene samples and reference samples.[1] Whereas DNA analysts look at "alleles" (or rather, bands, blots, graphic peaks, or other literary proxies for allele size and frequency), LPEs look at impressions of "friction ridge details." Historically, these details (also variously called

1. To avoid confusion concerning the many possible meanings of the term "fingerprint," we adopt Champod & Evett's (2001) convention of using the term "mark" for a putative impression of friction ridge skin found at a crime scene. "Fingerprint" is an inappropriate term, because a "mark" may derive from the friction ridge skin of a finger, palm, or sole, or even from another object, such as the spine of a leather-bound book. A "print" is an impression of friction ridge skin taken deliberately from the finger, palm, or sole of an individual.

"Galton details," "points," "points of similarity," "points of identification," "minutiae," or "ridge characteristics") were primarily ridge endings
and bifurcations. LPEs seek to determine whether the ridge details are
in the same topographic relationship in the mark and the print (fig. E.1).
Even in the early days, however, Edmond Locard pointed out that other
features, such as the shapes of the ridges themselves or the locations of
skin pores, might also provide identifying information (fig. E.2). David
Ashbaugh (1999) has coined the terms "second-level detail," referring to
ridge endings and bifurcations, and "third-level detail," referring to finer
features of the ridges (first-level details are the overall patterns of loops,
arches, and whorls).

It might be thought that an LPE could simply match up ridge details,
much in the way a DNA analyst matches up autoradiographic bands
or STR peaks, and then assign a probability figure to measure the rarity of each matching detail. The situation is not so simple, however. In
currently used STR analysis, the alleles are assumed to be "discrete"
(though as described in chapter 8, in practice the graphic display of possible "alleles" can be ambiguous, because of possibilities of contamination and sequence degradation). In principle, forensic statisticians are
able to calculate the frequency with which a particular allele occurs in a

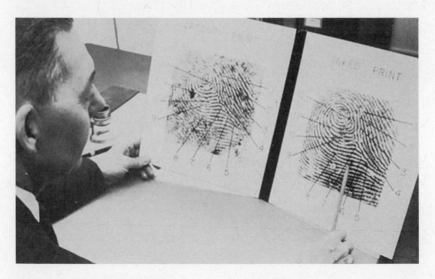

FIGURE E.1. Latent print examination

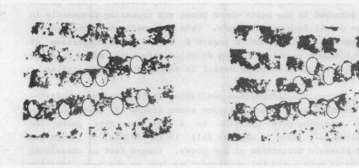

FIGURE E.2. Were the pores not visible, these two areas of friction ridge skin would appear to "match." However, with the pores visible it is clear that they derive from different sources. Source: Ashbaugh, 1999.

given population or subpopulation. The same is not true of the friction ridge details that LPEs analyze. Theoretically, each detail is visibly different from each other detail. Even two bifurcations in the same relative location differ from one another in some way. "Third-level detail" can make this apparent (fig. E.2).

Each impression "inscribed" by the same ridge detail visibly differs from other such impressions. Material differences between impressions can be attributed to pressure distortion, contamination, substrate noise, and so on. Note that such differences can be cited in order to *save* the inference of a match as well as to challenge it (fig. E.3); differences can be attributed to artifacts or taken as signs that the prints originate from different sources. Which explanation is preferred is a matter of interpretation for the analyst.

One simple way to characterize the difference between fingerprints and DNA profiles is to say that the former are "analog" while the latter are "digital." Digital information is easier to count, manipulate, reproduce, share, discuss, and analyze.[2] However, this easy distinction breaks down when the production of DNA evidence is examined in detail. As we saw in chapters 5 and 6, single-locus and multilocus probe techniques

2. However, as many audiophiles have claimed, information loss may be associated with digitizing information, which might be conceived as the "sounds 'between the knobs'" (Pinch & Trocco, 2002, 319) or "the information that falls between the digits" (Perlman, 2004). Analog, though more difficult to work with, is in some sense truer to the original product since it reproduces information in continuous, rather than discrete fashion.

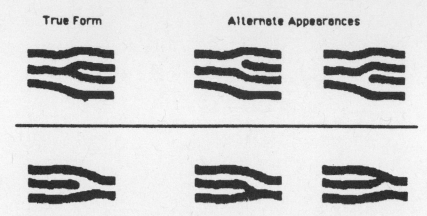

FIGURE E.3. A pixel of ridge that either fails to reproduce or reproduces spuriously can have a significant impact on the impression, changing an ending into a bifurcation or vice versa. Source: Stoney and Thornton, 1986; © ASTM International.

were not digital, because bands in parallel lanes of an autoradiograph do not always line up discretely, and a single band may be blurred, requiring the creation of a "bin" that includes molecules within a size interval. The STR system apparently fixed these problems with laser reading of exact molecular weight, graphic displays of analytical results, and also by the selection and color-coding of discrete chromosomal loci, but even with that system it is necessary to make judgments about the thresholds at which to count a given peak as an "allele" included in the profile, rather than an artifact due to degraded material, contaminant DNA, or incomplete hybridization.

According to Christophe Champod, a Swiss forensic statistician working at the time of the interview with the British Forensic Science Service, the differences between STR and fingerprint evidence are more apparent than real. In part, the "digital" properties of STR profiles are due to what Champod calls the "preprocessing" of graphic data before they are quantified and presented publicly.[3] What Champod seems to suggest is that the "subjective" or judgmental aspects of STR analysis become hidden, because they are "black boxed" by delegating visual inspection and analysis to machines (Latour, 1987).

3. Interview of Christophe Champod by M. Lynch, Forensic Science Service, Lambeth, London, July 2003.

The Meaning of a "Match"

As the relative credibility of fingerprinting and DNA underwent inversion, the nonquantitative character of fingerprint evidence went from being an advantage to becoming, potentially, its Achilles heel. The public's widespread perception of science as characterized by numbers had the potential to undermine the credibility of fingerprinting. The fact that proponents of DNA evidence had worked so vigorously to establish the precise details of random match probabilities had the potential to make fingerprint examiners look rather lax for evading the issue of probability altogether.

One expression of the danger may be found in the first of the two *Llera Plaza* decisions made in January 2002. Judge Pollak concluded that latent fingerprint evidence was deficient in terms of almost all of the *Daubert* factors, but he also recognized that such evidence was strongly established and heavily relied upon by criminal justice organizations. Consequently, he reasoned, blanket exclusion of such evidence would be "unwarrantably heavy-handed," and he proposed a compromise worthy of Solomon.[4] According to Pollak, latent print examiners would be permitted to testify about similarities and differences between a mark and a print, but they would not be permitted to "present 'evaluation' testimony as to their 'opinion' that a particular latent print is in fact the print of a particular person." Specifically, they would not be permitted to testify that the mark and the print matched.

Though Pollak did not mention DNA evidence explicitly, it is possible to make sense of his reasoning by drawing an analogy with the "two stage process" by which forensic analysts present DNA profile evidence (Aitken, 2005). An expert witness (or series of expert witnesses) describes two separate determinations: first, that a crime scene sample and reference sample (for example, the suspect's sample) are consistent; and, second, that the matching evidence has an estimated random match probability (RMP) in the relevant population. The first determination (of matching profiles) is done through visual inspection and interpretation of autoradiograph bands or graphic displays of STR peaks, and the second involves statistical calculations based on studies of particular al-

---

4. He adapted his solution from *United States v. Hines* (1999), a federal district court decision on the admissibility of handwriting evidence, which in turn derived this solution from an unpublished ruling in *United States v. McVeigh* (1997).

lele frequencies in human populations. Latent print analysts perform the equivalent of the first determination when they declare a match, but lacking data with which to perform the second step, they adopt the convention that the RMP is zero. Pollak's ruling against declaring matches rejected the convention of treating the determination of a match as a proxy for the calculation of an RMP of zero.

Distrust in Numbers: Resisting Quantification

Although Pollak eventually was persuaded to reverse his ruling, the continuing absence of quantification leaves fingerprinting vulnerable. One potential escape from the dilemma is, of course, to somehow "quantify" latent print matches, a strategy which is being pursued most aggressively by Champod and his colleagues (Neumann et al., 2006).

Champod's proposal to treat fingerprint evidence probabilistically has not been greeted enthusiastically by the LPE profession.[5] Paradoxically, quantification would actually increase the utility of latent print evidence by salvaging "gray" evidence that is currently discarded as inconclusive, though at the expense of decreasing the probative value of all latent print evidence below the current claimed level of absolute certainty.[6]

However, any approach to presenting fingerprint evidence necessarily countenances a loss of information. For example, the bifurcations in figure E.2 would necessarily have to be treated as identical, even though, according to Ashbaugh, we have clear evidence from the skin pore (level 3) traces that they are not from the same source. It is not difficult to understand why LPEs abhor the prospect of "going digital" (and probabilistic). To do so would squander what historically has been viewed as the great strength of latent print analysis: the seemingly fractal individuality of friction ridge skin itself (i.e., individuality at every level of magnification). The latent print community not only does not know how to calcu-

5. The initial reaction to Champod was fairly hostile at times. His presentation at the annual meeting of the Fingerprint Society in the United Kingdom was greeted with "stony silence" (Specter, 2002). His publications also provoked a hostile reaction (McKasson, 2001; but see also Crispino, 2001). More recently, however, he has been invited to join the Scientific Working Group on Friction Ridge Analysis, Study, and Technology (SWG-FAST), the standard-setting body for fingerprint analysis in the United States; http://www.swgfast.org/SWGFAST_members_feb06.pdf.

6. Budowle et al. (2006) make a similar argument to the effect that "exclusion" is underutilized in fingerprint practice.

late an RMP for a print, it does not want to calculate one. Instead, examiners believe the focus should remain on the individuality of friction ridge skin. This is why, when asked for information about the validity, accuracy, or error rate of latent print analysis, LPEs frequently cite the individuality of friction ridge skin as though it provided such a measure (Cole, 2006b).

CHAPTER NINE

Fingerprinting: An Inversion of Credibility

W hat we have called the "postclosure" phase of the techno-legal controversy over DNA evidence involved a reciprocal development: "DNA" was heralded as a new "paradigm" of forensic evidence, while many other forms of criminal evidence became suspect in light of new standards of "scientific" certainty. The model of scientific evidence provided by DNA profiling, coupled with the Supreme Court's *Daubert* and *Kumho Tire* rulings, threatened the secure status of handwriting evidence, firearms and tool marks, bite marks, and other forms of trace evidence. In this chapter, we focus on the most prominent example of that development: an inversion of relations between DNA "fingerprinting" and latent print identification, commonly known as "fingerprinting." Within a mere decade or so, DNA profiling went from measuring itself against fingerprinting to becoming the standard against which fingerprinting was measured. DNA "fingerprinting" supplanted its older namesake as the gold standard of forensic science. The process of achieving closure around forensic DNA profiling, we argue, ruptured the closure that for many decades had surrounded fingerprinting and other established forms of comparison evidence. This occurred primarily because the achievement of closure around forensic DNA profiling was accomplished by valorizing certain features of the law's construction of science—quantification, validation, connections to "high" science—that these older techniques conspicuously lacked.

This chapter traces how this rupture led proponents of older forensic identification techniques to fight back, seeking to repair the damage done by the ascendance of DNA profiling. In particular, fingerprint proponents engaged in a process we call "Daubertizing," in which they sought to remake the century-old technique in the idealized image of science promulgated by the U.S. Supreme Court's revision of admissibility standards.

Finally, we will examine how the courts responded to this growing controversy. "*Daubert* challenges" to fingerprint evidence (and other older forms of trace evidence) put the courts in a bind (Saks, 1998). Viewed in terms of *Daubert,* fingerprinting seemed to fall short of DNA profiling's scientific pedigree and degree of validation. And yet courts continued to believe that fingerprint evidence was highly accurate. Encouraged by the FBI and other proponents of fingerprinting, U.S. courts undertook rather tortured readings of *Daubert* to uphold the "scientific" (and thus legal) status of fingerprinting.

The Challenge to Fingerprinting

By 1990, fingerprint identification had reigned for nearly a century as the gold standard of forensic techniques. Fingerprinting enjoyed a widespread reputation as foolproof or "infallible" (Federal Bureau of Investigation 1985: iv). Fingerprinting held such cultural authority that "fingerprint" was frequently used as a metaphor signaling absolute certainty and unique identity, as with terms like "chemical fingerprint" or "electronic fingerprint." Fingerprinting was not just the gold standard of forensic science; it became a popular icon of unique personal identity (see fig. 9.1 for an example in a antiabortion billboard). Proponents of new forensic techniques have long noted the credibility of fingerprinting and sought to use the term when devising techniques such as "ear printing" and "lip printing." In some cases, the connection with fingerprinting was more tenuous, as, for example, with a technique called "brain fingerprinting."[1]

In this milieu, proponents of the new, untested technique of DNA profiling explicitly used the analogy with fingerprinting to bolster its credibility. Alec Jeffreys drew a direct analogy with fingerprinting in 1986 when he dubbed the multilocus probe technique "DNA fingerprinting."

---

1. This analogy is an obvious and rather far-fetched attempt to harness the credibility of fingerprinting, because it is not claimed that the brain is imprinted on anything. The claim, rather, is that memories (of the crime) are indelibly imprinted on the brain.

FIGURE 9.1. Minneapolis Billboard, c. 1998. Reprinted with permission from Prolife Across America.

He and other proponents of the new technique claimed that DNA pro-files were unique identifiers, just like fingerprints, though with the exception of identical twins.[2] During the most heated period of controversy in the early 1990s, the analogy with fingerprinting was questioned by proponents and critics alike: DNA matches were acknowledged to be merely probabilistic, rather than absolute, and the process of collecting and analyzing DNA evidence was deemed far more complicated and subject to error than the relatively simple practices associated with fingerprinting. The nominal analogy with fingerprinting was downplayed and scare-quoted, and other names (DNA profiling; DNA typing) became commonplace. Proponents of forensic DNA evidence invoked a hierarchy of credibility, in which they willingly accepted second billing to fingerprinting. The idea was that if they could convince people that DNA typing was nearly as trustworthy as fingerprinting, they would have accomplished their goal. Lander and Budowle invoked this hierarchy in their

2. See Moss (1988) on "the new fingerprints." The fingerprint analogy (and also the bar-code analogy that was used at the time) worked best with the early multilocus probe technique developed by Jeffreys, which produced a complex profile that was (arguably) unique to an individual, identical twins excepted, consisting of an indefinite number of al-leles. It was soon replaced by the single-locus probe technique, for which probability estimation was an explicit requirement.

famous "closure-inducing" editorial in *Nature:* they referred to DNA as "perhaps the greatest advance in forensic science since the development of ordinary fingerprints in 1892" (1994: 735).

The court's assessment of the relative value of fingerprint and DNA evidence in the trial *New York v. Rush* (1995) is typical of the period. Rush asked the court to reverse his rape conviction on the grounds that DNA evidence alone should be insufficient to support a conviction when contradicted by eyewitness evidence. The court noted that fingerprint evidence alone was considered sufficient to support a criminal conviction under New York's and many other states' laws, but added,

> a fingerprint expert will testify that the fingerprints left at the scene "match" those of the defendant. The DNA expert here was clear, however, to distinguish his finding from a fingerprint match:
>
> Q. Agent, recently in the news there's been a lot of the use of the term DNA fingerprinting suggesting the possibility of absolute identification. Can the DNA comparisons you described result in such an absolute identification?
>
> A. No, they cannot. The DNA tests that I do and I've described to you do not result in an absolute identification, as a fingerprint comparison results in an absolute identification.

The DNA analyst's deference to the greater probative value of fingerprint evidence was not particularly damaging to the prosecution's case, as the court denied Rush's demand for reversal and the jury convicted him. The probability figures assigned to DNA evidence, though less than absolute, were nevertheless impressive. In fact, within a few years those figures would become emblematic of credibility, as they conveyed the impression that the possibility of error was precisely and objectively measured, and also under strict control, given the very low random match probabilities (RMPs) assigned to DNA profiles.

Today, one would not expect to find explicit deference to fingerprinting in a DNA analyst's testimony. Within less than a decade, the relationship between DNA evidence and fingerprinting almost completely reversed. By the beginning of the twenty-first century, it was fingerprinting whose legitimacy was being questioned, while the controversy over DNA typing was, as we have noted, in a state of postclosure. Consequently, proponents (and also critics) of fingerprinting appealed to the credibility of DNA, much in the way proponents of DNA had used fingerprinting only a few years earlier. We refer to this stunning reversal as an *inversion of credibility*. The closure of the DNA fingerprinting controversy did not represent a general closing of controversy about forensic identification science more generally. Instead, it appears that it may have opened up a far broader, and perhaps less tractable, controversy. Achieving closure in one epistemic space only resulted in displacement and possible rupture elsewhere.

By the mid-1990s, various legal commentators were noting that latent print identification would probably not withstand the sort of rigorous scrutiny being applied to DNA profiling. The disconnect was all the more glaring because latent print examiners advanced stronger claims than DNA experts: absolute rather than probabilistic identifications. This apparent strength gradually became a sign of "subjective" weakness. DNA profiling, with its connections to biomedical technique and its probabilistic formulae, became emblematic of science, while latent print identification became associated with "merely" experiential expertise.

In 1995, in light of a recently publicized error by the fingerprint bureau of Scotland Yard, the *Times* quoted Adrian Clarke, a leading solicitor, as saying that "[f]ingerprint comparison is all quite Neanderthal and an innately subjective process." In the same article, Michael Mansfield, QC, was quoted as saying, "It is time that fingerprint techniques were subjected to the same rigour of questioning [in court] as DNA comparisons" (Grey, 1997). Similarly, David Stoney (1997) remarked:

> The criteria for absolute identification are wholly dependent on the subjective professional judgment of the fingerprint examiner. When a fingerprint examiner determines that there is *enough* corresponding detail to warrant the conclusion of absolute identification, then the criteria have been met. Period.
>
> Efforts to assess the individuality of DNA blood typing make an excellent contrast. There has been intense debate over which statistical models are to be applied, and how one should quantify increasingly rare events. To many, the absence of adequate statistical modeling, or the controversy regarding calculations, brings the admissibility of the evidence into question. Woe to fingerprint practice were such criteria applied!

Michael Saks (1998) followed Stoney's assessment with a prediction:

> Fingerprint evidence may present courts applying *Daubert* with their most extreme dilemma. By conventional scientific standards, any serious search for

evidence of the validity of fingerprint identification is going to be disappointing. Yet the intuitions that underlie fingerprint examination, and the subjective judgments on which specific case opinions are based, are powerful. When and if a court agrees to revisit the admissibility of fingerprint identification evidence under *Daubert,* the *Daubert* approach—that courts may admit scientific evidence only if it meets contemporary standards of what constitutes science—is likely to meet its most demanding test: A vote to admit fingerprints is a rejection of conventional science as the criterion for admission. A vote for science is a vote to exclude fingerprint expert opinions.

In 1999, Saks's prediction came true: in *United States v. Mitchell* (1999), the defendant filed a motion to exclude fingerprint evidence under *Daubert.* The adversaries mobilized for battle, expert witnesses were recruited, studies were hastily prepared, two by the FBI and one by the defendant, and a five-day admissibility hearing was held in July 1999. Notable at this hearing was the appearance of Bruce Budowle as an expert witness for the government. Budowle, a PhD-holding FBI scientist known primarily for his role in the development of forensic DNA profiling, was not a latent print examiner. Nevertheless, at the hearing, he vouched for the validity of latent print identification. Budowle did not explicitly compare fingerprinting with DNA profiling, but instead testified *as a scientist* that, with fingerprinting, there is "nothing odd or different than any other kind of forensic fields." Given his association with "DNA," Budowle could be heard to suggest that fingerprinting was a normal forensic science, not essentially different from DNA profiling. In essence, Budowle deployed the credibility of DNA typing in defense of fingerprinting, a stunning reversal of the situation in *Rush* only four years earlier.

Fingerprinting survived this first challenge, but the very idea that fingerprint evidence could be challenged was not lost on the defense bar, and admissibility challenges proliferated. Because of the nature of legal record-keeping, it is difficult to know exactly how many such challenges were mounted, but a nonofficial count lists more than forty as of this writing.[3]

In most of these cases, the court simply dismissed the *Daubert* challenge out of hand and refused to hold an admissibility hearing on the

3. A tally of "Legal Challenges to Fingerprinting" is maintained on the site <http:// onin.com/fp/daubert_links.html>. The current figure of more than forty challenges most likely does not include some unrecorded cases in lower courts.

matter (Faigman et al., 2002). In a substantial number of the cases besides *Mitchell,* however, the court agreed to hold a "*Daubert* hearing" on
the admissibility of latent print evidence. In addition, the record of the
admissibility hearing in *Mitchell* (1999) was recycled in a number of later
cases. Although this hearing generated an extensive record (five days of
testimony), the court did not issue a written opinion, but instead held in
a bench (oral) ruling that the latent print evidence was admissible. Subsequent courts, however, did issue written rulings. The first of these was
United States v. Havvard (2001).

Fingerprinting fared well in these *Daubert* hearings until receiving
a temporary setback in 2002, when a distinguished federal judge, Louis
Pollak (a former law school dean at Yale University and the University of
Pennsylvania), issued a lengthy ruling in *United States v. Llera Plaza* that
precluded examiners from offering opinions as to who was the source of
a latent print. This decision generated a firestorm of publicity, and the
government filed a motion for reconsideration. Judge Pollak agreed to
reconsider, and ten weeks later he issued a second, equally lengthy, opinion, known as *Llera Plaza II,* reversing his earlier decision. With a couple of technical exceptions,[4] only one court (*Maryland v. Rose,* 2007) has
ruled fingerprint inadmissible, although some individual judges have dis-
sented from majority opinions in favor of admissibility.[5] In 2004, Mitchell's appeal of his conviction finally reached the Third Circuit Court of
Appeals, after a slow passage through the court system. The Third Circuit ruled fingerprint evidence admissible in a lengthy opinion and also
noted that although

> this case does not announce a categorical rule that latent fingerprint identi-
> fication evidence is admissible in this Circuit . . . we trust that the foregoing

4. In *Virgin Islands v. Jacobs* (2001), the court ruled fingerprinting inadmissible after
the government put forward no evidence whatsoever in response to a defendant's *Daubert*
challenge. In *Commonwealth v. Patterson* (2005), the Supreme Judicial Court of Massachusetts ruled inadmissible a particular application of fingerprinting called "simultaneous impressions." In *New Hampshire v. Langill* (2007) the court excluded the latent print
evidence when the examiner failed to follow state protocols mandating contemporaneous
documentation, but was then overruled.

5. Judge Michael of the Fourth Circuit Court of Appeals dissenting in *United States v.
Crisp* (2003). In addition, Judge Thorne of the Utah Court of Appeals wrote a concurring
opinion in *Utah v. Quintana* (2004) that held that the defendant was entitled to a jury instruction to overcome the "cultural assumption" of the "infallibility" of fingerprinting.

discussion provides strong guidance.... Thus a district court would not abuse its discretion by limiting, in a proper case, the scope of a Daubert hearing to novel challenges to the admissibility of latent fingerprint identification evidence—or even dispensing with the hearing altogether if no novel challenge was raised (246).

Thus, the Third Circuit sought to bring closure to the fingerprint controversy, much as the courts had brought closure to the DNA wars (Cole 2005a).

Although the government prevailed in all but one of these challenges through the time of this writing, the edifice of fingerprinting has sustained a great deal of collateral damage. The criticisms of fingerprint evidence have not resulted in legal rulings barring the use of the evidence, but the argument that latent print examiners' knowledge claims have not been validated has been supported in the scholarly community (Kennedy, 2003; Siegel et al., 2006), widely reported in the media, and even acknowledged in many of the judicial opinions that nonetheless upheld the legal admissibility of fingerprint evidence. Although fingerprinting remains admissible evidence, its claims to being infallible, error-free, validated, and even scientific have been greatly diminished. Whereas in early cases fingerprinting was upheld as "the very archetype of reliable expert testimony" (*United States v. Havvard* 2001: 855), in later cases prosecutors argued merely that it was expert testimony based on "good grounds" (Cole, 2004b).

Responsibility for this stunning reversal of fortune for fingerprinting may to a large extent be laid at the doorstep of the twin developments of *Daubert* and DNA profiling. One, surely unintended, outcome of the vigorous battles over the proper statistical characterizations of DNA inclusions was the surprising realization that fingerprinting had been rendered essentially exempt from these debates because of its longstanding claim to be "non-probabilistic" evidence (Vanderkolk, 1993; Grieve, 1996; Champod, 1995). How, DNA analysts wondered, were latent print examiners able to make stronger evidentiary claims based on less data? How, moreover, were they able to defy forensic scientists' widely shared assumption that all evidence must necessarily be probabilistic (Stoney, 1991)? In addition to lacking quantification, fingerprinting lacked the connections to high science that characterized DNA profiling. Finally, although there was room to debate about which particular DNA assays had been validated, and prosecutors continued to resist reporting proficiency test error rates to juries, fingerprinting did not have a record of validation comparable to that of DNA typing.

Daubertizing Fingerprinting

The challenges to fingerprint evidence under *Daubert* that began with the 1999 *Mitchell* appeal, presented latent print examiners, prosecutors, and courts with a dilemma. *Daubert* demanded that the evidence must be reliable in order to be admissible. Although legal actors shared a widespread assumption that fingerprint evidence was highly reliable, there were in fact no empirical measures of the accuracy of latent print evidence. Consequently, the examiner community attempted hastily to "Daubertize" fingerprinting, a process of reconstructing the rhetorical practices of latent print identification in order to appear consistent with the idealized model of reliable science that *Daubert* and *Kumho Tire* appeared to require.

Even before *Daubert,* the fingerprint professional literature had contained references to impending scrutiny from the defense bar. Examiners were urged to avoid complacency and to get their house in order before the coming storm. As early as 1972, André Moenssens said, "Let's face it, we've had an easy ride for a long time. We had a few hard cases, but on the whole most defense attorneys did not know anything about fingerprints and accepted almost everything we said at face value without question. And they cross-examined only on peripheral matters" (Moenssens 1972: 15). It took *Daubert* and the establishment of the DNA gold standard however, to actually bring the anticipated defense scrutiny to pass.

SWGFAST

One such action was the convening of the Technical Working Group on Friction Ridge Analysis Study and Technology (TWGFAST) in 1995. This group aimed to develop "consensus standards which would preserve and improve the quality of service provided by the latent print community nationwide" (Scientific Working Group on Friction Ridge Analysis Study and Technology, 1999). This group was modeled after earlier quality assurance programs, particularly the Technical Working Group on DNA Analysis Methods (TWGDAM, 1991). The titles of these groups changed from TWG to SWG (Scientific Working Group . . .) in 1999, when the FBI changed the nomenclature of all its working groups (Grieve, 1999). Defense attorneys have sought to portray this change in nomenclature as a blatant attempt to turn latent print identification into "science" through linguistic fiat. Founding members of SWG-FAST claim the impetus for the change was merely administrative, but the creation of an administrative body represented a clear attempt to preempt any argument that latent print practice lacked the standards required by *Daubert*.[6]

ACE-V

Perhaps the most significant Daubertizing move was the coining of the term "ACE-V" (an acronym for analysis, comparison, evaluation, and verification) as the purported "methodology" of latent print analysis. Until *Daubert*, it had never been thought necessary to formalize such a protocol. As late as 1991, and even again in 2001 after the *Mitchell* hearing, Olsen (1991; Olsen and Lee, 2001) had delineated various "methods" for analyzing latent prints, but he saw no need to identify "the dominant method" other than to call it "the conventional method."

In the early 1990s, David Ashbaugh, a member of the Royal Canadian Mounted Police, coined the inelegant term "ridgeology" to convey the idea that the analysis was of ridges and all their attributes, and not just of endings and bifurcations (Ashbaugh, 1991). Later, Ashbaugh (1999) adapted the term ACE-V, which had been used by R. A. Huber (1959–60) and Harold Tuthill (1994) for many years to describe other areas of trace evidence analysis, for latent fingerprint analysis. He also used the term "quantitative-qualitative" to convey the idea that the amount of information available in a latent print was a function not merely of the number of "points" but also of the clarity. By characterizing latent print analysis as "quantitative," he also invoked the authority of numbers (Porter, 1986) that fingerprinting had conspicuously lacked. The ACE-V acronym became widely disseminated, however, only through its use in

6. This again reflects the curious politics of the word "science" in this area. It is the agency with the *least* claim to being a scientific institution (the FBI) that uses the term SWG, whereas other agencies, such as the National Institute of Justice and NIST, seem more comfortable with the term TWG.

the court for defending *Daubert* challenges, first in *Mitchell* and then in other cases (Triplett and Cooney, 2006).

To academic critics, the attempts by a craft profession to articulate a "methodology" seemed an amateurish effort to mimic "real" science: Zabell (2005) wrote, "ACE-V is an acronym, not a methodology,"[7] and Denbeaux and Risinger (2003) have called ACE-V "nearly hilarious." Defense lawyers also have sought to highlight its supposed meaningless-ness by cataloging aspects of everyday life, such as grocery shopping, that also use the ACE-V process, and asking who, in life, does *not* use ACE-V. As with SWGFAST, critics view ACE-V as an attempt to ap-pear scientific by superimposing an impressive sounding but ultimately vacuous terminology on latent print practice. However, as we have seen, these efforts paid dividends in the courts, as ACE-V was used to frame judicial evaluations of "reliability" under *Daubert*. Despite critics' and defense attorneys' efforts to dismiss ACE-V as vacuous, it has proven to be a spectacularly successful rhetorical gambit. Whatever courts have ultimately said about latent print individualization, they have generally treated ACE-V as a "methodology" and accepted the debatable claim that it governs latent print practice in the United States (Haber and Haber, 2008).

Graphs and Charts

Other efforts to make the practice appear more scientific included the conspicuous use of visual apparatus typical of science, such as graphs and charts. For example, Vanderkolk (1999; 2001; 2004) published a chart that purported to illustrate the threshold for "individualization" (fig. 9.2). In addition to being published in a latent print examiners' jour-nal, the chart has appeared at a national conference on scientific evi-dence (Vanderkolk, 2002) and *Daubert* hearings on latent print evidence (e.g., *United States v. Mitchell,* 1999; *Commonwealth v. Patterson,* 2005). The diagram essentially indicates that there is a tradeoff between quan-tity (the amount of friction ridge detail "in agreement") and quality (the clarity of the information in the mark), and that at some point the ex-

7. Indeed, it is not altogether clear that "methodology," which usually connotes an ap-proach to an open-ended research question, is an appropriate way to characterize what LPEs do. Again, LPEs' perception that they must have a methodology would seem to de-rive from the law, in this case the *General Electric v. Joiner* (1997) opinion, the second in the "*Daubert*" trilogy," which contains extensive discussion of "methodology."

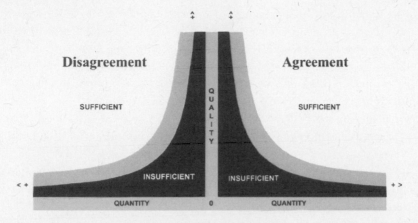

FIGURE 9.2. Illustration of ACE-V. From Vanderkolk, 2001.

aminer makes a determination that there is "sufficient" corresponding ridge detail to exclude every possible area of friction ridge skin in the universe other than the source of the known print. Vanderkolk is specifically vague about how much detail counts as "sufficient." Such illustrations are akin to the sorts of conceptual diagrams favored by social theorists which borrow the iconic form of graphic data superimposed on Cartesian coordinates in order to put forward arguments that are otherwise stated in the text (Lynch, 1991). Like ACE-V, these devices use "qualitative-quantitative" formats to enhance their appearance as a science. But while the graphic devices take the form, they do not play the role, of independent evidence in the visual language game.

Reliability Studies

Daubert also prompted efforts to conduct studies that would support LPEs' knowledge claims. These studies were a focus of extended testimony and discussion in the *Mitchell* and *Llera Plaza* cases. Indeed, in some cases, the studies were done in anticipation of such testimony. The FBI conducted two studies in the run-up to the *Mitchell* admissibility hearing. The first study, a survey of fifty-three crime laboratories (the fifty state laboratories, the District of Columbia, Puerto Rico, and the United Kingdom) was conducted by Stephen Meagher, the point man for defending against Mitchell's *Daubert* challenge. The FBI conveyed its urgency in a cover letter: "The FBI needs your immediate help! The

FBI Laboratory is preparing for a *Daubert Hearing* on the scientific basis for fingerprints as a means of identification."[8] The first part of the
survey consisted of a series of questions on how long latent prints had
been analyzed in that laboratory, on the use of AFIS (Automated Fingerprint Identification System), and so forth. In hindsight, the crucial
survey item was question 19: "Have you ever determined that two different person's [*sic*] fingerprints have resulted in an identification with
the [same?] latent fingerprint?" The question itself is poorly worded and
confusing. Not surprisingly, all respondents answered "no" to this question; to respond "yes" would violate the central dogma of friction ridge
individuality. Nonetheless, some courts have found this response to be
persuasive evidence that, after thousands and thousands of fingerprint
comparisons, examiners have never found prints from two different persons to match.[9]

In the second part of the study, Meagher sent the actual marks from
the *Mitchell* case to the respondents. First, participants were asked to
enter the prints into their AFIS databases and report any matches. The
only respondent that reported that the marks matched a set of prints on
their database was the Pennsylvania State Laboratory, which reported
that they matched Byron Mitchell's prints.[10] Trouble began when respondents were asked to compare the marks to a single ten-print card containing the print of Byron Mitchell. The agencies were asked to report
whether or not they thought the source of the prints was the source of
the marks. For latent A, twenty-seven agencies reported "yes," while
seven reported "no." For latent B, twenty-nine agencies reported "yes,"
and five reported "no." This was a potentially disastrous result, because
it revealed a lack of "reliability" (consistency) among LPEs, something
that Evett and Williams (1996) had already documented.

A second set of packets was mailed to the respondents. For agencies

8. FBI Laboratory, Survey of Law Enforcement Agency's [*sic*] Fingerprint Operations
in Support of a *Daubert Hearing,* 1999.

9. These courts have failed to grasp that, because latent print examiners do not believe
that any latent print can properly be attributed to two different areas of friction ridge skin,
they would never make the determination the question suggests. Faced with two candidate
areas that appear to "match," they would always either choose the best available "match"
or call the comparison "inconclusive" (Cole, 2005a).

10. It was not clear whether this result meant that Mitchell was the only candidate produced by an AFIS search, or that an AFIS search producing multiple candidates had been
conducted, followed by a manual review of the candidates, which resulted in a human conclusion that attributed the latent print to Mitchell.

that had not responded to the first mailing, the new cover letter empha-sized the importance of the study. Enlarged photographs of the marks were provided, and they were packaged within plastic sleeves. Red dots were marked on the sleeves to indicate ridge characteristics that purport-edly corresponded to ridge characteristics on Mitchell's ten-print card. The agencies that had failed to attribute one or more of the marks to Mitchell were instructed to "test your prior conclusions" against the new enlargements. This time, 100 percent agreement on the attribution of the marks to Mitchell was obtained. During the *Mitchell* hearing, and over defense objections, the government sought to substitute the new survey results for those from the earlier round that had already been submit-ted into evidence. The court denied this motion, and the results of both rounds of the survey were entered into evidence.

In the face of criticisms from the defense, Meagher defended the methodology used in the second round by analogy with training an examiner:

> [I]n terms of the prints in question, we wanted to offer the opportunity for the agencies to—just as I would if an examiner in my operation had not affected [*sic*] an identification that we believed he should have, we would have brought it to his attention, shown him the points in terms of correspondence, let him reassess and reevaluate and come back with another evaluation of that com-parison process. (*United States v. Mitchell*, 1999: 212)

Not surprisingly, the defense contended that the first round of the sur-vey demonstrated the inconsistent and subjective nature of latent print analysis (Epstein, 2002). Meagher blamed the inconsistency on the poor quality of the reproductions of the prints or on other contingent factors, such as it being "late in the day" or not taking the survey seriously, or on having "just screwed up" (*United States v. Mitchell*, 1999: 138–50). Five years later, however, in a further appeal by Mitchell, the Third Circuit Court would find the survey, in its totality, convincing evidence of accu-racy, construing it as an experiment in which no false positives occurred (*United States v. Mitchell*, 2004).[11]

The second FBI study was more elaborate. Working with Lockheed

11. This conclusion, of course, entails a regress, because the court assumed that Mitch-ell was the source of the marks, thus resolving the very dispute over the accuracy of the evi-dence by assuming the accuracy of that evidence.

Martin, which had helped build the IAFIS (Integrated Automated Fingerprint Identification System computerized fingerprinting searching system for the FBI, Meagher and Bruce Budowle selected from the FBI database 50,000 left loop patterns derived from index fingers of white males. Each of these images in turn was searched against the entire database of 50,000 prints, using two different automated fingerprint search systems. The scores generated by the two systems were combined to create a single similarity score for each comparison. The FBI found that all of the comparisons of prints to themselves generated higher similarity scores than the comparisons to other prints.

The FBI used these data to estimate the probability that two duplicate friction ridge patterns would occur. Assuming a bell-shaped curve, this estimate was 1 in 10^{97}. The FBI then performed the same exercise using the same images cropped to around 20 percent of their size, as a proxy for a typical mark (latent fingerprint). This generated a probability of exact duplicate friction ridge skin patterns of that size as 1 in 10^{27}. Meagher (2004) would later deny that these impressive numbers represented an error rate, insisting that they only estimated the probability of finding an exact duplicate of an area of friction ridge skin.

The study has been heavily criticized, both for using comparisons of an image with itself as a measure of similarity (rather than comparing it with an image of another print or mark generated by the same source), and for generating probability figures out of all proportion to the data on which the extrapolation was based (Wayman, 2000; Stoney, 2001; Champod and Evett, 2001; Pankanti, Prabhakar, and Jain, 2002; Kaye, 2003; Zabell, 2005). Again, however, despite "academic" criticisms, in the legal arena the studies have been successful. Latent print proponents continue to present the 50K study in court, and courts generally have been persuaded that it proves something relevant to the reliability of latent print identification. Indeed, the numerous published criticisms of the 50K study have yet to be cited by any published legal opinion.

Models of Method

Another example of the Daubertizing of fingerprinting is a series of attempts by proponents of fingerprinting to show that it conforms to "the scientific method." Proponents attempting to do so have turned to popular, idealized models of the scientific method and sought to demonstrate how latent print individualization practice is consistent with them.[12] Ashbaugh (2002) describes forensic fingerprint identification as a seven-step process that, he claims, is indistinguishable from the scientific method used by Newton himself. The steps are:

1. Statement of Problem
2. Hypothesis
3. Observation
4. Experimentation
5. Conclusion
6. Repetition
7. Report

Similarly, Wertheim (2000: 6–7) conceptually translated ACE-V into a five (rather than three) step process:

1. Examination of the Unknown
2. Formulation of a Hypothesis and Counterhypothesis
4. Experimentation
4. Formation of a Tentative Conclusion
5. Testing the Conclusion

Such formulations are, of course, commonplace in textbook treatments of science, especially in fields with insecure scientific status such as psychology and sociology (Lynch & Bogen, 1997), and they are frequently dismissed for having little relation to laboratory practice. However, special difficulties emerge in mapping a routine process like latent print analysis onto idealized versions of classical experimental method. Wertheim (2000: 6) concedes that, in practice, LPEs always hold the same hypothesis: that the latent print was made by the same person as the inked prints. Similarly, Wertheim (7) conceptualizes "experimentation" as follows: "The examiner goes back and forth between the two

12. Interestingly, some proponents of fingerprint evidence have made a Kuhnian turn by rejecting standard textbook models of hypothesis testing and falsificationism. Just as (popularized) Popperian arguments are used both by critics and by proponents of latent fingerprint evidence (as they are for debunkers and defenders of creationism and intelligent design), so can arguments about the disunity of science and local-historical conditions of credibility be mobilized to counteract those arguments in adversary discourse. Not surprisingly, given the context, one can find invocations of the ubiquitous notion of "paradigm" that would set Kuhn spinning in his grave (for example, Howard, 2002; Acree, 1998).

prints, first finding features in the unknown print, then examining the known print for the same features within tolerance." Leaving aside for the moment the fact that "tolerance" remains unspecified in latent print practice, this process would seem to stretch the concept of classical experimentation to the breaking point.

A more recent effort is forced into rather tortured understandings of "the scientific method." For example, it is claimed that "data" (rather than hypotheses) must be "falsifiable"; that Newton's Law of Gravity "was accepted by most people as infallible for more than 200 years"; that "hypothesis testing" has an error rate; and that "ACE-V is synonymous with hypothesis testing" (Triplett & Cooney, 2006: 350–53). These statements suggest a certain conflation of the notions of validation, hypothesis testing, and scientific method. How else to explain the extraordinary declaration "ACE-V is a valid scientific method *if it is used as a valid scientific method*" (Triplett & Cooney, 2006, original emphasis)? It would appear that "valid" is being used here as an honorific (see Haack [2005] on the use of "scientific" and "scientific method" as honorifics in legal discourse).

The Turn to Biology

Fingerprint proponents—and, in particular, the team preparing for the *Mitchell* admissibility hearing—adopted a tactic that we shall call "the turn to biology." They argued that biology—specifically the anatomy and embryology of the formation of friction ridge skin—constituted the scientific knowledge upon which latent print individualization rested. The groundwork for this approach was set by Ashbaugh. Continuing a longstanding tradition of grounding claims about the individuality of human ridge formation in anatomical study,[13] Ashbaugh argued that variations in temperature and pressure in the womb are responsible for the individuality of friction ridge details. Although the exact basis for this claim remains somewhat unclear, it allows fingerprint proponents to say, "It's been well documented in scientific literature that the process of prenatal

13. Ashbaugh's research renewed a tradition of anatomical studies of fingerprint patterns, starting with Harris Hawthorne Wilder and Inez Whipple Wilder, which was continued decades later by Kristine Bonnevie, Harold Cummins, and Sarah Holt (Cole, 2001a). Such research investigated the embryology, heredity, and evolution of friction ridge patterns, and has been used to support claims about the individuality of friction ridge skin. Harris Wilder, for example, claimed to have demonstrated "the impossibility of the duplication of a finger print, or even the small part of one" (Wilder & Wentworth, 1918: 328).

development causes *an infinite variation* of individual friction ridge details" (Moenssens, 2003: 32, emphasis added).

Whether or not anatomical research clearly supports the central dogma about fingerprints, it has no direct bearing on the reliability of latent print individualization (Cole, 2004a). Nevertheless, it offers certain advantages in the courts. First, to the extent that "the reliability question" and "the science question" are confounded, it enables proponents to argue that their practice is "rooted in science" or is "applied science."[14] Second, it enables the proponent of fingerprint evidence to appeal to academic journals in fields such as human anatomy and embryology in order to meet the "peer review and publication" prong of *Daubert*. (Alternatively, many courts and practitioners would adopt the position that "verification" constitutes "peer review.") It also allows proponents of fingerprint evidence to refer to an extensive literature. In *Mitchell*, for example, the government submitted a bibliography of more than one hundred publications, containing a number of the foundational articles and books of fingerprint practice, but also numerous biological and medical texts concerning the formation of friction ridge skin. Third, it became possible to proffer the authors of such texts as expert witnesses. In *Mitchell*, the government called William Babler, a professor of dentistry at Marquette University. Babler had done extensive research on the embryological formation of friction ridge skin, and he possessed precisely the kind of academic credentials that latent print practitioners almost always lack. The turn to biology thus enabled LPEs to answer questions about the accuracy of latent print individualization or the vexing question of what constitutes "sufficiency" by referring to "biological uniqueness." Consider, for example, the following cross-examination of Stephen Meagher at the *Mitchell* hearing:

Q. Could you please refer me to any testimony we have heard here . . . as to any scientific experimentation or controlled studies that [go] to the issue of what is a "sufficient" basis to make a latent print identification?

14. We borrow the phrase "the science question," from Sandra Harding's book title *The Science Question in Feminism* (1986). Though Harding examines "the science question" in a very different context, and we are mainly concerned with the way U.S. courts treat the matter, in either case the question is whether some field or practice meets abstract epistemological or methodological criteria that supposedly distinguish a science from other, often regarded as lesser, forms of knowledge production. And, like Harding, we do not merely ask the question; we reflect upon its salience and critically review the elevated epistemic status frequently associated with science.

A. The scientific basis lies with the biophysiology of the friction ridge skin to create unique formation[s]. The basis of comparing the fingerprints to sufficiency is finding the correspondence of quantity with quality in order to determine the arrangement of those unique formations of that skin are in agreement. (Day 3, 13–14)

Similarly, in their published literature, fingerprint proponents consistently answer challenges to the validity of latent print individualization by reference to "biological uniqueness" (Cole, 2006b).

Credentials

Latent print examiners have made concerted efforts to enhance the visibility of their professional standards, credentials, and publications. Historically, there have been no formal educational requirements for latent print examiners (Grieve, 1990). LPEs typically were (and still are) drawn from the ranks of both sworn police officers and civilians, and were trained primarily in an apprentice system, supplemented by some formal classes offered by mail, by independent instructors, or by the FBI. Efforts to develop certification standards and exams in recent decades had mixed success, and LPEs who failed to pass certification examinations were free to continue practicing and testifying.[15] There are no records of any U.S. case barring a latent print examiner from testifying for a lack of being certified or having failed the certification examination.

This situation began to change during the 1990s, and efforts were taken to strengthen the profession's credentialing and quality control

15. In 1977, the International Association for Identification (IAI), a professional organization, instituted a certification program (Certification for Latent Fingerprint Examiners, 1977). To become a "Certified Latent Print Examiner," an individual must pass a series of requirements including written and practical examinations involving latent print matching tasks. However, the certification program has been weakened by several factors. First, the IAI disavowed any claim that failure to be certified rendered an individual unqualified to practice or testify in court (Wertheim, 1992). Second, the IAI "grandfathered" some experienced latent print examiners into the certification, rendering ambiguous whether any individual examiner had attained certification by passing the examination or by being grandfathered (Certification for Latent Fingerprint Examiners, 1977). Third, many FBI examiners do not appear to participate in the IAI certification program, opting instead for "FBI certification," a credential that the FBI claims is more rigorous. Fourth, the IAI reports that around half of all test-takers fail the certification examination, and that the majority of failures are of the practical examination. (See Newman, 2001.)

mechanisms. SWGFAST instituted nominal loose training requirements and required annual proficiency tests and "testimony reviews," as well as reports of any erroneous conclusions (SWGFAST, 2002).[16] More latent print units sought accreditation from the American Society of Crime Laboratory Directors' Laboratory Accreditation Board. And the professional literature exhorted more examiners to seek certification.

The FBI was especially aggressive in Daubertizing its credentialing procedures. It instituted a new policy in which each LPE trainee was required to be a college graduate, preferably with some training in a physical science. This policy marked a shift away from the traditional emphasis on experiential expertise, usually developed through ten-print work, as a prerequisite for latent work. In an article about the future of the "latent print discipline," Meagher (2001) compared an examiner with ten-print experience but minimal educational background to a college graduate with no fingerprint experience, noting that the latter "[b]rings a skill set better equipped to address secondary issues, e.g. research, formal writing, public speaking". This new policy proved persuasive in the 2002 *Llera Plaza* case, in which Judge Pollak wrote:

> Whatever may be the case for other law enforcement agencies, the standards prescribed for qualification as an FBI fingerprint examiner are clear. . . . The uniformity and rigor of these FBI requirements provide substantial assurance that, with respect to certified FBI fingerprint examiners, properly controlling qualification standards are in place and are in force. (*Llera Plaza II* 2002: 566)[17]

The irony, of course, was that the very witness who informed Judge Pollak of these qualification standards lacked these qualifications himself because his employment predated the implementation of the standards.[18]

16. Read closely, the recommendations in the SWGFAST guidelines are far from ironclad, and they allow more traditionally trained examiners to fulfill the education requirements without the science degree. Moreover, the recommendations are "strictly voluntary" (Grieve, 1999).

17. Note that these standards appear to have been devised in order to "Daubertize" fingerprinting, and that few of the FBI's most experienced and trusted examiners would be able to meet them.

18. This also raises the interesting question of whether formal scientific training actually does render individuals more competent at analyzing latent prints than long years of experience.

Quality Control/Quality Assurance

Some have appealed to "quality control," "quality assurance," or "quality management" to vouch for the accuracy of latent print identification. Champod (2000), for example, advocates "total quality management" as "the way forward." Similar to the administrative fixes that helped settle the legal-technical controversy about DNA profiling (chapter 7), fingerprint proponents sometimes point to quality control measures when challenged about the validity and reliability of their practice. The formal existence of quality control measures stands proxy for their reliable enactment, a familiar strategy in an "audit society" (Power, 1999). This effort is most salient in defending fingerprinting on the "standards" prong of *Daubert*. By interpreting "standards controlling the technique's operations" to refer to such things as quality control standards and training standards, fingerprint proponents try to avoid the issue of whether the standard used for declaring a match is a specific number of points or a case-by-case experiential judgment. Defense attorneys, meanwhile, attempt to keep the focus on match standards.

Translating the Daubert List in Relation to Fingerprint Practice

"Daubertizing" fingerprinting involved more than an effort by proponents of latent fingerprint examination to bring the practice into apparent alignment with legal conceptions of reliable evidence. Equally important was a reciprocal effort that involved the complicity of particular courts to *translate* admissibility standards to be congruent with the practice they were used to evaluate.[19] The records of the admissibility hear-

19. "Translation" is a key concept in the science and technology studies literature. Actor-network theorists such as Callon (1985; 1986) and Latour (1987) treat translation as a multifaceted process through which scientific and technical innovations enroll users as they travel through society. The theme of translation covers material as well as verbal means to articulate an innovation so that potential recipients take an interest in adopting it, and adapting it to their local interests and activities. Star & Griesemer (1989) coin the term "boundary object" to supplement this conception of translation with a reciprocal focus on how the things that transact the relations among producers, users, and other interested parties both facilitate and are modified by those relations. In this case we describe how the "flexible standards" articulated by the U.S. Supreme Court's *Daubert* ruling have been translated in accordance with the practice (fingerprinting) they are used to evaluate,

ings and judicial opinions in *Mitchell* and *Llera Plaza* vividly document latent print examiners', prosecutors', and judges' efforts to articulate how fingerprint evidence satisfied *Daubert's* reliability requirement, as well as the decision's specifications: testability, peer review and publication, standards, error rate, and general acceptance.[20] The case records also provide access to the efforts of defense attorneys and their experts to rebut those arguments. By following the way these actors systematically translate each of the *Daubert* "factors" in light of fingerprint examiners practices, we can appreciate the extent to which the "flexible guidelines" articulated by the *Daubert* ruling are indeed subject to "interpretive flexibility" (Pinch & Bijker, 1984). More than that, we can appreciate that Daubertizing not only involves a rhetorical and organizational effort to bring a practice in line with a set of legal standards, it also involves a continuing effort to renegotiate what those standards mean in order to salvage the admissibility of an existing and highly valued practice.

Testing

In the *Mitchell* hearing, the government argued that the use of fingerprint evidence in casework over the past century constituted the requisite testing of the claim that latent print examiners could attribute latent prints accurately. Thus, they argued that fingerprinting had passed "100 years of empirical testing." The *Havvard* court accepted this argument, concluding that "the methods of latent print identification can be and have been tested. They have been tested for roughly 100 years. They have been tested in adversarial proceedings with the highest possible stakes—liberty and sometimes life" (*U.S. v. Havvard*, 2000: 854).

In *Llera Plaza I*, Judge Pollak rejected such reasoning, declaring, "'[A]dversarial' testing in court is not, however, what the Supreme Court meant when it discussed testing as an admissibility factor." Judge Pollak stuck to this view in *Llera Plaza II*, but he found the evidence admissible

---

and how the surface form of the practice, in turn, has been reconfigured in terms of those standards.

20. As noted in interlude B (also see Edmond, 2002), the *Daubert* and *Kumho Tire* rulings and their various glosses give various characterizations of the list of factors, standards, guidelines, and so forth. Depending on how the list is characterized, it can seem as though it is a strong set of criteria requiring adherence, or an optional set of considerations for discretionary judgments.

even without any testing. In subsequent cases, courts dropped the "adversarial testing" argument. Apparently confused by the *Daubert* court's reference to Popper, falsifiability, and testability, they argued instead that *Daubert* required for admissibility only that the claim be "testable" rather than actually tested. Thus, in *United States v. Sullivan* (2003) the court found that latent print examiners' claim to be able to correctly attribute the source of latent prints was testable, but had never been tested (after almost a century of use). It nonetheless ruled the evidence admissible. In *Mitchell* (2004) the Third Circuit used similar reasoning when it found that fingerprinting satisfied the testing prong while positing the curious notion of "implicit testing" to revive the argument that casework constituted testing (Cole, 2006a).

Peer Review and Publication

The peer review and publication factor is one of the most obviously "scientizing" of the *Daubert* factors. It posed difficulties for fingerprint proponents because, while they did have the trappings of a scientific literature (a journal, an annual conference, etc.), the vast majority of their publications dealt with such matters as the detection and visualization of latent prints, rather than with the accuracy of the attribution process. In *Mitchell* and *Havvard,* the government's tactic was to posit a novel, literal reading of the term "peer review" by saying that the process of "verification" in the ACE-V scheme—by which an examiner would pass a purported fingerprint match to a colleague for double-checking— constituted "peer review."

Again, Judge Pollak rejected this reasoning, writing that that was not what the Supreme Court meant by "peer review." Referring to the government claim that fingerprint "theories and techniques have been published and peer reviewed during a period of over 100 years" Pollak observed:

> [I]t is not apparent that their publication constitutes "submission to the scrutiny of the scientific community" in the *Daubert* sense. Even those who stand at the top of the fingerprint identification field—people like David Ashbaugh and Stephen Meagher—tend to be skilled professionals who have learned their craft on the job and without any concomitant advanced academic training. It would thus be a misnomer to call fingerprint examiners a "scientific community" in the *Daubert* sense. (*Llera Plaza I,* 30)

In *Llera Plaza II,* however, Judge Pollak decided that since fingerprint identification was not science, it was essentially exempt from the peer review requirement.

In *Mitchell,* the court acknowledged that the "cultural mystique attached to fingerprint identification may infect the peer review process" (2004: 239), that "the ACE-V verification step may not be peer review in its best form," and that "the publication facet of peer review is not a strong factor." Nevertheless, the court weighed in on the side of the government's claims about peer review. Other courts found various ways in which fingerprinting satisfied the peer review requirement. Most extraordinary was a New Hampshire trial court opinion which held that the published criticisms of the unpublished "50K study" (see above), which the court conceded did not address the relevant hypothesis necessary to assess the reliability of latent print identification, were sufficient to satisfy the peer review factor (*New Hampshire v. Sullivan,* 2005).

Standards

Defendants and the government waged a long battle over the meaning of the *Daubert* court's language concerning "standards controlling the technique's operation." The government presented evidence about training and quality control standards, which the defense countered by insisting that, in the context of fingerprinting, "standards" must refer to "match standards": how many corresponding ridge characteristics were necessary to declare a fingerprint match. Most courts adopted the government's view until, again, Judge Pollak rejected this reasoning, concluding, because of the government's witnesses' insistence that there could be no required number of corresponding ridge details to constitute a match, "With such a high degree of subjectivity, it is difficult to see how fingerprint identification—the matching of a latent print to a known fingerprint—is controlled by any clearly describable set of standards to which most examiners subscribe" (*Llera Plaza I,* 2002: 514).

It was at this point that the cross-national character of the controversy became salient. While considering his initial ruling, Pollak apparently made inquiries with his "good friend . . . for some thirty years," Lord Lester of Herne Hill, a British barrister and member of the House of Lords. Pollak characterized his conversations with Lord Lester as "a method of legal research to which I could cheerfully become accustomed" and described as "gratifying" the feeling of presenting the fruits of his "research" (*Llera Plaza II,* 2002: note 9). In a move that can only be described as atypical of U.S. criminal cases, and which no doubt astonished the attorneys in Philadelphia to whom it was presented before being incorporated into the text of his second ruling, Pollak quoted a dialogue in the House of Lords on 25 February 2002. The interchange consisted of a series of questions by Lord Lester concerning latent print evidence, and answers by Lord Rooker on behalf of Her Majesty's Government (*Llera Plaza II,* 2002: 42–44).

No fewer than four times, Pollak declared that a system that is accepted in Britain ought to be accepted in the United States. His ostensible justification for these assertions was "the primacy of English endeavors" in developing the fingerprint system. Pollak quoted at length from "the fabled eleventh edition of the *Encyclopedia Britannica*" to support his view of fingerprinting as a British creation, an Anglocentric view of history that is unsupported by current scholarship (Sengoopta, 2002; Cole, 2001a).

Legal commentators have been perplexed by the opinion's extensive reliance on British law (which has no *Daubert*-like admissibility standard for expert evidence) and proceedings on the floor of the emblematically undemocratic House of Lords (Sombat, 2002).[21] Britain's supposed significance, however, lay in the fact that Scotland Yard had switched philosophies from "point counting" to the non-numeric system. Britain had (at least generally) adhered to a 16-point minimum standard for latent print attribution since 1924, and British examiners had been among the staunchest defenders of point standards in the debates over minimum standards (Lambourne, 1984). On 11 June 2001, in part because of fallout from two high-profile misidentifications, the British officially abandoned the 16-point standard and switched to the North American philosophy of adhering to no minimum standard. In other words, the U.K. switchover fell between the *Mitchell* admissibility hearing in 1999 and the *Llera Plaza* hearing in 2002. This allowed Pollak to argue that something had changed: while "lack of harmony" over "standards" may have existed in 1999, it no longer existed in 2002. Choosing his words carefully, Pollak noted "that there is sufficient uniformity within the principal common law jurisdictions to satisfy *Daubert*." The "common law" restriction allowed Pollak to gloss over the continued use of point

21. Such critiques are particularly interesting in light of recent debates within the U.S. Supreme Court between Justices Scalia and Kennedy over whether any heed should be paid to the state of the law in foreign (principally European) countries with regard to the death penalty. See *Roper v. Simmons* (2005).

standards in continental Europe (European Fingerprint Standards, 2002), while the modifier "principal" presumably did the work of excluding common law jurisdictions, such as Western Australia, that did use point standards at that time.

Pollak's Anglophilia provides an unusually clear illustration of the cross-national character of legal-scientific controversies over forensic evidence. The dramatic abandonment of the points system by Scotland Yard during the short interval between the 1999 *Mitchell* and 2002 *Llera Plaza* hearings appears to have had a very significant impact on events in the United States. Of course, had the United Kingdom not changed systems, Pollak might have found other ways to justify his reversal of opinion in *Llera Plaza II,* but it gave him a useful way to frame the opinion.

In *Mitchell* (2004), the Third Circuit, interestingly, ruled that fingerprinting failed the standards prong because fingerprint standards "are insubstantial in comparison to the elaborate and exhaustively refined standards found in many scientific and technical disciplines," but found the evidence admissible nonetheless.

Error Rate

In the *Mitchell* hearing, the government approached the error rate prong with a novel concept: the bifurcation of "error rate" into two types of error: methodological and practitioner. Thus, the government developed a novel typology of methodological error rate, which was said to be zero, in contrast to practitioner error rate, which was not known, but was said to be small. This distinction appears to have a similar logic to that of "mechanical" versus "pilot" error in aviation disasters, but under closer scrutiny the analogy breaks down, because it becomes clear that *all* exposed errors are assigned to the practitioner category (Cole, 2005b).[22]

Most courts, while hesitant to endorse the "zero methodological error rate" concept, readily accepted—as echoed in their written opinions—

22. In his testimony as an expert witness for the prosecution in numerous *Daubert* challenges to fingerprint evidence, FBI analyst Stephen Meagher uses an analogy with elementary arithmetic to support the claim that practitioner errors do not impact the truth of the underlying claim. He argues that an elementary error (for example, $2 + 2 = 5$) does not count against adequacy of the technique, but is presumed to be the product of a mistake. The problem with Meagher's argument is that the rate of practitioner error, not the underlying truthfulness of the claim (whatever that claim may be in this context), is the information that a legal fact-finder needs to know in order to evaluate the persuasive value of the evidence.

latent print examiners' testimonial characterizations of the practitioner error rate as "vanishingly small," "essentially zero," "negligible," or "microscopic." None of these characterizations were based on any actual attempt to estimate the error rate of latent print attributions; they were simply expressions of latent print examiners' confidence in their own practice.

Many courts acknowledged that the error rate for fingerprint identification was not known, but turned this point against the defendants, arguing that they had failed to show that the error rate was too high. In *Llera Plaza II,* Judge Pollak wrote:

> Mr. Meagher knew of no erroneous identifications attributable to FBI examiners. Defense counsel contended that such non-knowledge does not constitute proof that there have been no FBI examiner errors. That is true, but nothing in the record suggests that the obverse is true. It has been open to defense counsel to present examples of erroneous identifications attributable to FBI examiners, and no such examples have been forthcoming. I conclude, therefore, on the basis of the limited information in the record as expanded, that there is no evidence that the error rate of certified FBI fingerprint examiners is unacceptably high. (566)

Lawyers calls this "burden shifting": in a *Daubert* challenge, the burden is on the proponent of the evidence to show that the error rate is known (or estimated) and that it is reasonably low.

General Acceptance

In most challenges, little attention was paid to the legacy-*Frye* "general acceptance" prong because in most cases the defendant conceded that thousands of latent print examiners constituted "general acceptance in the particular field in which" the proposition "belongs." Even Judge Pollak, in *Llera Plaza I,* found that fingerprinting satisfied the general acceptance prong, alone among the *Daubert* factors.

However, general acceptance hinges crucially on what Gieryn (1983) calls "boundary work"—how the "relevant scientific community" is defined. Indeed, one of the principal criticisms of *Frye* has long been that defining the community determines the outcome of a *Frye* ruling (Schwartz, 1997: 214). This ambiguity came up during an early challenge to the admissibility of DNA profiling (*New York v. Wesley,* 1988), in which the court ruled that the forensic science community alone was not sufficient to establish admissibility (Aronson, 2007). Although for fingerprinting there is no equivalent to the academic molecular biologists and population geneticists who weighed in on the general acceptance of forensic DNA evidence, in recent years more members of the academic community have testified on the side of defendants that latent print individualization has not been validated. In *Commonwealth v. Patterson* (2005), sixteen scientists and scholars (including the second author of this book) signed an *amicus curiae* brief in support of this position (Siegel et al., 2006). The courts are now faced with the question of whether the "relevant" community is constituted by a group of practitioners, almost none of whom have formal scientific training, or by representatives from scientific and scholarly fields, almost none of whose members have practical training in fingerprint examination.[23] In *Patterson,* the court opted for the former, ruling not only that latent print examiners could constitute a community (as presumably astrologers, though numerous, could not) but also that nonpractitioner scientists and scholars could be excluded from that community for general acceptance evaluation purposes.

Inversion of the Inversion

Fingerprint evidence thus far has survived its legal tribulations, while enduring increased skepticism about whether or not it is science. Its claims to scientific standing continue to be treated with skepticism (Kennedy, 2003; Mnookin, 2003), but the courts have put the question of its scientific status out of play—at least on the surface. As we have argued, the science question is still in the picture, given the way the *Daubert* ruling defines reliability, and especially now that "DNA"—with its iconic status as scientific evidence and its association with molecular biology and diagnostics—is regarded as the gold standard for criminal forensics. And, certainly, as we have seen, whether or not the science question is

---

23. The categories are not exclusive, though. The *amicus* signers included a former FBI latent print examiner, Mark Acree. And, some latent print examiners, like Glenn Langenburg, have master's degrees in natural sciences like chemistry. The *Patterson* court's definition of the relevant community is, therefore, actually based on ideology (the position that the actors take) rather than qualifications.

in play for admissibility, the latent print examiner community treats it as important for establishing its professional credibility.

In addition to questions about latent print identification's status as science, the long-standing claim of infallibility became increasingly unsustainable (Cole, 2005b). Although latent print misattributions have been known since 1920, two more recent high-profile misattributions have been particularly damaging. In 1997, Stephan Cowans was convicted and sentenced to thirty to forty-five years in prison for nonfatally shooting a police officer, based on eyewitness identifications by the victim and a bystander and a latent print recovered from a coffee mug in a house where the perpetrator had sought refuge (*Commonwealth v. Cowans*, 2001). After Cowans had served six years and worked biohazard duty in order to earn the money to pay for a postconviction DNA test, his lawyers convinced the court to allow DNA testing to be performed on the mug and two items of clothing discarded by the perpetrator. DNA profiles from these items were consistent with one another, but inconsistent with Cowans's sample. Any expectations for a showdown between fingerprint and DNA evidence were quickly dashed, however, when the Boston and State Police reexamined the print and declared that it did not match. Cowans was quickly freed with an apology in 2004, and the Boston Police Department's latent print unit was eventually completely disbanded. Although it would have been interesting to know what would have happened had the latent print examiners stayed with their initial conclusion, the DNA evidence's quick trumping of the fingerprint evidence poses an interesting contrast to the earlier *Thomas* case (see chapter 8). Further, the case highlighted the danger of exaggerated trust in both latent print and eyewitness evidence. Had the true perpetrator not obligingly left his DNA at the crime scene, Cowans would presumably still be in prison today. Indeed, it was Cowans himself who most eloquently summed up the continued cultural authority of fingerprint and eyewitness evidence, stating that "the evidence against him was so overwhelming that if he had been on the jury, he would have voted to convict himself" (Thomas, 2004). Tragically, Cowans was murdered shortly after receiving a settlement from the state for his wrongful conviction.

The second error, in the 2004 Madrid terrorist bombing case, made it impossible for latent print examiners to use their usual strategy of blaming all exposed errors on incompetence (Cole, 1998), because it was committed by three highly qualified FBI examiners and then corroborated by a highly regarded independent examiner retained on the suspect's behalf.

Such high-profile misattributions (Cole, 2005b) raised questions about latent print identification's status as science, and made it more difficult to sustain the long-standing claim of infallibility by blaming all exposed errors on incompetence (Cole, 1998). Such cases also raised concerns about Judge Pollak's apparent credulity, expressed in his second ruling, about the FBI being error-free. The Madrid train bombing case has now generated numerous reports by various investigatory bodies (Stacey, 2004; Fine, 2006; Cole, 2005b, 2006c). One of these reports was generated by an FBI review committee and first-authored by Budowle (Budowle et al., 2006). Interestingly, in this report Budowle no longer takes the position he testified to in *Mitchell,* that fingerprinting is essentially the same as DNA typing. Instead, he now distinguishes between the techniques, noting that "[t]he latent print ACE-V process has a greater component of subjectivity than, for example, chemical analysis or DNA typing." Yet, he argues, "this does not in itself call into question the reliability of the latent print analysis methodology."[24] Thus, fingerprinting appears now to be riding on the coattails of DNA rather than the other way around.

The widespread acceptance of DNA profiling in the 1990s coincided with "the rise of statistical thinking" (Porter, 1986), particularly Bayesian statistical thinking, in forensic science generally. Today, forensic scientists, and increasingly the courts, tend to believe that all forensic evidence can be presented in probabilistic terms. Indeed, in the case of DNA evidence, many courts in the United States insist that DNA evidence should be quantified, paraphrasing Lord Kelvin with statements, such as, "A match is virtually meaningless without a statistical probability expressing the frequency with which a match would occur."[25] Conse-

---

24. We should note that Budowle et al. do not actually point to any evidence that affirmatively demonstrates the reliability of latent print analysis, although they claim that "[t] here is indisputable evidence supporting that such practices *can* be carried out reliably and that the general process should not be rejected" (emphasis added). The use of the word "can," essentially defangs the proposition.

25. *New Hampshire v. Vandebogart* (1992). Also see *Iowa v. Brown* (1991): "Without statistical evidence, the ultimate results of DNA testing would become a matter of speculation." Lord Kelvin (William Thomson) famously stated in 1883, "When you cannot measure it, when you cannot express it in numbers, your knowledge is of a meager and unsatisfactory kind" (*Popular Lectures and Addresses,* vol. 1, "Electrical Units of Measurement," 5 March 1883; available at http://zapatopi.net/kelvin/quotes/).

quently, the avoidance of statistical probability in declarations of finger-print matches can seem quaint and dubious:

> The traditional role of expertise in forensic science is well recognized and it is something of a stereotype to visualise the distinguished, greying individual on the stand saying, "my opinion is based on my many years of experience in this field." Whereas we do not for one moment deny the value of experience, we claim, as a matter of principle, that the scientist should, as far as possible, support his/her opinion by reference to logical reasoning and an established corpus of scientific knowledge. This is what we mean by "transparency": the former "in my experience" justification we refer to as "obscurity." (Champod & Evett, 2001: 1-6–7)

What is remarkable is the way the notion of "transparency" has changed. Not very long ago, a latent print expert testifying in court held an advantage in terms of credibility over a DNA typing expert. Fingerprint evidence seemed transparent because of its surface visibility, familiarity, and demonstrability to lay jurors (Cole, 1998; Lynch & Jasanoff, 1998).[26] The images of the mark and print, regardless of what chemical and photographic alterations they had undergone, retained an apparent transparency ("apparency"), as though the evidence spoke for

---

26. Mark Twain's (1894) novella *Pudd'nhead Wilson* provides a mythic story of the transparency of fingerprint evidence (Mnookin, 2001: 14ff.; Cole, 2007). In a dramatic trial episode, Pudd'nhead Wilson, a lawyer with a reputation as a local oddball, vindicates himself with a successful defense of twin brothers accused of a murder. Using a bloody fingerprint recovered from the crime scene, which he characterizes as a signature or "physiological autograph" unique to a single individual, he enables the jurors to see for themselves that the twins' fingerprints do not match the bloody print. He then shows them the print of the true perpetrator—a young man who was assumed to be the murder victim's nephew. Twain then has Wilson demonstrate that the "nephew" was actually unrelated to the victim. According to the story's fatalistic treatment of racial and biological destiny, the boy was born to an African American slave family in 1830. Although he was only 1/32 black, this qualified him as a slave in Missouri. The boy's mother had switched him in the cradle with another infant—the son of a prosperous white merchant. Subsequently, he was raised in the prosperous family, only to be found out through Wilson's sleuthing. As Paul Rabinow (1996b) chronicles in his essay "Galton's Regret," the idea that fingerprints would allow detection of genetic relationships among persons was a fantasy that Francis Galton initially entertained but later abandoned when he investigated and codified the properties of fingerprints. Although other researchers were able to establish weak racial and hereditary correlations for general fingerprint pattern types, this research was largely obscured as the idea that fingerprints were an individually unique "signature" caught on and became established in criminal justice systems worldwide.

itself or jurors could see for themselves what the evidence said. In contrast, proponents of DNA evidence were compelled to work with an image that was several degrees removed from the original biological material and, crucially, no longer looked like biological material. And, even worse, the weight of the evidence was expressed in the dreaded form of statistics.

Today, it is latent print evidence that seems obscure, and forensic DNA evidence appears transparent. "Transparency" now refers to clarity about how the probative value of the evidence is derived from recognized scientific procedures—a variant on the theme of traceability.[27] Rather than DNA being viewed as "merely" probabilistic, as it was in the early 1990s, probability has become a marker of a higher claim to scientific status. This new view of transparency radically differs from the apparency that once supported the credibility of fingerprinting. The new transparency is grounded in procedures and a body of knowledge that is available to members of an expert field. Although Champod and Evett label opinions grounded "in my experience" as "obscurity"—obscure, because such opinions lack a public rational foundation—their conception of transparency runs its own risks of obscurity, because there is no assurance that the juror will be able to follow the expert's chain of reasoning or have independent access to what is or is not included in the relevant corpus of knowledge.

The story does not end at this point, however. In something of a double-reverse in what we have called the inversion of credibility between DNA "fingerprinting" and its older namesake, a move is afoot to present DNA evidence as individually identifying, and without the confusing probability statistics. For example, Budowle et al. (2000) spell out the circumstances under which it would be possible to characterize a DNA inclusion simply as a "source attribution": "It's the defendant's DNA." Indeed, a year earlier in his testimony in *Mitchell,* Budowle was able to resist the defense attorneys' attempt to draw invidious comparisons between fingerprint and DNA evidence by stating that, like the fingerprint expert, the DNA analyst can simply declare an expert opinion of the individual source of the evidence:

27. The quoted passage from Champod and Evett (2001) identifies traceability with deductive links to accepted scientific axioms. A different conception of traceability exemplified in our discussion of chains of custody (chapter 4) includes bureaucratic means for assuring accountability—paper trails, identification codes, signatures, and other tracking and tracing devices (also see Latour, 1995; Lezaun, 2006).

Q. So when you go to court in DNA cases, you give that conclusion as a statistical basis, correct?

A. Not always. When we have sufficient information we say this individual is
the source of the material in my opinion. I have to say this is opinion, based
in our opinion given a sufficient amount of information, this individual is the
source of the information. (*United States v. Mitchell,* 1999*[2004 in refs.]:* 157)

This proposal has been fairly widely criticized (Weir, 2001; Inman and
Rudin, 2001; Evett, 2000). Buckleton (2005), after reviewing all the commentaries, concluded, "I cannot see much positive coming from a policy of declaring a common source." Thus, just as latent prints were being
pilloried for their lack of statistics, forensic DNA scientists were trying
to do away with statistics. In an inversion of the inversion, just as statistically minded forensic scientists seem to be trying to turn fingerprinting
into something more like DNA, some forensic DNA analysts seem to be
trying to turn DNA into something more like fingerprinting.

CHAPTER TEN

Finality?

It may be that in more than one sense [cross-examination] takes the place in our system which torture occupied in the mediaeval system of the civilians. Nevertheless, it is beyond any doubt the greatest legal engine ever invented for the discovery of truth. However difficult it may be for the layman, the scientist, or the foreign jurist to appreciate this its wonderful power, there has probably never been a moment's doubt upon this point in the mind of a lawyer of experience. — John Henry Wigmore, *Evidence in Trials at Common Law.*

DNA testing is to justice what the telescope is for the stars: not a lesson in biochemistry, not a display of the wonders of magnifying optical glass, but a way to see things as they really are. It is a revelation machine. — Barry Scheck, Peter Neufeld, and Jim Dwyer, in *Actual Innocence*

Slowly, in the course of the classical age, we see the construction of those "observatories" of human multiplicity for which the history of the sciences has so little good to say. Side by side with the major technology of the telescope, the lens and the light beam . . . there were the minor techniques of multiple and intersecting observations, of eyes that must see without being seen; using techniques of subjection and methods of exploitation, an obscure art of light and the visible was secretly preparing a new knowledge of man. — Foucault, *Discipline and Punish*

Writing a century apart, Barry Scheck and John Henry Wigmore liken legal instruments to machineries for revealing the truth. Wigmore suggests that cross-examination occupies a space vacated by the instruments of torture (a space that, as we are well aware in the twenty-first century, was vacated selectively and only in principle rather than in practice). According to Wigmore, a kinder, gentler, and more transparent "engine" for revealing confessional truth replaces the cruel and secretive handiwork of the torturer.[1] Scheck and Neufeld deploy a different analogy.

1. For a brilliant exposition of the techno-politics of "engine science," see Patrick Carroll (2006). Carroll derives the idea from Shapin & Schaffer's (1985) elucidation of Robert

They liken DNA testing to the telescope, an instrument that epitomizes light, transparency, and revelation. Both quotations remind us of Foucault's (1979) famous analysis of the birth of the modern system of criminal justice characterized by the replacement of the spectacle of corporal punishment with the quieter, more efficient means of extracting truth through the examination. For Foucault, the transformation marks a shift in the mechanisms of power from the spectacular, but inefficient, exercise of a despotic sovereignty to the quiet and measured efficiency of administrative discipline.

As Paul Rabinow (1992) has argued, "DNA" has become a scientific instrument and popular icon in a new regime of biopower he dubs "biosociality." Increasingly, "DNA" articulates identities, prospects, fates, and liabilities, and in popular discourse it becomes an instrument of justice— a sure means for determining guilt and innocence. Wigmore's faith in cross-examination may now seem misplaced. Together with its dialogical partner eyewitness testimony, cross-examination is now denigrated for being prescientific and error-prone, rather than being trusted as the best means possible for exposing contradictions and inconsistencies. However, as we have seen, cross-examination also can expose the fact that forensic evidence is delivered through testimony: verbal and written accounts of what various specialists saw, said, and did as they worked with evidence samples. "DNA" points the way to guilt or innocence only when it is surrounded by a story containing other claims and evidences. The "genetic witness" (OTA, 1990) speaks for itself only when presented in the form of expert *testimony,* and as we have seen, interrogation of that "voice" points to an extended, indefinitely complicated, series of fallible practices through which evidence is collected, transported, analyzed, and quantified. The embeddedness of "DNA" in stories and sense making has not deterred courtroom participants and legal scholars from making millenarian pronouncements about an ascendant "truth machine" in a new era of criminal justice. In this final chapter, we reflect upon some implications of the exceptional and extraordinary power now granted to "DNA."

Boyle's nascent experimental program in the seventeenth century, in which the mechanical air pump was the central figure in the experiment.

Revisiting the New Paradigm

In chapter 1, we noted that DNA fingerprinting has been likened to a
Kuhnian paradigm in forensic science. Accordingly, it is treated as a
ground truth discovering erroneous convictions, and a "model for a scientifically sound identification science" (Saks & Koehler, 2005: 892).
The prescientific state of the art—what Saks and Koehler (ibid.) call
"traditional forensic individualization sciences"—involved comparisons
between "pairs of marks (handwriting, fingerprints, tool marks, hair,
tire marks, bite marks, etc.), [from which forensic scientists] intuited
whether the marks matched, and testified in court that whoever or whatever made one made the others." Such comparisons aim for "discernible
uniqueness" between the traces. Fingerprinting is the exemplary case of
a traditional forensic individualization science. Very soon after fingerprint evidence was admitted in criminal trials, proponents "attempted
to distinguish their knowledge from other forms of testimony by declaring that they offered not opinion but fact" (Mnookin, 2003: 50). Such
claims to certainty are rapidly being displaced by the ascendancy of a
new claimant to transcendent evidentiary status.

Consigning fingerprinting to the status of the old paradigm does not
simply herald the arrival of a new technique that is better than fingerprinting for discerning uniqueness. Instead, as Saks and Koehler make
clear, DNA typing is an exemplar for an entirely different, *statistical approach* to identification:

> One of the great strengths of DNA typing is that it uses a statistical approach
> based on population genetics theory and empirical testing. Experts evaluate
> matches between suspects and crime scene DNA evidence in terms of the
> probability of random matches across different reference populations (e.g.,
> different ethnicities). These probabilities are derived from databases that
> identify the frequency with which various alleles occur at different locations
> on the DNA strand. The traditional forensic sciences could and should emu-
> late this approach. (Saks & Koehler, 2005: 893)

"Fingerprinting," they suggest, "could be one of the first areas to make
the transition," because databases already exist (ibid.). The challenge
will be to develop "workable measures" of "complex images presented
by fingerprints, tool marks, bite marks, handwriting, etc." For this pur-

pose, they add, forensic scientists may need to enlist the help of "experts in differential geometry, topology, or other fields" (ibid.).

In many respects the two techniques do not compete: both continue to be used; fingerprint evidence is available in many cases in which no DNA evidence is collected, and vice versa. The arrival of DNA testing has not annulled the dogma that fingerprints are unique or the claim that such uniqueness is discernible. As chapter 9 elaborates, latent print examination thus far has withstood a series of challenges in U.S. admissibility hearings. Whether or not we agree that it is appropriate to speak of a paradigm shift, a profound shift has occurred with the ascendancy of DNA typing.

Saks and Koehler acknowledge that they use the notion of paradigm shift "not as a literal application of Thomas Kuhn's concept, but as a metaphor highlighting the transformation involved in moving from a pre-science to an empirically grounded science" (2005: 892). Perhaps the major difference between their usage and Kuhn's is that Kuhn used the term historically and analytically, while Saks and Koehler use it prospectively and tendentiously. In their view, DNA typing not only is an exemplar of a highly credible forensic technique, it also offers strong leverage for making broad reforms in the administration of criminal justice.

The contrast between DNA typing and other forensic identification methods lends rhetorical force to the argument that other forensic methods can and should be upgraded to normal scientific status. One can imagine that criminal defense lawyers will rehash the contrasts Saks and Koehler outline when interrogating prosecution witnesses who present "traditional" comparison evidence. What their argument glosses over, however, is the extent to which forensic DNA typing still falls short of their idealized conception of a normal science: as with other forms of evidence, most forensic DNA analysis is done by scientists who do not hold PhD degrees; proficiency testing falls short of the recommended standard of "blind, external proficiency tests using realistic samples" (ibid.); and the "cultural differences" between forensic science and (idealized) normal science pertain as much to forensic DNA typing as they do to other forms of forensic analysis (often performed by the same organizations). Moreover, it remains to be seen if it is even possible to develop meaningful probability measures for fingerprint, hair, and ballistic comparisons comparable to those used for calculating allele frequencies. Although we may want to applaud any effort to encourage forensic organizations to upgrade the quality and integrity of criminological practice, Saks and Koehler's polemical use of DNA typing as a paradigmatic normal science encourages the assumption that DNA evidence is a certain, scientific, and even infallible form of criminal evidence. Although they do not discuss the implications of their paradigm shift for ordinary evidence (e.g., eyewitness testimony, confessions, etc.), an implication would seem to be that eyewitness testimony, confessions, and the like also should be subjected to probabilistic analysis, and compared to the model of DNA typing.[2]

Forensic science occupies an intersection between law, science, and police practice, and in the courtroom DNA evidence is considered together with eyewitness testimony, confessions, circumstantial evidence, and other forms of forensic evidence. Despite efforts to expand its use, DNA evidence continues to be collected in a relatively small proportion of criminal cases. And, as we saw in chapters 5 and 6, even when DNA evidence is available, its value and significance depend upon other evidence. Such other evidence may or may not be quantifiable in terms of a Bayesian procedure such as the one attempted by the defense in the *Adams* case discussed in chapter 6. Aside from the appeal court's disapproval of that procedure in that case, it is unclear that eyewitness evidence and other forms of forensic comparison evidence could ever be quantified in a way comparable to DNA evidence, with its population databases, frequency statistics for different alleles, quality control and quality assurance regimes, and so forth. In addition, DNA evidence is strongly associated with *science* and an exceptional degree of ontological certainty has been attributed to it. Moreover, such certainty is of a different order than that claimed by fingerprint examiners. In cases in which convictions have been overturned, the mathematical certainty attributed to DNA evidence trumps the practical or moral finality associated with legal decisions. The paradigm shift thus implicates traditional legal concepts as well as other forms of "scientific" and "nonscientific" evidence. Adjustments have been made to existing conceptions of criminal suspects and of suspect rights; and statutes of limitation have been redefined to accommodate the potential for DNA reanalysis.

2. Saks and Koehler (2005: 892, fig. 1) present figures on eighty-six DNA exoneration cases, in which they classify types of evidence responsible for the errors exposed by DNA testing. Eyewitness errors led all types, with a 71 percent rate, closely followed by forensic testing errors.

Closure and Finality

In the postclosure phases of its history in the criminal justice systems of the United States, United Kingdom, and other nations, DNA evidence has been granted exceptional evidentiary status—to the point of becoming "a scientific arbiter of truth" that exposes the fallibility of all other forms of criminal evidence (Findley, 2002: 336). It is as though "DNA" has transcended the mortal realm and moral certainty of criminal evidence and taken its place in a dream world of unassailable scientific evidence and mathematical certainty. As noted in the preface of this book, we are referring here not only to views expressed in popular culture and echoed by ambitious politicians, but also to opinions expressed by judges, criminal lawyers, forensic scientists, and many academic analysts. A strong expression of such an opinion can be found in an article "Lessons from DNA" by Margaret Berger of Brooklyn Law School, who served on the Committee on DNA Forensic Science that authored the second NRC report (1996).[3]

Berger observes that the advent of DNA testing has challenged the principle of *finality* in law. Finality is a legal concept of closure. Unlike closure in scientific controversies, which marks consensus about previously uncertain or disputed matters of fact or theory in a particular field, finality is imposed by legal authority—it involves time limits and procedural barriers against challenging verdicts after appeals have been exhausted.[4] An extreme—almost apocryphal—example in which finality was invoked was expressed in a dialogue from a Missouri Supreme Court hearing quoted in a 2003 *New York Times* article:

> Judge Laura Denvir Stith seemed not to believe what she was hearing. A prosecutor was trying to block a death row inmate from having his conviction reopened on the basis of new evidence, and Judge Stith, of the Missouri Supreme Court, was getting exasperated. "Are you suggesting," she asked the prosecutor, that "even if we find Mr. Amrine is actually innocent, he should

3. An earlier version of Berger's chapter is criticized by Johnson & Williams (2004: 81). Our discussion of the influence of postconviction DNA testing on the legal concept of "finality" also is informed by Aronson (2005b).

4. Cases resulting in not-guilty verdicts have finality because of rights against double jeopardy, although as defendants in the Rodney King police trial and the O. J. Simpson trial learned, it is possible in the U.S. justice system to follow a not-guilty verdict in a criminal trial with a conviction in a civil trial.

be executed?" Frank A. Jung, an assistant state attorney general, replied, "That's correct, your honor." That exchange was, legal experts say, unusual only for its frankness.[5]

The contrast could not be clearer between Mr. Jung's admittedly extreme invocation of finality and the concept of closure in a scientific field. In theory at least, communities of natural scientists remain open to the possibility that novel evidence will lead to revisions of even the most settled theories and facts. Although, in practice, it may require extraordinary effort to gain a hearing for a challenge to deeply held convictions in a scientific field, one can imagine the howls of protest that would occur if leading figures in a professional association were to stipulate administrative barriers to introducing novel factual evidence that might reopen a closed controversy. In law, however, as Berger points out, there are several good reasons for imposing finality. Some of these reasons have long been recognized in the Anglo-American tradition: physical evidence tends to degrade over time; eyewitnesses die or become difficult to locate; memories become fogged and biased with the passage of time; victims and/or their families need closure so that they can go on with their lives; and trust in the system of criminal justice would break down if verdicts were constantly exposed to reversal. Consequently, as proponents of the Innocence Project are well aware, courts and prosecutors often impose temporal and procedural restrictions on efforts to reopen settled cases, even when new evidence that was not introduced in the original trial might persuade a jury to reach a different verdict.[6]

Berger, in line with proponents of the Innocence Project, argues that DNA evidence is exceptional because it does not degrade over time, and newer techniques (particularly systems incorporating PCR) enhance the ability to examine or reexamine old evidence.[7] Using shorthand expres-

5. Liptak, 2003.

6. See *Herrera v. Collins* (1993), in which the Supreme Court quotes from *Wainwright v. Sykes* (1977) to support the finality of the original determination of factual guilt or innocence: "Society's resources have been concentrated at that time and place in order to decide, within the limits of human fallibility, the question of guilt or innocence of one of its citizens."

7. Berger cites the fact that DNA typing has been used in archaeological investigations of Egyptian mummies. However, as William Thompson mentioned in an interview with two of the authors (October 2005), DNA can degrade when samples are exposed to the elements or become infected with bacteria. A completely degraded sample would very likely show a null result, but a partially degraded sample may suppress the visibility of longer

sions that have become commonplace in the postclosure era, Berger repeatedly assigns "DNA" with unassailable status and causal agency: *it* provides "irrefutable proof of the fallibility of eyewitness testimony"; *it* exposes errors and inaccuracies in traditional forensic comparison methods; and *it* "has had revolutionary impact on forensic laboratory practices" (112).[8] In her account, "DNA does not fade away" (113); and "technological advances have steadily enhanced the discriminating power of DNA to identify correctly the source of biological samples" (113).

There are very good rhetorical and political reasons for treating DNA evidence as unassailable: it has been used effectively as a source of leverage to overturn convictions based on other evidence, and has given new life to efforts by civil libertarians to challenge the constitutionality of the death penalty in the United States. Postconviction DNA testing has resulted in more than a dozen exonerations of death row inmates. The death penalty, of course, compounds legal finality with *life* finality. Although the principle of finality is a matter of moral certainty, and is not meant to endow criminal verdicts with absolute certainty, in a death penalty case special considerations come into play in order to assure the public that the verdict is as near to certainty as the criminal justice process can get.[9] As William Rehnquist argued for the Supreme Court majority in *Herrera v. Collins* (1993):

> We may assume, for the sake of argument in deciding this case, that in a capital case a truly persuasive demonstration of "actual innocence" made after trial would render the execution of a defendant unconstitutional, and warrant federal habeas relief if there were no state avenue open to process such a claim. But because of the very disruptive effect that entertaining claims of

---

STR sequences, which break down into shorter sequences, which show up as artifactual peaks in a graphic STR profile.

8. Like many commentators, Berger deploys the expression "DNA" as shorthand for various forms of forensic DNA evidence. Such usage conflates approximate and contingent measures with the totality of genomic DNA—a totality that is not measured with existing techniques. In the last quotation above, she also assigns "DNA" with responsibility for the quality assurance/quality control regimes recommended by the NRC and other review agencies. In a note, Berger (2004: 125, note 24) mentions a number of cases of U.S. forensic laboratories in which systematic errors have been exposed. These include laboratories responsible for DNA testing, and so her assertion about the impact of "DNA" is contradictory unless read as shorthand for ongoing efforts to regulate forensic laboratories that were initially spurred by evaluations of DNA testing in the 1990s.

9. See Laudan (2003) and Cole & Dioso-Villa (2007) for discussions of the classic distinction between moral certainty and mathematical certainty.

actual innocence would have on the need for finality in capital cases, and the enormous burden that having to retry cases based on often stale evidence would place on the States, the threshold showing for such an assumed right would necessarily be extraordinarily high.[10]

The *Herrera* ruling left unclear where the balance was to be struck between the demonstrability of "actual innocence" and the demand of finality, and courts and legal commentators have struggled with that issue ever since. It appears, however, that "DNA" is on the verge of being raised above the "extraordinarily high" threshold that Rehnquist mentions.

Although DNA evidence is available in only a small proportion of capital crimes, its significance for challenging finality extends well beyond its actual utility in particular cases. Berger and others (for example, Findley, 2002; Saks & Koehler, 2005) credit postconviction DNA testing with having exposed the extent to which other forms of evidence are fallible. Such exposure is especially dramatic in death penalty cases, given the assumption by the Supreme Court majority in *Herrera* that death penalty sentences meet a stringent standard of "reliability" as demanded by the Eighth Amendment, and that the availability of appeals or clemency ensures against wrongful executions. This assumption was directly challenged in a 2002 preliminary ruling by U.S. District Court judge Jed Rakoff in the case *United States v. Quinones* (2002). Citing the exoneration of at least twelve death row inmates through DNA testing,[11] Rakoff concluded that existing procedures were insufficient to prevent a substantial number of actually innocent persons from being convicted and sentenced to the death penalty:

> Ironically, it was only a year or so after *Herrera* was decided that the new availability of DNA testing began to supply the kind of "truly persuasive demonstration" of actual innocence to which Chief Justice Rehnquist had hypothetical[ly] alluded. Thus, not only did *Herrera* not reach the issue here presented, but also it was premised on a series of factual assumptions about the unlikelihood that proof of actual innocence would emerge long after con-

10. *Herrera v. Collins,* 506 U.S. 390, 408 (1993).
11. Rakoff recognized that a larger number of exonerations had been based on reviews that identified procedural flaws and introduced other forms of evidence besides DNA, but he singled out the DNA exonerations as having been based on evidence with "remarkably high reliability."

viction that no longer seem sustainable. More generally, as already discussed, it implicitly premised a degree of unlikelihood of wrongful capital convictions that no longer seems tenable. (*U.S. v. Quinones,* 2002)

Rakoff's ruling was reversed upon appeal by the U.S. Court of Appeals, Second Circuit (*United States v. Quinones,* 2002), partly on the ground that the argument advanced "is not new" and that the Supreme Court had anticipated such arguments in earlier rulings on the constitutionality of the Federal Death Penalty Act.

Positivism about Science, Skepticism about Law

Although Berger (2004) acknowledges that the meaning and significance of matching and nonmatching DNA evidence depend upon the circumstances of the case, her shorthand usage ("DNA proves . . ."), and her characterization of DNA evidence as "irrefutable" express a view that we can characterize as *positivism about science, skepticism about law.* The problem that Berger and many other commentators overlook is that "DNA" is *not* "irrefutable," and should not be exempted from the fallibility that "it" reveals to be a property of all other forms of evidence. Ironically, proponents of "DNA" increasingly endow it with mathematical certainty and infallibility, when fallibility and probability were once deemed to be its signal strengths as scientific evidence. Recall that the establishment of DNA evidence as the gold standard of forensic science was set up, in part, by the 1993 U.S. Supreme Court ruling in *Daubert v. Merrell Dow Pharmaceuticals, Inc.* In that ruling, the court majority emphasized that science was distinguished by fallibility: its supposed openness to revision on the basis of novel evidence.[12] Moreover, probability (in the form of measured error rates) was another distinguishing strength. To define evidence derived from a particular technique (or, rather, a family of techniques) as irrefutable and uniquely identifying is to violate the very characteristic of science that supposedly endowed DNA evidence with its exceptional probative value.

The transition from fallibility to irrefutability was bridged by the use of probability figures to estimate the probative value of DNA evidence.

---

12. This loosely Popperian criterion also was invoked as a basis for separating science from religion in the notable "creation science" ruling in *McLean v. Arkansas* (1982).

As we chronicled in reference to fingerprint evidence, matching DNA "fingerprints" initially were deemed *merely* probable—less certain than categorically declared matches between latent marks and rolled prints—but the apparently precise measures of uncertainty provided by probability figures became a source of credibility. Then, with the multiplication of markers in currently used STR systems, random match probabilities approached a vanishing point, and match declarations effectively implied certainty. However, the fact that such declarations were supported by quantitative measures encouraged commentators to distinguish the *mathematical* certainty associated with "DNA" from the "subjective" or *moral* certainty expressed by the fingerprint expert. Nevertheless, as challenges to DNA probability figures have demonstrated, the information value of DNA evidence and the probability estimates associated with such evidence continue to be based on judgments in fields of *possibility*.[13] Such fields of possibility are invoked by rival stories of a case and mundane questions about how to make sense of the evidence; questions about how many people might have worn a hat, or how a suspect's DNA could have found its way to the scene of a burglary. Police and prosecutorial motives and practices also come into play when DNA evidence is treated less as an abstract source of "truth" than as material that is collected, handled, labeled, and possibly planted.

Throughout this book we have emphasized that "DNA" is deeply imbued with the very complexes of testimony, evidence, and criminal justice practice that its proponents hope to reform. While we recognize that "DNA" has been a strong source of leverage for exoneration, and a rhetorical resource for general arguments against the death penalty, it also has been used to support and even promote the use of the death penalty, and more generally its expansive use enhances the state's ability to monitor and control the detailed conduct of individuals. As Berger observes, with the passage of time and the stabilization of DNA profiling techniques, postconviction exonerations based on such techniques will become less frequent.[14] Consequently, convictions attributed to "DNA" are

13. Insight into a concept of information that requires one to take into account *possibilities* through which relevance, meaning, and the very value *as* information are implied, was gained from reading an early paper by Harold Garfinkel (1952) on information theory.

14. Surprisingly, the number of "DNA exonerations" has continued to increase as more lawyers get involved, and courts and prosecutors offer less resistance to postconviction testing (Willing, 2007).

likely to seem irrefutable, ignoring the extent to which the criminological relevance and significance of DNA evidence is, like other evidence,
circumstantial. The lesson that "DNA" does not transcend mundane organizational practices or the possibilities residing in stories of a crime
becomes increasingly important as "DNA" becomes reified as a machinery of truth for determining guilt and innocence.

Cases

Andrews v. State, 533 So.2d 841 (Fla. App. 1988).
Caldwell v. State, 393 S.E.2d 436 (Ga. 1990).
California v. Moffett, no. CR103094 (Cal. Super. Ct. 1991).
California v. Simpson, no. BA097211 (Cal. Super. Ct. 1995).
Commonwealth v. Cowans, 756 N.E.2d 622 (Mass. App. Ct. 2001).
Commonwealth v. Patterson, 840 N.E.2d 12 (Mass. 2005).
Daubert v. Merrell Dow Pharmaceuticals, Inc., 509 U.S. 579 (U.S. 1993).
Daubert v. Merrell Dow Pharmaceuticals, Inc., 43 F.3d 1311 (9th Cir. 1995).
Frye v. United States, 293 F. 1013 (D.C. Cir. 1923).
Furman v. Georgia, 408 U.S. 153 (U.S. 1972).
General Electric Co. v. Joiner, 522 U.S. 136 (U.S. 1997).
Gregg v. Georgia, 428 U.S. 153 (U.S. 1976).
Herrera v. Collins, 506 U.S. 390 (U.S. 1993).
Iowa v. Brown, 470 N.W. 2d 30 (Iowa 1991).
Iowa v. Steffen, 230 N.W. 536 (Iowa 1930).
Jurek v. Texas, 428 U.S. 262 (U.S. 1976).
Kumho Tire Co. v. Carmichael, 526 U.S. 137 (U.S. 1999).
Maine v. McLeod, CR-89–62 (Cumberland Cty. 18 July 1989)
Maryland v. Rose, K06–0545 (Cir. Ct. Baltimore Cty. 2007).
McLean v. Arkansas Board of Education, 529 F. Supp. 1255 (E.D. Ark. 1982).
New Hampshire v. Langill, 05-S-1129 (Rockingham Sup. Ct. 2007).
New Hampshire v. Sullivan, no. 03-S-1635 (Hillsborough City Sup. Ct. 2005).
New Hampshire v. Vandebogart, 616 A.2d 483, 494 (N.H. 1992).
New Jersey v. Williams, 599 A.2d 960 (N.J. Sup. Ct. Law Div. 1991).
New York v. Castro, 545 N.Y.S.2d 985 (N.Y. Sup. Ct. 1989).
New York v. Hyatt, 2001 N.Y. Slip Op. 50155 (U) (N.Y. Sup. Ct. 2001).
New York v. Lee, 750 N.E.2d 63 (N.Y. 2001).
New York v. Rush, 165 Misc. 2d 821 (N.Y. Sup. Ct. 1995).
New York v. Wesley, 140 Misc. 2d 306 (Albany Cty. Ct. 1988).

People v. Halik, VA00843 (Los Angeles Co., Cal. Super. Ct. 26 September 1991).

People v. Robinson, GA044716 (Los Angeles Co. Cal. Super. Ct. 2001).

Powel v. Chaminade College Preparatory, Inc., 197 S.W.3d 576 (Mo. 2006).

Proffitt v. Florida, 428 U.S. 242 (U.S. 1976).

Regina v. Adams (Denis John), 1 Court of Appeal R 369 (C.A. 1996).

Regina v. Adams (Denis John), 1 Court of Appeal R 377 (C.A. 1997).

Regina v. Chief Constable of Yorkshire Police (ex parte LS and Marper, consolidated appeals). UK House of Lords (22 July 2004).

Regina v. Deen, The Times, 10 January 1994 (C.A. 1993).

Regina v. Doheny and Adams (Gary Andrew), 1 Court of Appeal R 369 (C.A. 1997).

Regina v. Gordon, 1 Cr App Rep 290 (C.A. 1995).

Regina v. Lashley, CA 9903890 Y3 (8 February 2000).

Regina v. Shirley, Court of Appeal (Criminal Division), London, 29 July 2003.

Roper v. Simmons, 543 U.S. 551 (U.S. 2005).

State v. Josiah Sutton, District Court of Harris County, Texas, Case no. 800450 (1999).

Stogner v. California, 539 U.S. 607 (U.S. 2003).

Thomas v. Trent, 108 F.3d 1373 (4th Cir. 1997).

United States v. Crisp, 324 F.3d 261, 266 (4th Cir. 2003).

United States v. Havvard, 260 F.3d 597 (7th Cir. 2001).

United States v. Hines, 55 F.Supp.2d 62 (D. Mass. 1999).

United States v. Jenkins, 887 A. 2d 1013 (D.C. Cir. 2005).

United States v. Llera Plaza, 179 F. Supp.2d 492 (E.D.Pa. 2002) [Llera Plaza I— later redacted from the Federal Supplement].

United States v. Llera Plaza, 188 F. Supp.2d 549 (E.D.Pa. 2002) [Llera Plaza II].

United States v. McVeigh, 1997 WL 47724 (D. Colo. 1997).

United States v. Mitchell, Cr. no. 96–407 Tr. trans. (E.D. Pa. 1999).

United States v. Mitchell, 365 F. 3d 215 (3d Cir. 2004).

United States v. Quinones, 205 F. Supp. 2d 256 (S.D.N.Y. 2002).

United States v. Quinones, 313 F.3d 49 (2nd Cir. 2002).

United States v. Sullivan, 246 F.Supp.2d 700 (E.D. Ky. 2003).

Utah v. Quintana, 103 P.3d 168 (Ct. App. Utah 2004).

Vermont v. Passino, 640 A.2d 547 (Vt. 1991).

Virgin Islands v. Jacobs, 2001 U.S. Dist. LEXIS 22511 (D.V.I. 2001).

Wainwright v. Sykes, 433 U.S. 72, 90 (1977).

References

Acree, M. A. (1998). What is science? The dilemma of fingerprint science revisited. *The Print* 14 (4): 4–5.

Agger, B. (1994). Derrida for sociology? A comment on Fuchs and Ward. *American Sociological Review* 59: 501–5.

Aitken, C. G. G. (2000). Statistical interpretation of evidence: Bayesian analysis. *Encyclopedia of Forensic Science* 2: 717–24.

Aitken, C. G. G. (2005). The evaluation and interpretation of scientific evidence. Paper presented at "Forensic Science: The Nexus of Science and the Law," Arthur M. Sackler Colloquia of the National Academy of Sciences (Washington, DC, 16–18 November).

Allard, J. E. (1992). Series sexual crimes identified by a DNA computerised database. *Advances in Forensic Haemogenetics* 4: 295–97.

Allen, G. (2004). DNA and human behavior genetics: Implications for the criminal justice system. In D. Lazer (ed.), *DNA and the Criminal Justice System: The Technology of Justice.* Cambridge, MA: MIT Press: 287–314.

Alsdorf, M. (1999). Do statutes of limitations apply to DNA cases? *Slate* (12 October); available at http://slate.msn.com/id/1003808/.

Angell, M. & Relman, A. S. (2002). Patents, profits & American medicine. *Daedalus* 131 (2): 102–11.

Aronson, J. (2004). The introduction, contestation and regulation of forensic DNA analysis in the American legal system (1984–1994). PhD diss., University of Minnesota.

Aronson, J. (2005a). DNA fingerprinting on trial: The dramatic early history of a new forensic technique. *Endeavour* 29 (3): 126–31.

Aronson, J. (2005b). Post-conviction DNA testing in the American criminal justice system. International Society for the History, Philosophy, and Social Studies of Biology, University of Guelph (14 July). Abstract available at http://www.ishpssb.org/ocs/viewabstract.php?id=83.

Aronson, J. (2006). The "starch wars" and the early history of DNA profiling. *Forensic Science Review* 18 (1): 59–72.

Aronson, J. (2007). *Genetic Witness: Science, Law, and Controversy in the Making of DNA Profiling.* New Brunswick, NJ: Rutgers University Press.

Ashbaugh, D. R. (1991). Ridgeology. *Journal of Forensic Identification* 41: 16–64.

Ashbaugh, D. R. (1999). *Quantitative-Qualitative Friction Ridge Analysis: An Introduction to Basic and Advanced Ridgeology.* Boca Raton, FL: CRC Press.

Ashbaugh, D. R. (2002). Ridgeology: Science and law. Paper read at Annual Professional Learning Conference of the Canadian Identification Society (Mississauga, ON, 6 June).

Atkinson, J. M. & Drew, P. (1979). *Order in Court: The Organisation of Verbal Interaction in Judicial Settings.* London: Macmillan.

Atkinson, J. M. & Heritage, J. C. (eds.) (1984). *Structures of Social Action: Studies in Conversation Analysis.* Cambridge, UK: Cambridge University Press.

Austin, J. L. (1962). *How to Do Things with Words.* Oxford: Clarendon.

Balding D. J. (2002). The DNA database search controversy. *Biometrics* 58: 241–44.

Balding, D. J. & Donnelly, P. (1994). The prosecutor's fallacy and DNA evidence. *Criminal Law Review* (October): 711–21.

Balding, D. J. & Donnelly, P. (1995). Inference in forensic identification. *Journal of the Royal Statistical Society* 158 (part 1): 21–53.

Barnes. S. B. (1977). *Interests and the Growth of Knowledge.* London: Routledge & Kegan Paul.

Barnes, B. & Bloor, D. (1982). Relativism, rationalism, and the sociology of knowledge. In M. Hollis & S. Lukes (eds.), *Rationality and Relativism.* Cambridge, MA: MIT Press.

Barnes, B., Bloor, D. & Henry, J. (1996). *Scientific Knowledge: A Sociological Analysis.* Chicago: University of Chicago Press.

BBC News (2007). All UK "must be on DNA database." British Broadcasting Corporation (5 September), available at http://news.bbc.co.uk/1/hi/uk/6979138.stm.

Beck, U. (1992). *Risk Society: Towards a New Modernity.* London: Sage.

Bennett, W. L. (1978). Storytelling in criminal trials: A model of judgment. *Quarterly Journal of Speech* 64: 1–22.

Berg, M. & Timmermans, S. (2000). Orders and their others: On the constitution of universalities in medical work. *Configurations* 8: 31–61.

Berger, M. A. (2004). Lessons from DNA: Restriking the balance between finality and justice. In D. Lazer (ed.), *DNA and the Criminal Justice System: The Technology of Justice.* Cambridge, MA: MIT Press: 109–31.

Biagioli, M. & Galison, P. (eds.) (2003). *Scientific Authorship: Credit and Intellectual Property in Science*. London & New York: Routledge.

Bijker, W., Hughes, T. & Pinch, T. (eds.) (1987). *The Social Construction of Technical Systems*. Cambridge, MA: MIT Press.

Bittner, E. (1967). Police discretion in emergency apprehension of mentally ill persons. *Social Problems* 14: 278–92.

Bloor, D. (1976). *Knowledge and Social Imagery*. London: Routledge and Kegan Paul.

Blumer, H. (1969). *Symbolic Interactionism: Perspective and Method*. Englewood Cliffs, NJ: Prentice Hall.

Bobrow, D. & Whalen, J. (2002). The eureka story: Community knowledge sharing in practice. *Reflections* 4 (2): 47–59.

Bowker, G. (1994). *Science on the Run: Information Management and Industrial Geophysics at Schlumberger, 1920–1940*. Cambridge, MA: MIT Press.

Brannigan, A. & Lynch, M. (1987). On bearing false witness: Perjury and credibility as interactional accomplishments. *Journal of Contemporary Ethnography* 16: 115–46.

Brayley, F. A. (1909). *Brayley's Arrangement of Finger Prints: Identification and Their Uses*. Boston: Worchester Press.

Bromwich, M. R., et al. (2007). *Final Report of the Independent Investigator for the Houston Police Department Crime Laboratory and Property Room*. Washington, DC (13 June); available at http://www.hpdlabinvestigation.org.

Brown, P. G. (1998). Private matters. Editorial in *The Sciences* (November/December): 2.

Buckleton, J. (2005). Population genetic models. In J. Buckleton, C. M. Triggs and S. J. Walsh (eds.), *Forensic DNA Evidence Interpretation*. Boca Raton, FL: CRC Press.

Budowle, B., Buscaglia, J., & Perlman, R. S. (2006). Review of scientific basis for friction ridge comparisons as a means of identification: Committee findings and recommendations. *Forensic Science Communications* 8 (1); available at http://www.fbi.gov/hq/lab/fsc/current/research/2006_01_research02.htm.

Budowle, B., Chakraborty, R., Carmody, G. & Monson, K. (2000). Source attribution of a forensic DNA profile. *Forensic Science Communications* 2 (3) (July); available at http://www.fbi.gov/hq/lab/fsc/backissu/july2000/source.htm.

Bugliosi, V. (1996). *Outrage: The Five Reasons O. J. Simpson Got Away with Murder*. New York: W. W. Norton.

Burney, I. A. (2000). *Bodies of Evidence: Medicine and the Politics of the English Inquest, 1830–1926*. Baltimore: Johns Hopkins University Press.

Burns, S. (2001). "Think your blackest thoughts and darken them": Judicial mediation of large money damage disputes. *Human Studies* 24: 227–49.

Butterfield, H. (1931). *The Whig Interpretation of History.* London: G. Bell & Sons.

Callon, M. (1985). Some elements of a sociology of translation: Domestication of the scallops and the fishermen of St. Brieuc Bay. In J. Law (ed.), *Power, Action, and Belief.* London: Routledge & Kegan Paul: 196–230.

Callon, M. (1986). The sociology of an actor-network: The case of the electric vehicle. In M. Callon and A. Rip (eds.), *Mapping the Dynamics of Science and Technology.* Basingstoke, UK: Macmillan: 19–54.

Cambrosio, A. & Keating, P. (1988). "Going monoclonal": Art, science, and magic in the day-to-day use of hybridoma technology. *Social Problems* 35: 244–60.

Cambrosio, A., Keating, P. & MacKenzie, M. (1990). Scientific practice in the courtroom: The construction of sociotechnical identities in a biotechnology patent dispute. *Social Problems* 37: 275–93.

Cambrosio, A., Keating, P., Schlich, T. & Weisz, G. (2006). Regulatory objectivity and the generation and management of evidence in medicine. *Social Science & Medicine* 63: 189–99.

Carroll, P. (2006). *Science, Culture, and Modern State Formation.* Berkeley: University of California Press.

Caudill, D. & LaRue, L. (2006). *No Magic Wand: The Idealization of Science in Law.* Lanham, MD: Rowman & Littlefield.

Caudill, D. & Redding, R. (2000). Junk *philosophy* of science? The paradox of expertise and interdisciplinarity in federal courts. *Washington & Lee Law Review* 57: 685–766.

Certification for Latent Fingerprint Examiners (1977). *Identification News* 27 (8): 3–5.

Chakraborty, R. & Kidd, K. (1991). The utility of DNA typing in forensic work. *Science* 254: 1735–39.

Champod, C. (1995). Edmond Locard: Numerical standards and "probable" identifications. *Journal of Forensic Identification* 45: 136–63.

Champod, C. (2000). Fingerprints: Standards of proof. In J. A. Siegel, P. J. Saukko & G. C. Knupfer (eds.), *Encyclopedia of Forensic Sciences.* New York: Academic Press: 884–90.

Champod, C. & Evett, I. W. (2001). A probabilistic approach to fingerprint evidence. *Journal of Forensic Identification* 51 (2): 101–22.

Chapman, J. & J. Moult. DNA test blunder nearly landed me in jail. *Daily Mail* (London) (11 February), 23.

Cockcroft, L. (2007). Innocents fear DNA database errors. *Telegraph* (London, 26 November), available at http://www.telegraph.co.uk/news/main.jhtml ?xml=/news/2007/11/26/ndna126.xml.

Cole, S. A. (1997). The profession of truth: Knowledge and the fingerprint examiner. Paper, Department of Science and Technology Studies, Cornell University.

Cole, S. A. (1998). Witnessing identification: Latent fingerprint evidence and expert knowledge. *Social Studies of Science* 28 (5/6): 687–712.

Cole, S. A. (1999). What counts for identity? The historical origins of the methodology of latent fingerprint identification. *Science in Context* 12 (1): 139–72.

Cole, S. A. (2001a). *Suspect Identities: A History of Fingerprinting and Criminal Identification*. Cambridge, MA: Harvard University Press.

Cole, S. A. (2004a). Grandfathering evidence: Fingerprint admissibility ruling from *Jennings* to *Llera Plaza* and back again. *American Criminal Law Review* 41 (3): 1189–1276.

Cole, S. A. (2004b). Jackson Pollack, Judge Pollak, and the dilemma of fingerprint expertise. In G. Edmond (ed.), *Expertise in Regulation and Law*. Aldershot, UK: Ashgate: 98–120.

Cole, S. A. (2005a). Does "yes" really mean yes? The attempt to close debate on the admissibility of fingerprint testimony. *Jurimetrics* 45 (4): 449–64.

Cole, S. A. (2005b). More than zero: Accounting for error in latent fingerprint identification. *Journal of Criminal Law and Criminology* 95: 985–1078.

Cole, S. A. (2006a) "Implicit testing": Can casework validate forensic techniques? *Jurimetrics* 46 (2): 117–28.

Cole, S. A. (2006b). Is fingerprint identification valid? Rhetorics of reliability in fingerprint proponents' discourse. *Law and Policy* 28 (1): 109–35.

Cole, S. A. (2006c). The prevalence and potential causes of wrongful conviction by fingerprint. *Golden Gate University Law Review* 37: 39–105.

Cole, S. A. (2007). Twins, Twain, Galton and Gilman: Fingerprinting, individualization, brotherhood, and race in *Pudd'nhead Wilson*. *Configurations* 15(3).

Cole, S. A. & Dioso, R. (2005). Law and the lab. *Wall Street Journal,* May 13, W13.

Cole, S. A. & Dioso-Villa, R. (2007). *CSI* and its effects: Media, juries, and the burden of proof. *New England Law Review* 41: 701–35.

Cole, S. A. & Lynch, M. (2006). The social and legal construction of suspects. *Annual Review of Law and Social Science* 2: 39–60.

Collins, H. M. (1974). The TEA set: Tacit knowledge and scientific networks. *Science Studies* 4: 165–86.

Collins, H. M. (1975). The seven sexes: A study in the sociology of a phenomenon, or the replication of experiments in physics. *Sociology* 9: 205–24.

Collins, H. M. (1985). *Changing Order: Replication and Induction in Scientific Practice*. London: Sage.

Collins, H. M. (1987). Certainty and the public understanding of science: Science on television. *Social Studies of Science* 17: 689–713.

Collins, H. M. (1990). *Artificial Experts: Social Knowledge and Intelligent Machines.* Cambridge, MA: MIT Press.

Collins, H. M. (1999). Tantalus and the aliens: Publications, audiences, and the search for gravitational waves. *Social Studies of Science* 29: 163–97.

Collins, H. M. (2004). *Gravity's Shadow: The Search for Gravitational Waves.* Chicago: University of Chicago Press.

Collins, H. M. & Evans, R. J. (2002). The third wave of science studies: Studies of expertise and experience. *Social Studies of Science* 32: 235–96.

Collins, H. M. & Evans, R. J. (2003). King Canute meets the Beach Boys: Responses to the Third Wave. *Social Studies of Science* 33: 435–52.

Collins, H. & Pinch, T. (1998). *The Golem: What You Should Know about Science.* Cambridge: Cambridge University Press.

Coulter, J. (1975). Perceptual accounts and interpretive asymmetries. *Sociology* 9 (3): 385–96.

Coulter, J. & Parsons, E. D. (1991). The praxiology of perception: Visual orientations and practical action. *Inquiry* 33: 251–72.

Crispino, F. (2001). Comment on *JFI* 51 (3). *Journal of Forensic Identification* 51 (5): 449–56.

Cussins, A. (1992). Content, embodiment, and objectivity: The theory of cognitive trails. *Mind* 101: 651–88.

Dao, J. (2005). Errors force review of 150 DNA cases. *New York Times* (7 May), p. 1.

Daston, L., & Galison, P. (1992). The image of objectivity. *Representations* 40 (1): 81–128.

Daston, L., & Galison, P. (2007). *Objectivity.* New York: Zone Books.

Davey, M. (2005). Freed by DNA, now charged with new crime. *New York Times* (23 November).

Dawkins, R. (1998). Arresting evidence: DNA fingerprinting: Public servant or public menace? *The Sciences* (November/December): 20–25.

Dear, P. (2001). Science studies as epistemography. In J. Labinger and H. Collins (eds.), *The One Culture? A Conversation about Science.* Chicago: University of Chicago Press: 128–41.

Delsohn, G. (2001). Cracking an unsolved rape case makes history. *APF Reporter* 20(1); available at www.aliciapatterson.org/APF2001/Delsohn/Delsohn.html.

Denbeaux, M. & Risinger, D. M. (2003). *Kumho Tire* and expert reliability: How the question you ask gives the answer you get. *Seton Hall Law Review* 34: 15–70.

Derksen, L. (2000). Towards a sociology of measurement: Making subjectivity invisible and negotiating measurement in the case of DNA fingerprinting. *Social Studies of Science* 30 (6): 803–45.

Derksen, L. (2003). Toward a sociology of measurement: Objectivity and the erasure of subjectivity in the DNA typing controversy. PhD diss., Department of Sociology, University of California, San Diego.

Derrida, J. (1977). Signature, event, context. *Glyph* 1: 172–97.

Devlin, B. (2000). The evidentiary value of a DNA database search. *Biometrics* 56 (4): 1276–77.

Doing, P. (2004). "Lab hands" and the "Scarlet O": Epistemic politics and (scientific) labor. *Social Studies of Science* 34 (3): 299–323.

Donnelly, P. & Friedman, R. D. (1999). DNA database searches and the legal consumption of scientific evidence. *Michigan Law Review* 97: 931–84.

Dorf, M. C. (2001). How reliable is eyewitness testimony? A decision by New York State's highest court reveals unsettling truths about juries. *Findlaw, Legal News and Commentary;* online at http://writ.news.findlaw.com/dorf/20010516.html.

Douglas, M. (1966). *Purity and Danger: An Analysis of Concepts of Pollution and Taboo.* New York: Praeger.

Dovaston, D. F. (1994a). A DNA database (1) Report: For consideration by ACPO Crime Committee (28/29 September). London: Association of Chief Police Officers.

Dovaston, D. F. (1994b). A DNA database (3) Business case: For consideration by ACPO Crime Committee (28/29 September). London: Association of Chief Police Officers.

Dovaston, D. F. (1995). The national DNA database revised police user requirement: Revised by the ACPO national DNA database joint implementation group. London: Association of Chief Police Officers.

Drew, P. (1992). Contested evidence in courtroom cross-examination: The case of a trial of rape. In J. Heritage & P. Drew (eds.), *Talk at Work: Interaction in Institutional Settings.* Cambridge: Cambridge University Press: 470–520.

Dreyfus, H. (1979). *What Computers Can't Do.* New York: Harper and Row.

Duster T. (2004). Selective arrests, an ever-expanding DNA forensic database, and the specter of an early-twenty-first-century equivalent of phrenology. In D. Lazer (ed.), *DNA and the Criminal Justice System: The Technology of Justice.* Cambridge, MA: MIT Press: 315–34.

Edmond, G. (2001). The law set: The legal-scientific production of medical propriety. *Science, Technology, and Human Values* 26 (2): 191–226.

Edmond, G. (2002). Legal engineering: Contested representations of law, science (and non-science), and society. *Social Studies of Science* 32: 371–412.

Edmond, G. & Mercer, D. (1997). Scientific literacy and the jury: Reconsidering jury "competence." *Public Understanding of Science* 6: 329–57.

Edmond, G. & Mercer, D. (2000). Litigation life: Law-science knowledge construction in (Bendectin) mass toxic tort litigation. *Social Studies of Science* 30: 265–316.

Edmond, G. & Mercer, D. (2002). Conjectures and exhumations: Citations of history, philosophy, and sociology of science in US federal courts. *Law & Literature* 14: 309–66.

Edmond, G. & Mercer, D. (2004). *Daubert* and the exclusionary ethos: The convergence of corporate and judicial attitudes toward the admissibility of expert evidence in tort litigation. *Law and Policy* 26 (2): 231–57.

Emerson, R. M. (1969). *Judging Delinquents.* Chicago: Aldine.

Epstein, R. (2002). Fingerprints meet *Daubert:* The myth of fingerprint "science" is revealed. *Southern California Law Review* 75: 605–57.

Epstein, S. (1995). The construction of lay expertise. *Science, Technology, and Human Values* 20 (4): 408–37.

Epstein, S. (1996). *Impure Science: AIDS, Activism, and the Politics of Knowledge.* Berkeley: University of California Press.

European Fingerprint Standards (2002). *Fingerprint Whorld* 28 (107): 19.

Evett, I. W. (2000). DNA profiling: A discussion of issues relating to the reporting of very small match probabilities. *Criminal Law Review:* 341–55.

Evett, I. W. & Williams, R. L. (1996). A review of the sixteen points fingerprint standard in England and Wales. *Journal of Forensic Identification* 46 (1): 49–73.

Faigman, D. L., Kaye, D. H., Saks, M. J. & Sanders, J. (2002). *Science in the Law: Forensic Science Issues.* St. Paul, MN: West.

Federal Bureau of Investigation (1985). *The Science of Fingerprints: Classification and Uses.* Washington, DC: United States Government Printing Office.

Findley, K. (2002). Learning from our mistakes: A criminal justice commission to study wrongful convictions. *California Western Law Review* 38: 333–54.

Fine, G. A. (2006). A review of the FBI's handling of the Brandon Mayfield case. Washington, DC: U.S. Department of Justice Office of the Inspector General.

Finger Print & Identification Magazine (1925). Vol. 6 (10 April).

Foster, K. R. & P. Huber (1999). *Judging Science: Scientific Knowledge and the Federal Courts.* Cambridge, MA: MIT Press.

Foucault, M. (1979). *Discipline and Punish: The Birth of the Prison.* New York: Random House.

FSS (2005). The national DNA database (NDNAD). Fact sheet (3). The Forensic Science Service (5 December).

Fuchs, S. & Ward, S. (1994). What is deconstruction, and where and when does it take place? Making facts in science, building cases in law. *American Sociological Review* 59: 481–500.

Funtowicz, S. O. & Ravetz, J. R. (1992). Three types of risk assessment and the emergence of post-normal science. In S. Krimsky & D. Golding (eds.), *Social Theories of Risk*. Westport, CT: Praeger: 251–74.

Galbraith, J. K. (1967). *The New Industrial State*. Boston: Houghton Mifflin.

Garfinkel, H. (1952). Notes on a sociological theory of information. Paper, Department of Sociology, UCLA.

Garfinkel, H. (1963). A conception of, and experiments with, "trust" as a condition of stable concerted actions. In O. J. Harvey (ed.), *Motivation and Social Interaction*. New York: Ronald Press: 187–238.

Garfinkel, H. (1967). *Studies in Ethnomethodology*. Englewood Cliffs, NJ: Prentice Hall.

Garfinkel, H. (1991). Respecification: Evidence for locally produced, naturally accountable phenomena of order, logic, reason, meaning, method, etc., in and as of the essential haecceity of immortal ordinary society (I)—an announcement of studies. In G. Button (ed.), *Ethnomethodology and the Human Sciences*. Cambridge, UK, Cambridge University Press: 10–19.

Garfinkel, H. (2002). *Ethnomethodology's Program: Working Out Durkheim's Aphorism*. Lanham, MD: Rowman & Littlefield.

Garfinkel, H., Lynch, M. & Livingston, E. (1981). The work of a discovering science construed with materials from the optically discovered pulsar. *Philosophy of the Social Sciences* 11: 131–58.

Garfinkel, H. & Sacks, H. (1970). On formal structures of practical actions. In J. C. McKinney & E. A. Tiryakian (eds.), *Theoretical Sociology: Perspectives and Development*. New York: Appleton-Century-Crofts: 337–66.

Garrett, B. & Neufeld, P. (2009). Improper forensic science and wrongful convictions. *Virginia Law Review* 95 (forthcoming).

Gaughan P. A. (1996). *Delivery of DNA Evidence in the Criminal Justice System*. MSc Thesis in the Study of Security Management: University of Leicester.

GeneWatch and Action on Rights for Children (2007). Over 100,000 innocent young people now on the National DNA Database. Joint Press Release, 22 May.

GeneWatch UK (2005). The police national DNA database: Human rights and privacy. GeneWatch Briefing 31 (June).

GeneWatch UK (2007). Submission to the Nuffield Council on Bioethics' consultation on forensic use of bioinformation; available at http://www.genewatch .org/uploads/f03c6d66a9b354535738483c1c3d49e4/Respondents_form001 .doc.

Gieryn, T. (1983). Boundary-work and the demarcation of science from non-science. *American Sociological Review* 48: 781–95.

Gieryn T. (1992). The ballad of Pons and Fleischmann: Experiment and narrative in the (un)making of cold fusion. In E. McMullin (ed.), *The Social Dimensions of Science*. South Bend, IN: University Notre Dame Press.

Gieryn, T. (1995). Boundaries of science. In S. Jasanoff, G. Markle, J. Petersen &

T. J. Pinch (eds.), *Handbook of Science, Technology & Society*. Beverly Hills, CA: Sage: 393–443.

Gieryn, T. (1999). *Cultural Cartography: Credibility on the Line*. Chicago: University of Chicago Press.

Gigerenzer, G. & Hoffrage, U. (1995). How to improve Bayesian reasoning without instruction: Frequency formats. *Psychological Review* 102: 684–704.

Gigerenzer, G., Swijtink, Z., Porter, T., Daston, L., Beatty, J. & Krüger, L. (1989). *The Empire of Chance: How Probability Changed Science and Everyday Life*. Cambridge, UK: Cambridge University Press.

Gilbert, G. N. & Mulkay, M. (1984). *Opening Pandora's Box: An Analysis of Scientists' Discourse*. Cambridge: Cambridge University Press.

Gilbert, W. (1992). Visions of the grail. In D. J. Kevles & L. Hood (eds.), *The Code of Codes: Scientific and Social Issues in the Human Genome Project*. Cambridge, MA: Harvard University Press: 83–97.

Gillespie, T. (2007). *Wired Shut: Copyright and the Shape of Digital Culture*. Cambridge, MA: MIT Press.

Ginzburg, C. (1989). Clues: Roots of an evidential paradigm. In C. Ginzburg, *Clues, Myths, and the Historical Method*. Baltimore: Johns Hopkins University Press: 96–125.

Glaberson, W. (2001). Juries, their powers under siege, find their role being eroded. *New York Times* (2 March).

Gluckman, M. (1963). The reasonable man in Barotse law. In M. Gluckman, *Order and Rebellion in Tribal Africa*. New York: Free Press.

Goffman, E. (1959). *The Presentation of Self in Everyday Life*. Garden City, NY: Anchor Books.

Goffman, E. (1961). *Asylums: Essays on the Social Situation of Mental Patients and Other Inmates*. Garden City, NY: Doubleday-Anchor.

Goffman, E. (1971). The territories of self. In E. Goffman, *Relations in Public*. New York: Harper & Row: 28–61.

Golan, T. (1999). History of scientific expert testimony in the English courtroom. *Science in Context* 12: 5–34.

Golan, T. (2004a). *Laws of Men and Laws of Nature: A History of Scientific Expert Testimony*. Cambridge, MA: Harvard University Press.

Golan, T. (2004b). The legal and medical reception of x-rays in the USA. *Social Studies of Science* 34: 469–99.

Gonzalez, E. M. (2006). The interweaving of talk and text in a French criminal pretrial hearing. *Research on Language and Social Interaction* 39 (3): 229–61.

Goodwin, C. (1994). Professional vision. *American Anthropologist* 96: 606–33.

Grey, S. (1997). Yard in fingerprint blunder. *Sunday Times* (London, 6 April): 4.

Grieve, D. L. (1990). The identification process: Traditions in training. *Journal of Forensic Identification* 40: 195–213.

Grieve, D. L. (1996). Possession of truth. *Journal of Forensic Identification* 46 (5): 521–28.

Grieve, D. L. (1999). The identification process: TWGFAST and the search for science. *Fingerprint Whorld* 25 (98): 315–25.

Gross, P. R. & Levitt, N. (1994). *Higher Superstition*. Baltimore: Johns Hopkins University Press.

Gross, P. R., Levitt, N. & Lewis, M. W. (eds.) (1996). *The Flight from Science and Reason*. New York: New York Academy of Sciences.

Gross, S. R & Mnookin, J. L. (2003). Expert information and expert evidence: A preliminary taxonomy. *Seton Hall Law Review* 34 (1): 141–89.

Grove, V. (1989). Why women owe a debt to prof's curiosity. *Sunday Times* (London, 23 April).

Gurwitsch, A. (1964). *The Field of Consciousness*. Pittsburgh: Duquesne University Press.

Haack, S. (2005). Trial and error: The Supreme Court's philosophy of science. *American Journal of Public Health* 95 (S1): S66–S73.

Haber, L. & Haber, R. (2008). Scientific validation of fingerprint evidence under *Daubert*. *Law, Probability and Risk* 87(2): 87–109.

Hacking, I. (1975). *The Emergence of Probability: A Philosophical Study of Early Ideas about Probability, Induction, and Statistical Inference*. Cambridge: Cambridge University Press.

Hacking, I. (1995). *Rewriting the Soul: Multiple Personality and the Sciences of Memory*. Princeton: Princeton University Press.

Hacking, I. (1999). *The Social Construction of What?* Cambridge, MA: Harvard University Press.

Halfon, S. (1998). Collecting, testing, and convincing: DNA experts in the courts. *Social Studies of Science* 28 (5/6): 801–28.

Hand, L. (1901). Considerations regarding expert testimony. *Harvard Law Review* 15: 40–58.

Hanson, N. R. (1961). *Patterns of Discovery*. Cambridge, UK: Cambridge University Press.

Haraway, D. (1997). *Modest_Witness@Second_Millennium. FemaleMan©_Meets _OncoMouse™: Feminism and Technoscience*. New York and London: Routledge.

Harding, S. (1986). *The Science Question in Feminism*. Ithaca, NY: Cornell University Press.

Hartl, D. (1994). Forensic DNA typing dispute. *Nature* 372 (1 December): 398.

Healy, D. (2004). Shaping the intimate: Influences on the experience of everyday nerves. *Social Studies of Science* 34: 219–45.

Hilgartner, S. (2000). *Science on Stage.* Stanford: Stanford University Press.

Hine, C. (2000). *Virtual Ethnography.* London: Sage.

Holton, G. (1978). *The Scientific Imagination: Case Studies.* Cambridge: Cambridge University Press.

Home Office (2000). Prime Minister hails hi-tech drive against crime. Home Office Announcement 269 (31 August).

Home Office (2003). *Review of the Forensic Science Service* (July).

Home Office (2006). DNA Expansion Programme, 2000–2005: Reporting achievement. Home Office Forensic Science and Pathology Unit (4 January); available at http://police.homeoffice.gov.uk/news-and-publications/publication/operational-policing/DNAExpansion.pdf?view=Binary.

Houck, M. M. (2004). Forensic science, no consensus. *Issues in Science and Technology* 20 (2): 6–8.

Houck, M. M. (2006). CSI: Reality. *Scientific American* 295 (1) (July): 84–89.

Houck, M. M. & Budowle, B. (2002). Correlation of microscopic and mitochondrial DNA hair comparisons. *Journal of Forensic Sciences* 47 (5): 1–4.

House of Commons Science and Technology Committee (2005). Session 2004–5, 7th Report: *Forensic Science on Trial.* Norwich: The Stationary Office.

Howard, S. (2002). The business of paradigms. *Weekly Detail* 66.

Huber, P. (1991). *Galileo's Revenge: Junk Science in the Courtroom.* New York: Basic Books.

Huber, R. A. (1959–60). Expert Witness. *Criminal Law Quarterly* 2: 276–95.

Human Genetics Commission (2002). Balancing interests in the use of personal genetic data. *Law and the Human Genome Review* 16 (January–June): 260–62.

Imwinkelried, E. J. (2004). The relative priority that should be assigned to trial stage DNA issues. In D. Lazer (ed.) *DNA and the Criminal Justice System: The Technology of Justice.* Cambridge, MA: MIT Press: 91–107.

Inman, K. & Rudin, N. (2001). *Principles and Practice of Criminalistics: The Profession of Forensic Science.* Boca Raton, FL: CRC Press.

International Association for Identification (1920). Minutes of the Sixth Annual Convention, at Minneapolis.

International Association for Identification (1979). Resolution VII. *Identification News* 29: 1.

Irwin, A. & Wynne, B. (eds.) (1996). *Misunderstanding Science? The Public Reconstruction of Science and Technology.* Cambridge, UK: Cambridge University Press.

Jasanoff, S. (1990). *The Fifth Branch: Science Advisors as Policymakers.* Cambridge: Harvard University Press.

Jasanoff, S. (1991). Judicial construction of new scientific evidence. In P. Durbin (ed.), *Critical Perspectives in Nonacademic Science and Engineering.* Bethlehem, PA: Lehigh University Press: 215–38.

Jasanoff, S. (1992). What judges should know about the sociology of science. *Jurimetrics Journal* 32: 345–59.

Jasanoff, S. (1995). *Science at the Bar: Law, Science, and Technology in America*. Cambridge, MA: Harvard University Press.

Jasanoff, S. (1998). The eye of everyman: Witnessing DNA in the Simpson trial. *Social Studies of Science* 28 (5/6): 713–40.

Jasanoff, S. (2003). Breaking the waves in science studies: Comment on H. M. Collins and Robert Evans, "The third wave of science studies." *Social Studies of Science* 33: 389–400.

Jasanoff, S. (2004). The idiom of co-production. In S. Jasanoff (ed.), *States of Knowledge: The Co-Production of Science and the Social Order*. London: Routledge: 1–13.

Jasanoff, S. (2005). Law's knowledge: Science for justice in legal settings. *American Journal of Public Health* 95 (S1): S40–S58.

Jeffreys A. J. (1993a). DNA typing: Approaches and applications. *Journal of the Forensic Science Society* 33: 204–11.

Jeffreys, A. J., Wilson V. & Thein, S. L. (1985a). Hypervariable "minisatellite" regions in human DNA. *Nature* 314: 67–73.

Jeffreys, A. J., Wilson, V. & Thein, S. L. (1985b). Individual-specific "fingerprints" of human DNA. *Nature* 316: 76–78.

Jha, A. (2004). DNA fingerprinting "no longer foolproof": Pioneer of process calls for upgrade. *Guardian Unlimited* (9 September); available at http://www.guardian.co.uk.

Johns, A. (1998). *The Nature of the Book: Print and Knowledge in the Making*. Chicago and London: University of Chicago Press.

Johnson, P. & Williams, R. (2004). Post-conviction DNA testing: The UK's first "exoneration" case. *Science and Justice* 44: 77–82.

Johnston, P. (2006). Huge rise in juvenile DNA samples kept by the police. *Daily Telegraph* (9 January); available at http://www.mobile.telegraph.co.uk/news/main.jhtml?xml=/news/2006/01/09/ndna09.xml.

Jordan, K. (1997). *Sociological Investigations into the Mainstreaming of the Polymerase Chain Reaction*. PhD diss., Department of Sociology, Boston University.

Jordan, K. & M. Lynch (1992). The sociology of a genetic engineering technique: Ritual and rationality in the performance of the plasmid prep. In A. Clarke and J. Fujimura (eds.), *The Right Tools for the Job: At Work in Twentieth-Century Life Science*. Princeton, NJ: Princeton University Press: 77–114.

Jordan, K. & Lynch, M. (1993). The mainstreaming of a molecular biological tool: A case study of a new technique. In G. Button (ed.), *Technology in Working Order: Studies in Work, Interaction, and Technology*. London and New York: Routledge: 160–80.

Jordan, K. & Lynch, M. (1998). The dissemination, standardization, and routinization of a molecular biological technique. *Social Studies of Science* 28 (5/6): 773–800.

Joseph, A., & Winter, A. (1996). Making the match: The hunt for human traces, the scientific expert, and the public imagination. In F. Spufford and J. Uglow (eds.), *Cultural Babbage: Technology, Time, and Invention*. London: Faber and Faber: 193–214.

Kaiser, D. (2005). *Drawing Theories Apart: The Dispersion of Feynman Diagrams in Postwar Physics*. Chicago: University of Chicago Press.

Kalven, H. & Zeisel, H. (1970). *The American Jury*. Chicago: University of Chicago Press.

Kaye, D. H. (2003). Questioning a courtroom proof of the uniqueness of fingerprints. *International Statistical Review* 71 (3): 521–33.

Keller, E. F. (2000). *The Century of the Gene*. Cambridge, MA: Harvard University Press.

Kennedy, D. (2003). Forensic Science: Oxymoron? *Science* 302: 1625.

Kevles, D. (1998). *The Baltimore Case: A Trial of Politics, Science, and Character*. New York: Norton.

Kirk, P. L. (1963). The ontogeny of criminalistics. *Journal of Criminal Law, Criminology & Police Science* 54 (2): 235–38.

Klein, A. (2004). Art trips up life: TV crime shows influence jurors. *Baltimore Sun* (25 July): 1A–8A.

Knorr, K. (1981). *The Manufacture of Knowledge*. Oxford: Pergamon.

Knorr-Cetina, K. (1983). The ethnographic study of scientific work: Towards a constructivist sociology of science. In K. Knorr-Cetina & M. Mulkay (eds.), *Science Observed*. London: Sage: 115–40.

Knorr-Cetina, K. (1999). *Epistemic Cultures: How the Sciences Make Knowledge*. Cambridge, MA: Harvard University Press.

Koehler, J. J. (1996). On conveying the probative value of DNA evidence: Frequencies, likelihood ratios, and error rates. *University of Colorado Law Review*, 67: 859–86.

Koehler, J. J. (1997). Why DNA likelihood ratios should account for error (even when a National Research Council report says they should not). *Jurimetrics Journal* 37: 425–37.

Krawczak, M. & Schmidtke, J. (1994). *DNA Fingerprinting*. Oxford: BIOS.

Kuhn, T. S. (1970 [1962]). *The Structure of Scientific Revolutions*. 2nd ed. Chicago: University of Chicago Press.

Kuhn, T. S. (1991). The road since "Structure." In A. Fine, M. Forbes & L. Wessels (eds.), *PSA 1990*. East Lansing, MI: Philosophy of Science Association: 3–13.

Kuhn, T. S. (2000). *The Road since "Structure": Philosophical Essays, 1970–1993, with an Autobiographical Interview*. Chicago: University of Chicago Press.

Krawczak, M. & Schmidtke, J. (1994). *DNA Fingerprinting*. Oxford: BIOS.

Labinger, J. & Collins, H. (eds.) (2001). *The One Culture? A Conversation about Science*. Chicago: University of Chicago Press.

Lakatos, I. (1970). Falsification and the methodology of scientific research programmes. In I. Lakatos & A. Musgrave (eds.), *Criticism and the Growth of Knowledge*. Cambridge: Cambridge University Press: 91–195.

Lambourne G. T. C. (1984). Fingerprint standards. *Medicine, Science, and the Law* 24: 227–29.

Lander, E. S. (1989). DNA fingerprinting on trial. *Nature* 339 (14 June): 501–5.

Lander, E. S. & Budowle, B. (1994). DNA fingerprinting dispute laid to rest. *Nature* 371 (27 October): 735–38.

Langmuir, I. (1989 [1968]). Pathological science. Transcribed & edited by R. N. Hall. *Physics Today* 42 (10): 36–48. [Originally: General Electric R&D Center Report no. 68-C-035, Schenectady, NY.]

Latour, B. (1987). *Science in Action*. Cambridge, MA: Harvard University Press.

Latour, B. (1990). Drawing things together. In M. Lynch & S. Woolgar (eds.), *Representation in Scientific Practice*. Cambridge, MA: MIT Press: 19–68.

Latour, B. (1992). Where are the missing masses? The sociology of a few mundane artifacts. In W. E. Bijker & J. Law (eds.), *Shaping Technology/Building Society: Studies in Sociotechnical Change*. Cambridge: MIT Press: 225–58.

Latour, B. (1995). The "pédofil" of Boa Vista: A photo-philosophical montage. *Common Knowledge* 4 (1): 144–87.

Latour, B. (1996). *Aramis: The Love of Technology*. Cambridge, MA: Harvard University Press.

Latour, B. (1999). *Pandora's Hope: Essays on the Reality of Science Studies*. Cambridge, MA: Harvard University Press.

Latour, B. (2004a). *Politics of Nature: How to Bring the Sciences into Democracy*. Cambridge, MA: Harvard University Press.

Latour B. (2004b). Scientific objects and legal objectivity. In A. Pottage (ed.), *Law, Anthropology and the Constitution of the Social*. Cambridge: Cambridge University Press.

Latour, B. & Woolgar, S. (1986 [1979]). *Laboratory Life: The Construction of Scientific Facts*. 2nd ed. Princeton, NJ: Princeton University Press [London: Sage, 1979].

Laudan, L. (1982). Science at the bar: Causes for concern. *Science, Technology, and Human Values* 7 (41): 16–19.

Laudan, L. (2003). Is reasonable doubt reasonable? *Legal Theory* 9: 295–331.

Law, J. (1986). On the methods of long-distance control: Vessels, navigation, and the Portuguese route to India. In J. Law (ed.), *Power, Action and Belief*. London: Routledge and Kegan Paul: 231–60.

Leapman, B. (2006). Three in four young black men on the DNA database. *Telegraph* (London) (5 November).

Lee, H. C., Ladd, C., Bourke, M. T., Pagliaro, E., & Tirnady, F. (2004). DNA typing in forensic science. *American Journal of Forensic Medicine and Psychiatry* 15 (4): 269–82.

Lerner, G. H. (1989). Notes on overlap management in conversation: The case of delayed completion. *Western Journal of Speech Communication* 53 (2): 167–77.

Levy, H. (1996). *And the Blood Cried Out.* New York: Avon Books.

Lewenstein, B. V. (1995). From fax to facts: Communication in the cold fusion saga. *Social Studies of Science* 25: 403–36.

Lewin, R. (1989). DNA typing on the witness stand. *Science* 244 (2 June): 1033–35.

Lewontin, R. C. (1992). The dream of the human genome. *New York Review of Books* (28 May): 31–40.

Lewontin, R. C. (1993). *Biology as Ideology: The Doctrine of DNA.* New York: Harper Collins.

Lewontin, R. C. (1994a). DNA fingerprinting: A review of the controversy; Comment: The use of DNA profiles in forensic contexts. *Statistical Science* 9 (2): 259–62.

Lewontin, R. C. (1994b). Forensic DNA typing dispute. *Nature* 372 (1 December): 398.

Lewontin, R. C. & Hartl, D. J. (1991). Population genetics in forensic DNA typing. *Science* 254: 1745–50.

Lewontin, R. C. & Hartl, D. J. (1992). Letters, DNA typing. *Science* 255: 1054–55.

Lezaun, J. (2006). Creating a new object of government: Making genetically modified organisms traceable. *Social Studies of Science* 36 (4): 499–531.

Lin, C. H., Liu, J. H., Osterburg, J. W. & Nicol, J. D. (1982). Fingerprint comparison I: Similarity of fingerprints. *Journal of Forensic Sciences* 27: 290–304.

Liptak, A. (2003a). Prosecutors see limits to doubt in death penalty cases. *New York Times* (24 February).

Liptak, A. (2003b). You think DNA evidence is foolproof? Try again. *New York Times* (16 March): 5.

Loftus, E. (1975). Leading questions and the eyewitness report. *Cognitive Psychology* 7, 560–72.

Loftus, E. & Doyle, J. M. (1997). *Eyewitness Testimony: Civil and Criminal.* Charlottesville, VA: Lexis Law Publishing.

Lowe, A. L., Urquhart, A., Foreman, L. A. & Evett, I. W. (2001). Inferring ethnic origin by means of an STR profile. *Forensic Science International* 119: 17–22.

Lynch, M. (1985a). *Art and Artifact in Laboratory Science*. London: Routledge & Kegan Paul.

Lynch, M. (1985b). Discipline and the material form of images: An analysis of scientific visibility. *Social Studies of Science* 15 (1): 37–66.

Lynch, M. (1991). Pictures of nothing? Visual construals in social theory. *Sociological Theory* 9 (1): 1–22.

Lynch, M. (1993). *Scientific Practice and Ordinary Action: Ethnomethodology and Social Studies of Science*. New York: Cambridge University Press.

Lynch, M. (1998). The discursive production of uncertainty: The OJ Simpson "Dream Team" and the sociology of knowledge machine. *Social Studies of Science* 28 (5/6): 829–68.

Lynch, M. (1999). Archives in-formation: Privileged spaces, popular archives, and paper trails. *History of the Human Sciences* 12 (2): 65–87.

Lynch, M. & Bogen, D. (1996). *The Spectacle of History: Speech, Text, and Memory at the Iran-Contra Hearings*. Durham, NC: Duke University Press.

Lynch, M. & Bogen, D. (1997). Sociology's asociological "core": An examination of textbook sociology in light of the sociology of scientific knowledge. *American Sociological Review* 62: 481–93.

Lynch, M. & Cole, S. (2005). Science and technology studies on trial: Dilemmas of expertise. *Social Studies of Science* 35 (3): 269–311.

Lynch, M., Hilgartner, S. & Berkowitz, C. (2005). Voting machinery, counting, and public proofs in the 2000 US presidential election. In B. Latour & P. Wiebel (eds.), *Making Things Public: Atmospheres of Democracy*. Cambridge, MA: MIT Press: 814–25.

Lynch, M. & Jasanoff, S. (guest eds.) (1998). *Science, Law, and Forensic Practice*. Special issue of *Social Studies of Science* 28 (5/6).

Lynch, M. & Jordan, K. (1995). Instructed actions in, of, and as molecular biology. *Human Studies* 18: 227–44.

Lynch, M., Livingston, E. & Garfinkel, H. (1983). Temporal order in laboratory work. In K. Knorr-Cetina and M. Mulkay (eds.), *Science Observed*. London: Sage: 205–38.

Lynch, M. & McNally, R. (2005). Chains of custody: Visualization, representation, and accountability in the processing of DNA evidence. *Communication & Cognition* 38 (3&4): 297–318.

Lynch, M. & Woolgar, S. (eds.) (1990). *Representation in Scientific Practice*. Cambridge, MA: MIT Press.

MacKenzie, D. (1981). *Statistics in Britain: 1865–1930; The Social Construction of Scientific Knowledge*. Edinburgh: Edinburgh University Press.

Maniatis, T., Fritsch, E. F. & Sambrook, J. (1982). *Molecular Cloning: A Laboratory Manual*. Cold Spring Harbor, NY: Cold Spring Harbor Laboratory.

Marcus, G. & Fischer, M. (1999). *Anthropology as Cultural Critique: An Experimental Moment in the Human Sciences*. Chicago: University of Chicago Press.

Martin, A. (2007). The chimera of liberal individualism in late 20th century America. In G. Eghigian, A. Killen & C. Leuenberger (eds.), *The Self as Project: Politics and the Human Sciences in the Twentieth Century. Osiris* 22: 205–22.

Maynard, D. & Manzo, J. (1993). On the sociology of justice: Theoretical notes from an actual jury deliberation. *Sociological Theory* 11: 171–93.

McCartney, C. (2006). *Forensic Identification and Criminal Justice*. Cullompton: Willan.

McKasson, S. (2001). I think therefore I *probably* am. *Journal of Forensic Identification* 51 (3): 217–21.

McLeod, N. (1991). English DNA evidence held inadmissible. *Criminal Law Review:* 583–90.

Meagher, S. (2001). What is the future of the latent print discipline? An FBI Perspective. Paper presented at International Association for Identification. Slide Presentation available online: http://www.onin.com/fp/iai_future_2001 .pdf.

Meagher, S. (2004). Letter to Randerson, J., *New Scientist* (28 January), available at http://www.newscientist.com/article.ns?id=dn7983.

Medawar, P. (1964). Is the scientific paper fraudulent? Yes; it misrepresents scientific thought. *Saturday Review* (1 August): 42–43.

Meehan, A. J., & Ponder, M. C. (2002). Race and place: The ecology of racial profiling African American motorists. *Justice Quarterly* 19 (3): 399–431.

Megill, A. (ed.) (1994). *Rethinking Objectivity*. Durham, NC: Duke University Press.

Melnick, R. L. (2005). A *Daubert* motion: A legal strategy to exclude essential scientific evidence in toxic tort litigation. *American Journal of Public Health* 95 (S1): S30–S34.

Merton, R. K. (1973). The normative structure of science. In R. K. Merton, *The Sociology of Science*. Chicago: University of Chicago Press: 267–78.

Mills, C. W. (1940). Situated actions and vocabularies of motive. *American Sociological Review* 5: 904–13.

Mirowski, P. & Van Horn, R. (2005). The contract research organization and the commercialization of scientific research. *Social Studies of Science* 35: 503–48.

Mnookin, J. (2001). Fingerprint evidence in the age of DNA profiling. *Brooklyn Law Review* 67 (Fall): 13–70.

Mnookin, J. (2003). Fingerprints not a gold standard. *Issues in Science and Technology* (Fall): 47–54; available at http://www.issues.org/issues/20.1/mnookin .html.

Mody, C. (2001). A little dirt never hurt anyone. Knowledge-making and contamination. *Social Studies of Science* 31 (1): 7–36.

Moenssens, A. A. (1972). Testifying as a fingerprint witness. *Finger Print and Identification Magazine:* 3–18.

Moenssens, A. A. (2003). Fingerprint identification: A valid, reliable "forensic science"? *Criminal Justice* 30 (September): 18.

Mopas, M. (2007). Examining the "CSI effect" through an ANT lens. *Crime, Media, Culture* 3 (1): 110–17.

Moss, D. C. (1988). DNA: The new fingerprints. *American Bar Association Journal* 74: 66.

Mullis, K. (1998). *Dancing Naked in the Mind Field.* New York: Pantheon.

Mullis, K., Faloona, F., Scharf, S., Saiki, R., Horn, G. & Erlich, H. (1986). Specific enzymatic amplification of DNA in vitro. *Cold Spring Harbor Symposium in Quantitative Biology* 51: 263–73.

Murphy, E. H. & Murphy, J. E. (1922). *Finger Prints for Commercial and Personal Identification.* Detroit: International Title and Recording and Identification Bureau.

Myers, G. (1995). From discovery to invention: The writing and rewriting of two patents. *Social Studies of Science* 25 (1): 57–105.

National DNA Database Annual Report (2003–4). Forensic Science Service (29 November 2004); available at http://www.forensic.gov.uk/forensic_t/inside/about/docs/NDNAD_AR_3_4.pdf.

National DNA Database Annual Report (2004–5). Forensic Science Service; available at http://www.acpo.police.uk/asp/policies/Data/NDNAD_AR_04_05.pdf.

National DNA Database Annual Report (2005–6). Forensic Science Service; available at http://www.homeoffice.gov.uk/documents/DNA-report2005–06.pdf?view=Binary.

Nelkin, D. & Lindee, M. S. (1995). *The DNA Mystique: The Gene as Cultural Icon.* New York: Freeman.

Nelson, A. (2006). Genealogical branches, genetic roots, and the pursuit of African ancestry. In B. Koenig & S. S.-J. Lee (eds.), *Revisiting Race in a Genomic Age.* New Brunswick: Rutgers University Press.

Nelson, B. (2003). The DNA revolution. *Newsday* (26 January 2003): A6.

Neufeld, P. (2005). The (near) irrelevance of *Daubert* to criminal justice and some suggestions for reform. *American Journal of Public Health* 95 (S1): S107–S113.

Neufeld, P., & Colman, N. (1990). When science takes the witness stand. *Scientific American* 262 (5) (May): 18–25.

Neufeld, P., & Scheck, B. (1989). Factors affecting the fallibility of DNA profiling: Is there less than meets the eye? *Expert Evidence Reporter* 1 (4).

Neumann, C., Champod, C., Puch-Solis, R., Egli, N., Anthonioz, A., Meuwly, D. &

Bromage-Griffiths, A. (2006). Computation of likelihood ratios in fingerprint identification for configurations of three minutia. *Journal of Forensic Sciences* 51 (6): 1255–66.

Newman, A. (2001). Fingerprinting's reliability draws growing court challenges," *New York Times* (7 April).

New York Times (2006). DNA's weight as evidence. Editorial (18 January).

Nolan, T. W. (2007). Depiction of the CSI in the popular culture: Portrait in domination and effective affectation. *New England Law Review* 41 (April).

Nordberg, P. (2003). *Daubert* on the Web; available at http://www.daubertonthe web.com/leader_board.htm.

NRC (National Research Council) (1992). *DNA Technology in Forensic Science.* Washington, DC: National Academy Press.

NRC (National Research Council) (1996). *The Evaluation of Forensic DNA Evidence.* Washington DC: National Academy Press.

Nuffield Council on Bioethics (2007). *The Forensic Use of Bioinformation: Ethical Issues.* Cambridge: Cambridge Publishers; available at http://www .nuffieldbioethics.org.

Olsen, R. D., Sr. (1991). Identification of latent prints. In H. C. Lee & R. E. Gaensslen (eds.), *Advances in Fingerprint Technology.* New York: Elsevier: 39–58.

Olsen, R. D., Sr. & Lee, H. C. (2001). Identification of latent prints. In H. C. Lee & R. E. Gaensslen (eds.), *Advances in Fingerprint Technology.* 2nd ed. Boca Raton: CRC Press: 41–61.

Oreskes, N. (2004). Beyond the ivory tower: The scientific consensus on climate change. *Science* 306 (5702) (3 December): 1686.

OTA (Office of Technology Assessment), U.S. Congress (1990). *Genetic Witness: Forensic Uses of DNA Tests, OTA-BA-438.* Washington, DC: U.S. Government Printing Office (July).

Oteri, J. S., Weinberg, M. G. & Pinales, M. S. (1982). Cross-examination in drug cases. In B. Barnes & D. Edge (eds.), *Science in Context: Readings in the Sociology of Science.* Milton Keynes: Open University Press: 250–59.

Packer, K. & Webster, A. (1996). Patenting culture in science: Reinventing the scientific wheel of credibility. *Science, Technology, and Human Values* 21 (4): 427–53.

Pankanti, S., Prabhakar, S. & Jain, A. K. (2002). On the individuality of fingerprints. *IEEE Transactions on PAMI* 24 (8): 1010–25.

Parliamentary Office of Science and Technology (2006). The National DNA Database. *POSTNote,* February, no. 258.

Parliamentary Question (2006a). House of Commons, Hansard Column 647W (28 November).

Parliamentary Question (2006b). House of Commons, Hansard Column 504W (6 December).

Parliamentary Question (2006c). House of Commons, Hansard Column 829W (11 December).

Parliamentary Question (2007). House of Commons, Hansard Column 1276W (30 October).

Parloff, R. (1989). How Barry Scheck and Peter Neufeld tripped up the DNA experts. *American Lawyer* (December): 1–2; 53–59.

Perlman, M. (2004). Golden ears and meter readers: The contest for epistemic authority in audiophilia. *Social Studies of Science* 34 (5): 783–807.

Perrow, C. (1984). *Normal Accidents: Living with High Risk Technologies.* New York: Basic Books.

Phillips, F. (2003). Science stressed on death penalty. *Boston Globe* (24 September): 1.

Pickering, A. (1984). *Constructing Quarks: A Sociological History of Particle Physics.* Chicago: University of Chicago Press.

Pickering, A. (1995). *The Mangle of Practice.* Chicago: University of Chicago Press.

Pinch, T. (1985). Towards an analysis of scientific observation: The externality and evidential significance of observation reports in physics. *Social Studies of Science* 15: 167–87.

Pinch, T., & Bijker, W. (1984). The social construction of facts and artifacts: Or how the sociology of science and the sociology of technology might benefit each other. *Social Studies of Science* 14 (1984): 399–441.

Pinch, T. & Trocco, F. (2002). *Analog Days: The Invention and Impact of the Moog Synthesizer.* Cambridge, MA: Harvard University Press.

Pitkin, H. F. (1967). *The Concept of Representation.* Berkeley: University of California Press.

Podlas, K. (2006). "The CSI effect": Exposing the media myth. *Fordham Intellectual Property, Media and Entertainment Law Journal* 16: 429–65.

Podlas, K. (2006–7). The "CSI effect" and other forensic fictions. *Loyola University of Los Angeles Entertainment Law Review* 27: 87–125.

Polanyi, M. (1964). *Personal Knowledge: Towards a Post-Critical Philosophy.* New York: Harper & Row.

Pollner, M. (1974). Mundane reasoning. *Philosophy of the Social Sciences* 4: 35–54.

Pollner, M. (1975). The very coinage of your brain: The anatomy of reality disjunctures. *Philosophy of the Social Sciences* 5: 411–30.

Pollner, M. (1987). *Mundane Reason: Reality in Everyday and Sociological Discourse.* Cambridge, UK: Cambridge University Press.

Pomerantz, A. M. (1987). Descriptions in legal settings. In G. Button & J. R. E. Lee (eds.), *Talk and Social Organization.* Clevedon, UK: Multilingual Matters: 226–43.

Popper, K. (1963). *Conjectures and Refutations.* London: Routledge & Kegan Paul.

Porter, T. M. (1986). *The Rise of Statistical Thinking, 1820–1900*. Princeton: Princeton University Press.

Porter, T. M. (1995). *Trust in Numbers: The Pursuit of Objectivity in Science and Public Life*. Princeton: Princeton University Press.

Power, M. (1999). *The Audit Society: Rituals of Verification*. 2nd ed. Oxford: Oxford University Press.

Prainsack, B., & Kitzberger, M. (2007). DNA behind bars: "Other" ways of knowing forensic DNA evidence. Paper presented at annual meeting of Society for Social Studies of Science (Montreal, 12 October).

Preston, J. (2006). In New York, power of DNA spurs call to abolish statute of limitations (2 January).

Purnick, J. (2005). Justice delayed (and delayed and delayed). *New York Times* (10 October): B1.

Quinn, P. (1984). The philosopher of science as expert witness. In J. T. Cushing et al. (eds.), *Science and Reality: Recent Work in the Philosophy of Science*. Notre Dame: University of Notre Dame Press: 32–53.

Rabeharisoa, V. & Callon, M. (2002). The involvement of patients' associations in research. *International Social Science Journal* 54 (171): 57–63.

Rabinow, P. (1992). Artificiality and enlightenment: From sociobiology to biosociology. In J. Crary (ed.), *Zone 6: Incorporations*. Cambridge, MA: MIT Press. Reprinted, in Mario Biagioli (ed.) (1999). *The Science Studies Reader*. NY and London: Routledge: 407–16.

Rabinow, P. (1996a). *Making PCR: A Story of Biotechnology*. Chicago: University of Chicago Press.

Rabinow, P. (1996b). Galton's regret: Of types and individuals. In Rabinow, *Essays on the Anthropology of Reason*. Princeton, NJ: Princeton University Press, 112–28.

Rampton, S. & Stauber, J. (2001). *Trust US, We're Experts!* New York: Penguin Putnam.

Randerson, J. (2006). Police DNA database holds 37 percent of black men. *Guardian* (London) (5 January): 6.

Ravetz, J. (2006). *A No-Nonsense Guide to Science*. Oxford: New Internationalist.

Raymond, J. J., Walsh, S. J., Van Oorchot, R. A., Gunn, P. R. & Roux, C. (2004). Trace DNA: An underutilized resource or Pandora's box? A review of the use of trace DNA analysis in the investigation of volume crime. *Journal of Forensic Identification* 54: 668–86.

Redmayne, M. (1995). Doubts and burdens: DNA evidence, probability, and the courts. *Criminal Law Review:* 464–82.

Riles, A. (ed.) (2006). *Documents: Artifacts of Modern Knowledge*. Ann Arbor: University of Michigan Press.

Rimer, S. (2002). Convict's DNA sways labs, not a determined prosecutor. *New York Times* (6 February).

Rip, A. (2003). Constructing expertise in a third wave of science studies? *Social Studies of Science* 33: 419–34.

Risinger, D. M. (2000a). Preliminary thoughts on a functional taxonomy of expertise for the post-Kumho world. *Seton Hall Law Review* 31: 508–36.

Risinger, D. M. (2000b). Navigating expert reliability: Are criminal standards of certainty being left on the dock? *Albany Law Review* 64: 99–152.

Risinger, D. M. & Saks, M. J. (2003). A house with no foundation. *Issues in Science and Technology* 20 (1) 35–39.

Risinger, D. M., Saks, M. J., Thompson, W. C. & Rosenthal, R. (2002). The *Daubert/Kumho* implications of observer effects in forensic science: Hidden problems of expectation and suggestion. *California Law Review* 90 (1): 1–56.

Roberts, L. (1991). Fight erupts over DNA fingerprinting. *Science* 254: 1721–23.

Roberts, L. (1992). Science in court: A culture clash. *Science* 257: 732–36.

Roeder, Kathryn (1994). DNA fingerprinting: A review of the controversy: Rejoinder. *Statistical Science* 9 (2): 267–78.

Rose, N. (2007). *The Politics of Life Itself: Biomedicine, Power, and Subjectivity in the Twenty-First Century.* Princeton, NJ: Princeton University Press.

Royal Commission on Criminal Justice (1993). *Report* CM 2263. London: HMSO.

Rubinstein, J. (1973). *City Police.* New York: Farrar, Straus & Giroux.

Rudin, N. & Inman, K. (2002). *An Introduction to Forensic DNA Analysis.* 2nd ed. Boca Raton: CRC Press.

Ruse, M. (1982). Response to the commentary: *Pro judice. Science, Technology, and Human Values* 7 (41): 19–23.

Ruse, M. (1986). Commentary: The academic as expert witness. *Science, Technology, and Human Values* 11 (2): 68–73.

Ryle, G. (1954). The world of science and the everyday world. In G. Ryle, *Dilemmas: The Tanner Lectures 1953.* Cambridge: Cambridge University Press: 68–81.

Sacks, H. (1992). *Lectures on Conversation.* 2 vols. Oxford: Blackwell.

Saks, M. J. (1998). Merlin and Solomon: Lessons from the law's formative encounters with forensic identification science. *Hastings Law Journal* 49: 1069–1141.

Saks, M. J. (2000). Banishing *Ipse Dixit:* The impact of *Kumho Tire* on forensic identification science. *Washington and Lee Law Review* 57: 879–900.

Saks, M. J. & J. J. Koehler (2005). The coming paradigm shift in forensic identification science. *Science* 309 (5 August): 892–95.

Santos, F. (2007). 'CSI effect': Evidence from bite marks, it turns out, is not so

elementary. *New York Times* (28 January); available at http://www.nytimes
.com/2007/01/28/weekinreview.

Schaffer, S. (1988). Astronomers mark time: Discipline and the personal equa-
tion. *Science in Context* 2: 115–45.

Schaffer, S. (1989). Glass works: Newton's prisms and the uses of experiment. In
D. Gooding, T. Pinch & S. Schaffer (eds.), *The Uses of Experiment: Studies in
the Natural Sciences.* Cambridge, UK: Cambridge University Press: 67–104.

Schanberg, S. H. (2002). A journey through the tangled case of the Central Park
jogger: When justice is a game. *Village Voice* (20–26 November); available at
http://www.villagevoice.com/news/0247,schanberg,39999,1.html.

Scheck, B., P. Neufeld & J. Dwyer (2000). *Actual Innocence: Five Days to Execution, and Other Dispatches from the Wrongly Convicted.* New York:
Doubleday.

Schegloff, E. A. & Sacks, H. (1973). Opening up closings. *Semiotica* 7: 289–327.

Schutz, A. (1962). Concept and theory formation in the social sciences: Choosing among projects of action. In A. Schutz, *Collected Papers I: The Problem
of Social Reality.* The Hague: Martinus Nijhoff: 48–96.

Schutz, A. (1964a). The dimensions of the social world. In A. Schutz, *Collected
Papers II: Studies in Social Theory.* The Hague: Martinus Nijhoff: 20–63.

Schutz, A. (1964b). The problem of rationality in the social world. In A. Schutz,
Collected Papers II: Studies in Social Theory. The Hague: Martinus Nijhoff:
64–90.

Schutz, A. (1964c). Making music together: A study in social relationship. In
A. Schutz, *Collected Papers II: Studies in Social Theory.* The Hague: Martinus Nijhoff: 159–78.

Schwartz, A. (1997). A "dogma of empiricism" revisited: *Daubert v. Merrell
Dow Pharmaceuticals, Inc.* and the need to resurrect the philosophical insight of *Frye v. United States. Harvard Journal of Law and Technology* 10 (2):
149–237.

Scientific Working Group on Friction Ridge Analysis Study and Technology
(1999). *What and Who is SWGFAST* (13 Sept.); available at http://onin.com/
twgfast/twgfast.html.

Scientific Working Group on Friction Ridge Analysis Study and Technology
(2002). *Quality Assurance Guidelines for Latent Print Examiners.* http://www
.swgfast.org/Quality_Assurance_Guidelines_for_Latent_Print_Examiners
_2.11.pdf.

Scientific Working Group on Friction Ridge Analysis Study and Technology
(2003). *Standards for Conclusions.* http://www.swgfast.org/standards_for_
conclusions_ver_1_0.pdf.

Searle, J. (1995). *The Construction of Social Reality.* London: Penguin.

Sengoopta, C. (2002). *Imprint of the Raj: The Emergence of Fingerprinting in India and Its Voyage to Britain.* London: Macmillan.

Shapin, S. (1989). The invisible technician. *American Scientist* 77: 554–63.

Shapin, S. (1994). *The Social History of Truth.* Chicago: University of Chicago Press.

Shapin, S. (1995). Cordelia's love: Credibility and the social studies of science. *Perspectives on Science* 3: 255–75.

Shapin, S. (1996). *The Scientific Revolution.* Chicago: University of Chicago Press.

Shapin, S. (2004). Expertise, common sense, and the Atkins diet. American Sociological Association Annual Meeting (San Francisco, 15 August).

Shapin, S. & Schaffer, S. (1985). *Leviathan and the Air Pump.* Princeton, NJ: Princeton University Press.

Shapiro, B. J. (1983). *Probability and Certainty in Seventeenth-century England: A Study of the Relationships between Natural Science, Religion, History, Law, and Literature.* Princeton, NJ: Princeton University Press.

Shapiro, B. J. (1986). "To a moral certainty": Theories of knowledge and Anglo-American juries, 1600–1850. *Hastings Law Journal* 38: 153–93.

Sharrock, W. W. & Watson, D. R. (1984). What's the point of "rescuing motives"? *British Journal of Sociology* 34 (3), 435–51.

Shelton, D. E., Kim, Y. S. & Barak, G. (2006). A study of juror expectations and demands concerning scientific evidence: Does the "CSI effect" exist? *Vanderbilt Journal of Entertainment and Technology Law* 9: 331–68.

Siegel, D. M., et al. (2006). The reliability of latent print individualization: Brief of *amici curiae* submitted on behalf of scientists and scholars by the New England Innocence Project. *Commonwealth v. Patterson. Criminal Law Bulletin* 42 (1): 21–51.

Smith, R. & Wynne, B. (eds.) (1989). *Expert Evidence: Interpreting Science in the Law.* London: Routledge.

Sokal, A. & Bricmont, J. (1998). *Intellectual Impostures.* London: Profile.

Solomon, S. & Hackett, E. (1996). Setting boundaries between science and law: Lessons from *Daubert v. Merrell Dow Pharmaceuticals, Inc. Science, Technology, and Human Values* 21 (2): 131–56.

Sombat, J. M. (2002). *Note,* latent justice: Daubert's impact on the evaluation of fingerprint identification testimony. *Fordham Law Review* 70: 2819.

Specter, M. (2002). Do fingerprints lie? *New Yorker* (27 May 27): 96–105.

Stacey, R. B. (2004). A report on the erroneous fingerprint individualization in the Madrid train bombing case. *Journal of Forensic Identification* 54 (6): 706–18.

Star, S. L. & Griesemer, J. R. (1989). Institutional ecology, "translations," and boundary objects: Amateurs and professionals in Berkeley's Museum of Vertebrate Zoology, 1907–39. *Social Studies of Science* 19: 387–420.

Starrs, J. E. (2000). The John Doe DNA profile warrant. *Scientific Sleuthing Review* 24: 4.

Stoney, D. A. (1991). What made us ever think we could individualize using statistics? *Journal of the Forensic Science Society* 31 (2): 197–99.

Stoney, D. A. (1997). Fingerprint identification: Scientific status. In D. L. Faigman, D. H. Kaye, M. J. Saks & J. Sanders (eds.), *Modern Scientific Evidence: The Law and Science of Expert Testimony,* vol. 2. St. Paul: West: 55–78.

Stoney, D. A. (2001). Measurement of fingerprint individuality. In H. C. Lee and R. E. Gaensslen (eds.), *Advances in Fingerprint Technology.* Boca Raton, Fla.: CRC Press: 327–87.

Stoney, D. A. & Thornton, J. I. (1986). A critical analysis of quantitative fingerprint individuality models. *Journal of Forensic Sciences* 31 (4): 1187–1216.

Stove, D. C. (1984). *Popper and After: Four Modern Irrationalists.* New York: Pergamon.

Strathern, M. (ed.) (2000). *Audit Cultures: Anthropological Studies in Accountability, Ethics, and the Academy.* London: Routledge.

Suchman, L. (1987). *Plans and Situated Actions.* Cambridge: Cambridge University Press.

Sudnow, D. (1965). Normal crimes: Sociological features of the penal code in a public defender's office. *Social Problems* 12: 255–76.

Swanson, K. (2005). Certain guilt: The social construction of a scientific death penalty statute. Paper, Department of History of Science, Harvard University.

Swanson, K. (2007). Biotec in court: A legal lesson on the unity of science. *Social Studies of Science* 37 (3): 354–84.

Sykes, G. (1958). *The Society of Captives.* Princeton: Princeton University Press.

Tackett, J. (2003). The Chronicle interview: David Protess. *Northwestern Chronicle* (28 September); available at http://www.chron.org/tools/viewart.php?artid=677.

Taroni, F. & Aitken C. (1998). Probabilistic reasoning in the law, Part 1: Assessment of probabilities and explanation of the value of DNA evidence; Part 2: Assessment of probabilities and explanation of the value of trace evidence other than DNA. *Science and Justice* 38: 165–77; 179–88.

Thomas, J. (2004). Two police officers are put on leave. *Boston Globe* (24 April).

Thompson, W. C. (1993). Evaluating the admissibility of new genetic identification tests: Lessons from the "DNA war." *Journal of Criminal Law & Criminology* 84: 22–104.

Thompson, W. C. (1996). DNA evidence in the O. J. Simpson trial. *University of Colorado Law Review* 67: 827–57.

Thompson, W. C. (1997). A sociological perspective on the science of forensic DNA testing. *UC Davis Law Review* 30 (4) 1113–36.

Thompson, W. C. (2003). Houston has a problem: Bad DNA evidence sent the wrong man to prison at least once. How many more are there and what can be done about it? *Cornerstone* 25 (1): 16–17.

Thompson W. C. (2006). Understanding recent problems in forensic DNA testing. *Champion* 30 (1): 10–16.

Thompson, W. C. & Cole, S. (2007). Psychological aspects of forensic identification evidence. In M. Costanzo, D. Krauss & K. Pezdek (eds.), *Psychological Testimony for the Courts*. Mahwah, NJ: Erlbaum: 31–68.

Thompson, W. C. & Ford, S. (1988). DNA typing: Promising forensic technique needs additional validation. *Trial* (September): 56–64.

Thompson, W. C. & Ford, S. (1991). The meaning of a match: Sources of ambiguity in the interpretation of DNA prints. In J. Farley & J. Harrington (eds.), *Forensic DNA Technology*. New York: CRC Press.

Thompson, W. C., Ford, S., Doom, T., Raymer, M. & Krane, D. (2003). Evaluating forensic DNA evidence: Essential elements of a competent defense review, part 1. *Champion* (April): 16–25; available at http://www.engineering.wright.edu/itri/EVENTS/SUMMER-INST-2003/SIAC03-Krane2.PDF.

Thompson W. C. & Schumann E. L. (1987). Interpretation of statistical evidence in criminal trials. *Law and Human Behavior* 11: 167–87.

Thornton, J. I. & Peterson, J. L. (2002). The general assumptions and rationale of forensic identification. In D. L. Faigman, D. H. Kaye, M. J. Saks & J. Sanders (eds.), *Modern Scientific Evidence: Science in the Law: Forensic Science*. St. Paul, MN: West: 1–45.

Toobin, J. (2007). The *CSI* effect. *New Yorker* 83 (11) (7 May): 30–35.

Topping, R. (2000). One-time opponent now advocate of DNA testing. *Newsday* (20 December): A4.

Travis, A. (2003). Police DNA log now has 2m profiles. *Guardian* (London), 26 June.

Traweek, S. (1988). *Beamtimes and Lifetimes: The World of High Energy Physics*. Cambridge, MA: Harvard University Press.

Triplett, M. & Cooney, L. (2006). The etiology of ACE-V and its proper use: An exploration of the relationship between ACE-V and the scientific method of hypothesis testing. *Journal of Forensic Identification* 56 (3): 345–55.

Trombley, S. (1992). *Execution Protocol: Inside America's Capital Punishment Industry*. New York: Crown.

Tucker, C. (2002). Promote use of DNA tests for all suspects. *Atlanta Journal-Constitution*, (10 February): E10.

Turner, S. (1986). *The Search for a Methodology of Social Science*. Dordrecht: D. Reidel.

Turner, S. (2001). What is the problem with experts? *Social Studies of Science* 31: 123–49.

Turow, S. (2003). *Ultimate Punishment: A Lawyer's Reflections on Dealing with the Death Penalty*. New York: Farrar, Strauss and Giroux.

Tuthill, H. (1994). *Individualization: Principles and Procedures in Criminalistics*. Salem, OR: Lightning Powder Company.

TWGDAM (Technical Working Group on DNA Analysis Methods) (1991). TWGDAM guidelines on validation. *Crime Laboratory Digest* 22: 21–43.

Tyler, T. (2006). Viewing *CSI* and the threshold of guilt: Managing truth and justice in reality and fiction. *Yale Law Journal* 115: 1050–85.

Valverde, M. (2003a). *Law's Dream of a Common Knowledge*. Princeton: Princeton University Press.

Valverde, M. (2003b). Pragmatist and non-pragmatist knowledge practices in American law. Paper presented at Conference on Pragmatism and Law, Cornell University School of Law (Ithaca, NY, 28 March).

Valverde, M. (2003c). Pragmatist and non-pragmatist knowledge practices in American law. *PoLAR* 26 (2): 86–108.

Valverde, M. (2005). Authorizing the production of urban moral order: Appellate courts and their knowledge games. *Law & Society Review* 39 (2): 419–55.

Valverde, M. (2006). *Law & Order: Signs, Meanings, Myths*. New Brunswick, NJ: Rutgers University Press; Abingdon, UK: Routledge.

Vanderkolk, J. R. (1993). Class characteristics and "could be" results. *Journal of Forensic Identification* 43 (2): 119–25.

Vanderkolk, J. R. (1999). Forensic individualization of images using quality and quantity of information. *Journal of Forensic Identification* 49 (3): 246–56.

Vanderkolk, J. R. (2001). Levels of quality and quantity of detail. *Journal of Forensic Identification* 51 (5): 461–68.

Vanderkolk, J. R. (2002). Forensic individualization of images. Paper read at Fourth Annual Conference on Science and the Law (Miami, FL, 5 October).

Vanderkolk, J. R. (2004). ACE+V: A model. *Journal of Forensic Identification* 54 (1): 45–53.

Wagner, D. & DeFalco, B. (2002). DNA frees Arizona inmate. *Arizona Republic* (9 April).

Walsh, N. P. (2002). False result fear over DNA tests. *Observer* (London) 27 January.

Wambaugh, J. (1989). *The Blooding*. New York: Morrow.

Wayman, J. L. (2000). When bad science leads to good law: The disturbing irony of the *Daubert* hearing in the case of *U.S. v. Byron C. Mitchell*. *Biometrics in the Human Services User Group Newsletter* (January).

Weber, M. (1968). *Economy and Society: An Outline of Interpretive Sociology*. 2 vols. Berkeley: University of California Press.

Weir, B. S. (2001). DNA match and profile probabilities: Comment on Budowle et al. (2000) and Fund and Hu (2000). *Forensic Science Communications* 3; available at http://www.fbi.gov/programs/lab/fsc/current/weir.htm.

Werret, D. (1995). The development of forensic science: DNA and the future. *Police and Government Security Journal* 1: 46–47.

Wertheim, P. A. (1992). Re: Certification (to be or not to be). *Journal of Forensic Identification* 42: 280–81.

Wertheim, P. A. (2000). Scientific comparison and identification of fingerprint evidence. *The Print* 16 (5): 1–8.

Wigmore, J. H. (1940). *Wigmore's Code of the Rules of Evidence in Trials of Law*. 3rd ed. Boston: Little Brown.

Wilder, H. H. & Wentworth, B. (1918). *Personal Identification: Methods for the Identification of Individuals, Living or Dead*. Boston: R. G. Badger.

Wilgoren, J. (2002). Confession had his signature: DNA did not. *New York Times* (26 August): A-13.

Williams, R. & Johnson, P. (2005). Inclusiveness, effectiveness, and intrusiveness: Issues in the developing uses of DNA profiling in support of criminal investigations. *Journal of Law and Medical Ethics* 333: 545–58.

Williams, R., Johnson, P. & Martin, P. (2004). *Genetic Information and Crime Investigation: Social, Ethical and Public Policy Aspects of the Establishment, Expansion, and Police Use of the National DNA Database*. Report funded by the Wellcome Trust; available from School of Applied Social Science, University of Durham, UK, and online at http://www.dur.ac.uk/p.j.johnson.

Willing, R. (2000). Mismatch calls DNA tests into question. *USA Today* (8 February), 3A.

Willing, R. (2007). DNA to clear 200th person: Pace picks up on exonerations. *USA Today* (23 April): 1.

Wilson, A. & Ashplant, T. G. (1988a). Whig history and present-centred history. *Historical Journal* 31 (1): 1–16.

Wilson, A. & Ashplant, T. G. (1988b). Present-centred history and the problem of historical knowledge. *Historical Journal* 31 (2): 253–74.

Winch, P. (1958). *The Idea of a Social Science and Its Relation to Philosophy*. London: Routledge & Kegan Paul.

Winch, P. (1970). Understanding a primitive society. In B. Wilson (ed.), *Rationality*. Oxford: Basil Blackwell: 78–111.

Winner, L. (1977). *Autonomous Technology: Technics-Out-of-Control as a Theme in Political Thought*. Cambridge, MA: MIT Press.

Wittgenstein, L. (1958). *Philosophical Investigations*. Oxford: Blackwell.

Wittgenstein, L. (1969). *On Certainty*. Oxford: Blackwell.

Woolgar, S. (1981). Interests and explanations in the social study of science. *Social Studies of Science* 11: 365–94.

Woolgar, S. (1988). *Science: The Very Idea*. London, Tavistock.

Woolgar, S. & Pawluch, D. (1985). Ontological gerrymandering: The anatomy of social problems explanations. *Social Problems* 32: 214–27.

Wynne, B. (1989a). Establishing the rules of laws: Constructing expert authority.

In R. Smith & B. Wynne (eds.), *Expert Evidence: Interpreting Science in the Law*. London: Routledge: 23–55.

Wynne, B. (1989b). Sheepfarming after Chernobyl: A case study in communicating scientific information. *Environment* 31: 10–15; 33–39.

Wynne, B. (2003). Seasick on the third wave? Subverting the hegemony of propositionalism. *Social Studies of Science* 33: 401–17.

Zabell, S. L. (2005). Fingerprint evidence. *Journal of Law and Policy* 13 (1): 143–70.

Index